Guide to Telecommunications Transmission Systems

For a complete listing of the *Artech House Telecommunications Library*,
turn to the back of this book.

Guide to Telecommunications Transmission Systems

Anton A. Huurdeman

Artech House
Boston • London

Library of Congress Cataloging-in-Publication Data
Huurdeman, Anton A.
 Guide to telecommunications transmission systems / Anton A. Huurdeman.
  p. cm.
 Includes bibliographical references and index.
 ISBN 0-89006-978-6 (alk. paper)
 1. Telecommunication systems. I. Title.
 TK5105.H88   1997
 384'.042—dc21   97-15157
                                                      CIP

British Library Cataloguing in Publication Data
Huurdeman, Anton A.
 Guide to telecommunications transmission systems
 1. Telecommunication systems
 I. Title
 621.3'82

 ISBN 0-89006-978-6

Cover design by Jennifer Makower

© 1997 ARTECH HOUSE, INC.
685 Canton Street
Norwood, MA 02062

International Standard Book Number: 0-89006-978-6
Library of Congress Catalog Card Number: 97-15157

10 9 8 7 6 5 4 3 2 1

Contents

Foreword

The impressive development of key technologies such as optoelectronics, microelectronics, and memories as well as software technologies initiated the introduction of the digital "transswitching" (transmission and switching for telecommunication signals) in the early 1980s; although Reeves filed his patent on pulse code modulation in 1938.

Digital transmission is predestined to integrate all telecommunication services in one network and, in combination with optical transmission, telecommunication satellites, and high-speed microelectronics, to introduce "integrated broadband communication." These perspectives stimulated big efforts for optimum coding of digital signals for data reduction (source encoding), elimination of transmission errors (channel encoding), and optimum signal shaping for transmission (line encoding).

These facts prepared the evolution from dedicated analog transport networks for speech, data, sound, and video signals through the integrated services digital network (ISDN) toward the broadband-ISDN.

Furthermore, telecommunication satellites ideally complementing optical transmission systems, because of supporting/enabling fast global implementation of telecommunication services, appeared as a spin-off of the space flight activities in the United States and the previous Union of Soviet Socialist Republics, likewise supported by the key technologies of the time period. This opened the way to "global communication."

This progress of the key technologies in combination with the progress in digital signal processing paved the way for "mobile communication for everybody"—the telecommunication service with the highest growth in the last years. Mobile communications together with "asynchronous telecommunication satellites" (known as LEO and MEO satellites) make Graham Bell's vision "that everybody will have the possibility to call everybody from any place of this globe at any time" a reality.

The integration of telecommunication services in one network enables the combined usage of these services and opens the way for "multimedia" serving the emerging "information society."

The tremendous development of telecommunication technologies and services resulted in the specialization of dedicated expert groups in such fields as microelectronics, optoelectronics, software, and systems, that introduced their own "jargon" with abbreviations and acronyms, thus creating a Babylonian confusion even within the family of telecommunications experts. "All-round experts" in telecommunications are disappearing. Efficient work even in specialized disciplines, however, requires a general overview of the whole field of activities and a basic understanding of the commonly used abbreviations and acronyms. A book that comprises an introduction, an overview, and an easy reference for the complex field of telecommunications transmission systems, describing the present state of the art as well as summarizing planned systems, is missing in the existing literature. Fortunately, this encouraged Anton Huurdeman—backed by his 40 years of experience in the field of telecommunication systems—to produce this comprehensive work.

This book may become the "hand luggage" of all persons who are obliged to occupy themselves more intensively with telecommunications such as researchers, developers, manufacturers, and operators; students of concerned disciplines; persons occupied in planning, marketing, sales, and purchasing; and representatives of media, sociology, and politics because telecommunication is an essential part of our culture and effects many areas of our society.

This book, therefore, should find its place in all libraries of universities, telecommunication systems manufacturers and operators, as well as relevant administrations such as ministries of research and technologies, PTTs, and economics.

Referring to researchers and developers, it should be noted that this book *does not offer fundamentals* for the developments of components, subassemblies, and systems for telecommunications *but offers an overview, a guideline, and a reference* for existing and planned systems and helps to overcome the inundation of abbreviations and acronyms. Moreover, for the manufacturers and operators it is a valuable assistance for strategic decisions.

Telecommunications students are introduced by a plausible way in this complex field and find ample literature references for further study.

Persons occupied with planning, marketing, sales, and purchasing get a guideline and a "support of memory" for their whole activities.

For representatives of media, sociology, and politics this book is also recommended for its contribution toward objectivity in presenting opinions and the discussions about the challenges, the importance, and the dangers of telecommunications.

Principles, definitions, standards, and meanings of "transswitching" systems and technologies are worked out and described by the author without redundancy. The text is enhanced by numerous simplified graphics about principles and is completed by tables of important data. The text and the whole contents concentrate on the essentials.

The whole field of telecommunications transmission systems is presented as an overview in a concise and easily understandable manner beginning with the history and ending with an outlook on the planned developments and introductions of the next century's first decade, thus closing a gap in the large field of telecommunications literature.

Prof. Dr.-Ing. Horst Ohnsorge
Retired as Director Research and Technology of Alcatel, Paris
Presently Honorary Professor of Stuttgart University
Blaubeuren, Germany
December 1, 1996

Preface

Transmission is one of the major technologies that make multimedia happen. The information society of tomorrow will be served by a *global information infrastructure* (GII, the first of unfortunately many acronyms in this book), which will enable us "to obtain an answer to our questions within a matter of seconds, at affordable cost." The biggest developments in transmission took place in the last 40 years; and yet, the last comprehensive book in the English language covering the whole field of transmission, *Transmission Systems for Communications,* also called the "bible of transmission," was written by staff members of the Bell Telephone Laboratories in 1954, many times revised until the last edition in 1982.

A highlight of transmission in 1954 was the proud announcement that the first transatlantic coaxial cable (TAT-1) would be put into operation in 1956 for the transmission of 36 telephone channels between the United States and Europe using submarine repeaters every 60 km. Today we might read a little note in a newspaper that TAT 12/13 started operation with a capacity of over 300,000 telephone channels without any signal regeneration on the almost 8,000-km transatlantic route—thanks to a technology called soliton transmission through optical fiber.

With this book I endeavor to provide an easy, understandable overview of the comprehensive domain of transmission within its evolutionary context, covering its media, applications, market volume, and its future, thereby showing that the five transmission media—copper lines, optical fiber, radio relay, mobile radio, and satellites—are complementary, with different applications of maximum cost effectiveness for each of the media.

To appreciate and understand the importance of transmission within the scope of telecommunications, this book starts with an introduction to telecommunications in the context of its evolutionary development over the last 200 years. Before describing the transmission media, a chapter titled "Multiplex" describes the transmission equipment that interfaces between terminal and switching equipment on one side and transmission media on the other, such as multiplex, circuit multiplication, data transmission interface, and video and sound codecs equipment.

Chapters 3 to 7 cover in detail the five transmission media and include a summary of the specific applications of the concerned media at the end of each chapter. Chapter 8 underlines the complementary nature of the five media and is intended as an assistance to select the most cost-effective solution for each specific application. Chapter 9 shows that all five transmission media will continue to contribute to a further significant growth of the telecommunication market.

Chapter 10 presents an outlook on the development of each of the five media in the next five years. Special coverage is given in this chapter of the planned, most spectacular transmission development of this century: the operation of new satellite networks that will significantly contribute to the ultimate goal of transmission "that everybody on this planet will be able to obtain the right answer to her/his question in a matter of seconds, at affordable cost."

Acknowledgments

Transmission technology has become so increasingly complex and versatile that a single person can hardly describe the technology exclusively from his own knowledge. Even the aforementioned "bible of transmission" was written by a *group* of transmission experts. In writing this book I have drawn on the expertise of numerous specialists as identified at the end of each chapter. I have used my experience and knowledge on transmission to select, judge, evaluate, interpret, and quote or rewrite from those numerous experts' descriptions, views, and articles. For the principles and basic information on transmission systems I have extensively drawn from the *Telecommunications Technology Handbook* written by Daniel Minoli, *Cellular Radio Systems* edited by D. M. Balston and R. C. V. Macario, and my own *Radio-Relay Systems*—all three published by Artech House, respectively, in 1991, 1993, and 1995. To ensure that this book describes the state of the art and future of the transmission systems and technologies as known at the end of 1996, I have included much information on the latest transmission developments scrutinized in the last three years from the following technical magazines:

- *Alcatel Telecommunications Review*;
- *Communications International*;
- *Discovery* (Nokia);
- *Ericsson Review*;
- *ITU News*;
- *Iridium Today*;
- *Mobile Asia Pacific*;
- *Mobile Communications International*;
- *Mobile Europe*;
- *Telcom Report International* (Siemens);
- *Telecommunications*;
- *Transat* (Inmarsat);

- *VIA INTELSAT*;
- *Via Satellite.*

Information and photographs of the latest transmission products and systems are included thanks to the kind cooperation of

- Mrs. Gertrud Braune (Siemens);
- Mrs. Helen Jarvinen (Nokia);
- Mrs. Christiane Rausch (Bosch Telecom);
- Mrs. Anita Wohlfahrt (Alcatel SEL);
- Mr. David C. Benton (Globalstar);
- Mr. Bernard Pastor (Arianespace);
- Mr. John M. Windolph (Iridium).

Many thanks are also due to my previous boss, Dipl. Ing. Gerd Lupke, for proofreading the complete manuscript and his very useful suggestions for improvements in the interest of a balanced content. Similarly, many thanks to Dr. Ing. Heinrich Rupp for proofreading Chapter 7, for improvement suggestions, and for kindly providing the latest information on satellite systems. Finally my thanks go to Prof. Dr.-Ing. Horst Ohnsorge for his manuscript improvement suggestions and especially for writing the foreword in which he underlines the complexity of transmission and clearly defines the groups of persons who can gain from reading this book.

Anton A. Huurdeman
Todtnauberg, Germany
June 1997

Telecommunications 1

1.1 INTRODUCTION

In this first chapter we give an overview of the basics of telecommunications in the context of their evolutionary development.

The word *communications,* derived from the latin word *communicatio*—which stands for the social process of information exchange—covers the human need for direct contact and mutual understanding. The word *telecommunications*—thus adding tele=distance—was created by Edouard Estaunie, director of the Ecole Supérieure des Postes et Télégraphes de France in 1904, and defined as the "information exchange by means of electrical signals."

Estaunie thus limits telecommunication explicitly to electrical signals. The *International Telecommunication Union* (ITU), which was founded at a common meeting of the 13th International Telegraph Conference and the 3rd International Radiotelegraph Conference in Madrid in 1932, defined telecommunications at that constitutional meeting in a more comprehensive way: "any telegraph or telephone communication of signs, signals, writings, images, and sound of any nature, by wire, radio, or other system or processes of electric or visual (semaphore) signaling."

The ITU currently defines telecommunications as: "any transmission, emission, or reception of signs, signals, writings, images, and sounds; or intelligence of any nature by wire, radio, visual, or other electromagnetic systems."

In order to telecommunicate, local, regional, national, and worldwide international telecommunication networks are required. Figure 1.1 shows the basic configuration of a telecommunication network.

In local telecommunication networks, also called access networks, individual telecommunication users (the telecommunication originators as well as the telecommunication recipients) are all connected with one or more local switches (also called the local exchanges or central offices). The telecommunication users, such as the subscribers of public networks, are connected by their

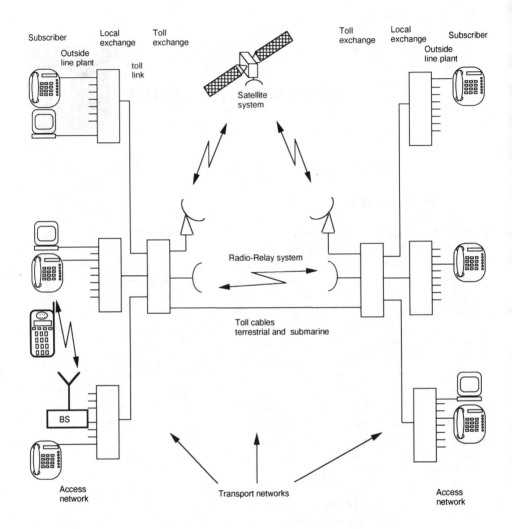

Figure 1.1 Typical telecommunication network.

local exchange—mainly by means of a single cable pair but previously also by open wire, at distant or isolated locations by radio, and in the near future increasingly by broadband optical fiber systems. In regional and national telecommunication networks a number of local exchanges are connected via transmission links with a tandem exchange (also called toll or trunk exchange); all the tandem exchanges of a region or a nation are interconnected by transmission links also. Such a transmission link can be by means of copper or optical fiber cable, radio relay, or satellite. In international telecommunication networks the telecommunication users are connected via their local exchange and one or more tandem exchanges with international exchanges of their country. The international exchanges worldwide are all interconnected by transmission links either directly or via one or more other international exchanges.

According to the first definition provided by the ITU that postulated "visual (semaphore) signaling" as a means of telecommunication, it can be stated that telecommunications started in the French Revolution with the optical telegraph developed by Claude Chappe.

There is no doubt that visual signaling had been applied long before Chappe constructed his optical telegraph. The Greeks, Persians, and Romans used smoke and fire signals for information transmission. Following the discovery of the telescope in 1608 by the Dutch optician Jan Lipperhey, the British astronomer Robert Hooke in 1684 presented to the Royal Society in London a plan for optical telegraphy on land and between ships upon the sea with a combination of a telescope and an optical telegraph. Hooke's plan was never tested. Similar proposals and experiments were made in France, Germany, and Scandinavia. On the new continent during the War of Independence from 1775 to 1783 a "revolution-telegraph" was temporarily used with a flag for daytime and a basket for nighttime signals.

It was during the French Revolution, however, with the creation of a new national republican state, that the merits of a permanently installed communication network were finally recognized and an optical telegraph network could be implemented with Claude Chappe's optical telegraph. Chappe, who thus constructed the first functional telecommunication device that was used successfully until succeeded by a superior solution (the electrical telegraph), deserves to be called the father of telecommunications.

Telecommunications, like so many technological developments, had thus been developed for military purposes, that is, to support Napoleon in the wars of France against its neighboring countries. In spite of this military origin and although telecommunications still plays a vital role in the military field, it is appropriate to state here that telecommunications is a peaceful technology. Telecommunications does not provoke aggressiveness as traffic does. Telecommunications helps to unite people without causing environmental damages. Instead of causing pollution, telecommunications can prevent negative environmental effects by reducing travel and improving emergency services. Teleworking can arguably save more natural resources than by using alternative renewable energy. Moreover, telecommunications can improve education and health services and can be vital in the event of a natural disaster. After the disastrous earthquake in Kobe (Japan) on January 17, 1995, less than 10% of a cellular network went out of service but was repaired in a few days. The fixed telecommunication infrastructure, although heavily damaged, took up limited operation within two hours, and within four days it was able to support an emergency network with open-air *phone pools* all over Kobe that provided around-the-clock free calls for several weeks.

Telecommunications, apart from reducing the influence of geographical obstacles in bringing people together, will significantly accelerate social and economic changes—hopefully toward better living conditions worldwide [1,2].

1.2 OPTICAL TELEGRAPH

Abbe Claude Chappe (1763–1805) made his first experiments with a device that he called *Tachygraphe* (Latin for quick writer) in Paris on March 2, 1791. The Tachygraphe had a very limited visibility. Chappe therefore constructed an improved version in 1793, then called optical telegraph (Figure 1.2), that consisted of an approximately 4.5-m-long *regulator* to which two approximately 2-m-long *indicators* were attached. The regulator could have four and each indicator seven different positions. Altogether $4 \times 7 \times 7 = 196$ different configurations were possible. The 92 best distinguishable configurations were allocated to characters of the alphabet, numbers, complete words, and sentences. The first message, which reported a success of Napoleon's troops, arrived at the French Revolutionary National Convent through the Paris-Lille semaphore line (230 km, with 23 stations) on August 15, 1794. Thus the era of telecommunications had begun!

Inspired by Chappe's success, similar optical telegraph lines were constructed in Spain, Italy, Algeria, England, Belgium, Holland, Germany, and Denmark along with a long line between Moscow and Warsaw with 220 stations. By 1844 France had approximately 534 stations connecting 29 cities covering 4,800 km. Since the early 1800s, the term *semaphore* (Greek for "sign bearer") was equally applied for optical telegraph.

Claude Chappe, regretfully, could not enjoy his success. People accused him of copying what they claimed to be their idea. The attacks affected Chappe very seriously. He became depressed and, aggravated by chronic bladder trouble, finally committed suicide by jumping into a well on January 23, 1805. A heavily damaged semaphore on a much uncared-for grave hidden in the famous *Pere Lachaise* cemetery in Paris (Figure 1.3) is all that remains as a remembrance of the genius who deserves to be called the father of telecommunications [2–4].

1.3 ELECTRICAL TELEGRAPH

The optical telegraph met the requirements of its era, being the best solution available. A drawback, however, was its dependence on weather and daylight. During the night; in fog, rain, or snow; or on hot hazy days with dusty air or strong winds, the semaphore could not be online. On the other hand, good visibility made the semaphore an easy military target.

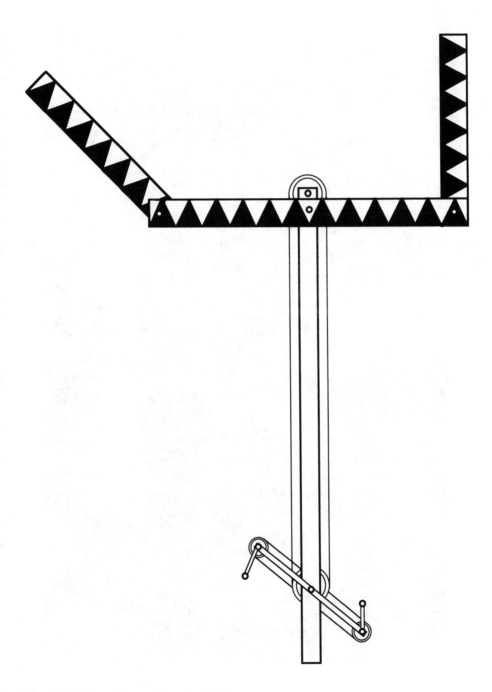

Figure 1.2 Chappe's semaphore in 1793.

Figure 1.3 The uncared-for grave of the father of telecommunications.

Located 100 km west of the neglected grave of Claude Chappe, in the splendid and modern American Museum at Giverny, a large painting shows an artist at work in front of *Gallery of the Louvre.* The artist who painted that gallery in 1831/32 would have noticed Chappe's semaphore on top of the Louvre. During his long voyage on the sailing ship *Sully* back to the New World in 1832, recollecting his impressions on art and the interesting news of electrical experiments performed by now-famous men such as Ampere, Ohm, Faraday, Gauß, Oersted, Steinheil, and others, that artist had an idea that six years later resulted in the beginning of electrical telecommunications. The artist, surely, was Samuel Finley Breese Morse (1791–1872), one of the fathers, or at least the most successful father, of electrical telegraphy.

Although Morse uncontestedly is the father of the electrical *writing* telegraph, there are many fathers of the electrical telegraph. In short, the story is that William Fothergill Cooke (1806–1879), as a student in anatomy, attended a lecture at the Heidelberg University in March 1836, where Wilhelm Muncke, a professor in physics, demonstrated a five-needle electrical telegraph. Professor Muncke had his demonstration model made after he attended demonstrations with a five-needle telegraph given at the Physics Society in Frankfurt in 1835 by Baron Pawel Lwowitsch Schilling (1786–1837). Baron Schilling saw a demonstration of an electrochemical telegraph given by S. T. Soemmering at the Munich Academy of Science on August 28, 1809. Baron Schilling, after many experiments, changed Soemmering's device from electrochemical to electromagnetic using the deflection of five needles as the information code. In 1837, shortly after Emperor Nicholas I of Russia appointed a commission to advise him on the installation of a Schilling telegraph between St. Petersburg and his imperial palace Peterhof, Schilling died. Cooke, fortunately having seen Schilling's five-needle telegraph in Heidelberg, built his first telegraph in Heidelberg and Frankfort and took it with him to England. He contacted railway companies and obtained a trial order for an electrical telegraph line from the Liverpool-Manchester Railway. Since he met with many difficulties, he formed a partnership with Charles Wheatstone (1802–1875), professor of natural philosophy at Kings College, London, who also had made experiments with electrical telegraph devices. Together they constructed a more reliable five-needle electrical telegraph; see Figure 1.4 showing the letter S (S being the only letter at an intersection point of two needles).

Cooke and Wheatstone's five-needle telegraph was patented in 1837 and became operational as the world's first electrical telegraph line on July 9, 1839.

All these needle telegraph versions are based upon the deflection of a magnetic needle in a magnetic field; Morse's merit was in using a direct writing device and applying a single transmission wire only (with the ground as return conductor) as compared to the five wires required for the five-needle telegraph.

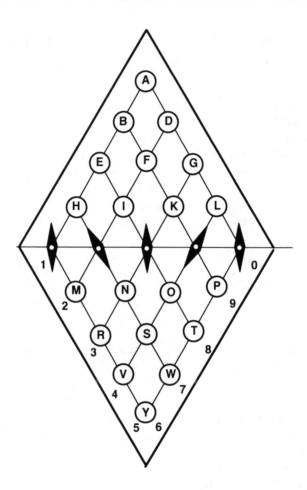

Figure 1.4 Front display of the Cooke-Wheatstone five-needle telegraph.

Morse, in the meantime, succeeded to get his electrical writing telegraph patented in France on August 18, 1838.

Cooke and Wheatstone continued to improve their five-needle telegraph, finally reducing the number of needles to one. Their system survived on British Railways into the 20th century. No other countries, apart from Spain for some time, adopted the Cooke and Wheatstone telegraph. For public services the Morse writing telegraph was preferred. As traffic grew rapidly, however, the Morse telegraph could not keep pace with the requirements. The Morse strip of paper with dots and dashes could not simply be passed on to the customer but had to be transcribed by hand into plain uncoded language. This transcribed plain language telegraphic message (a long description for a small piece of paper) got a short name *Telegram* proposed by E. P. Smith in the *Albany Evening Journal* of April 6, 1852.

The first step in the direction of plain language telegraphy came in 1855 when David E. Hughes (1831–1900)—a British professor of music who, like Morse, was a professor at the New York University—constructed an electric writing telegraph using a continuously rotating wheel with 28 letters of the alphabet and other signs. When the desired letter or sign was over a moving strip of paper, a clutch mechanism activated by an electromagnet brought the wheel momentarily to rest and pressed it onto the paper. Just as the artist Morse used an easel for his first electrical writing telegraph, the music professor Hughes used a piano keyboard to send messages with his telegraph. The Morse patent prevented the use of the Hughes telegraph in the United States, so Hughes successfully sought French cooperation. Together with his French partner Gustave Forment, a first operational *teleprinter* was manufactured in Paris and adopted by the French Telegraph Administration in 1860.

The next step forward came from Emile Baudot (1845–1903), who in 1869 as a newcomer to the French Telegraph Administration took a training course on Hughes' teleprinter. Baudot quickly improved the speed and reliability of the teleprinter and in 1874 introduced the still-in-use five-unit code in which each letter, symbol, and numeral is represented by a combination of five elements. Furthermore, he combined the use of the five-unit code with a time division multiplex system, thus allowing several telegraph communications to be transmitted over the same circuit. The piano keyboard was replaced by a keyboard of only five keys.

The Baudot teleprinter was officially adapted by the French Telegraph Administration in 1877, and with its many notable improvements—mainly by Creed in 1920 in England, and in Germany by Lorenz in 1926 and Siemens in 1928—it was adopted worldwide. In fact, both the electrical writing telegraph and the teleprinter are still in use: the telegraph for short, cost-effective messages and the teleprinter in the worldwide telex network mainly for administrations and business users. Telex provides both a permanent record of messages in directly written language rather than a text translated by a Morse operator, as well as evidence of delivery for the sender in the form of the answer back facility of the remote machine. In many countries a telex can be used as a legally binding document.

The invention of electrical telegraphy has, without doubt, provided a great stimulus to telecommunications; still, however, telegraphy is only an indirect way of communication, similar to the written letter but much quicker. Telex and telegram service presently amount to only a few percent of total telecommunication services. The worldwide number of telex subscriber lines stagnates at slightly more than one million, compared with to-date over 600 million telephone main lines with an annual increase of about 5%.

The major achievement of electrical telegraphy, apart from having substantially improved the security and reliability of railway transportation, has been the creation of an international telecommunication infrastructure, which is a prerequisite for the development of worldwide telecommunications. Telegraphy initially was a national affair. Telegraphic messages arriving at countries' borders needed to be terminated on paper and manually reentered in the telegraph network of the neighbor country. The first agreement for electrically passing messages across a country's border was signed on October 3, 1849 between Austria and Prussia. The agreement so successfully settled technical interface conditions, including an improved Morse-alphabet as a standard and rates and procedures, that it was used as an example for similar bilateral agreements between many other European countries and finally led to the foundation of the International Telegraph Union in Paris in 1865, which in 1932 merged with the International Radiotelegraph Convention [2–6].

1.4 TELEPHONE

The word *telephone* (Greek combination of *tele* for far off and *phone* for sound or voice) was first used in 1828 to describe a system for signaling with musical notes. Charles Wheatstone, the co-inventor of the telegraph, also applied the name to his *enchanted lyre,* used for transmitting sound from one room to another.

Philip Reis (1834–1874), a schoolmaster in Friedrichsdorf (near Frankfurt, Germany), on October 26, 1861 in Frankfurt demonstrated the first electrical reproduction and transmission of sound with an instrument, which he baptized *telephon.* He used an animal membrane stressed over a cone and connected with a platinum wire as the microphone. The platinum wire formed part of an electrical battery circuit; as the membrane vibrated, the platinum wire could make and break a contact in the electrical circuit. At the other end of the circuit was a coil wound around a knitting needle. The rapid magnetization and demagnetization of the knitting needle reproduced the sound that in his first experiments was amplified by a violin placed in front of the needle. Reis claimed that words could also be recognized.

Reis made various improvements to his *telephon.* A small quantity was produced in a workshop in Frankfurt and sent to various laboratories for further experimenting.

Reis, as a self-made man, unfortunately was not accepted by the German scientists, who did not consider his invention a serious matter. Disillusioned and ill, he died in 1874, justly convinced that he had given mankind a big invention. In the meantime, at least in Germany, he is considered to be the father of the telephone as documented on stamps of the German Federal Post of 1961 and 1986 commemorating 100 and 125 years of telephony; see Figure 1.5.

The successful vital step from telegraphy to telephony is documented in Letters Patent No. 174,465 issued by the United States Patent Office on March 7, 1876, in response to an application submitted February 14, 1876, at 2:00 p.m. by Alexander Graham Bell of Salem, Massachusetts on the subject of Improvement in Telegraphy.

There are, at least, two remarkable peculiarities concerning this document that is now generally accepted as the beginning of the telephone era. First, the patent application was filed at 2:00 p.m., which is exactly two hours before another application was submitted by Elisha Gray (1835–1901), cofounder of Western Electric Manufacturing Company, for a device similar to that constructed by Reis and also capable of transmitting the human voice. Gray submitted his application as a so-called *caveat*, which at that time was a declaration of an invention for which a one-year protection was requested to gain time for further investigation and submission of a final patent application. After a long

Figure 1.5 Stamps commemorating Philip Reis as the inventor of telephony.

legal battle backed by Bell's father-in-law, G. G. Hubbard, a rich and powerful lawyer, Gray's patent application was rejected.

Second, the six-page patent application from Bell explicitly refers to telegraphy only and does not even mention the word telephone or speech. In his application, Bell describes a method of simultaneous operation on a single line of a number of telegraph instruments, each trimmed to a different resonance frequency. Figure 1.6 gives a reproduction of the sixth page of Bell's patent application illustrating the parallel operation.

Alexander Graham Bell (1847–1922), a Scotsman, emigrated to Canada in 1870 and moved to Boston, Massachusetts in 1872 where he became professor of vocal physiology at Boston University in 1873. Bell had thus far devoted most of his time to educating the deaf and thereby acquired considerable knowledge of the physiology of human speech and hearing, but in Boston he started to experiment with the multiple transmission of telegrams over a single wire. In 1875 he obtained his first patent (No. 161,739) for simultaneously transmitting two or more telegraphic signals differing in pulse rate. Also, after obtaining his famous patent on March 7, 1876, he continued with his telegraphic experiments together with assistant Thomas A. Watson. During such experiments on March 10, with Watson in another room, an accident happened. Bell knocked down an acid container and must have automatically asked for help, calling, "Mr. Watson, come here, I want you," a phrase heard by Watson not via the corridor, ceiling, or wall between the two rooms but via the experimental telegraph arrangement now suddenly enhanced to a telephone experiment. From that very moment Bell concentrated his experiments on the improvement of his telephone in order to show a working model at the Philadelphia Centennial Exhibition (commemorating the Declaration of Independence of the United States of America on July 4, 1776 in Philadelphia, Pennsylvania). At that exhibition Bell's telephone did not attract much attention until Dom. Pedro II, Emperor of Brazil (from 1840 to 1889), who had met Bell previously at the Institute for Deafs in Boston, recognized Bell, who then gave him a demonstration of the telephone upon which the highly surprised emperor cried, "Good Lord, it speaks!" Now after imperial appreciation, Bell got the deserved attention. Sir William Thomson (from 1892 Lord Kelvin, 1824–1907), one of the judges at the exhibition, wrote: "With somewhat more advanced plans and more powerful apparatus, we may confidently expect that Mr. Bell will give us the means of making voice and word audible through the electric wire to an ear at hundreds of miles distant." A mayor of an American city was so impressed by Bell's telephone that he predicted, "I can see the time when every city will have one!" Quite an underestimation, and yet, for many Third World villages still a dream.

Bell installed the first telephone line across a two-mile stretch between Boston and Cambridge in 1876. On July 9, 1877, Alexander Graham Bell founded

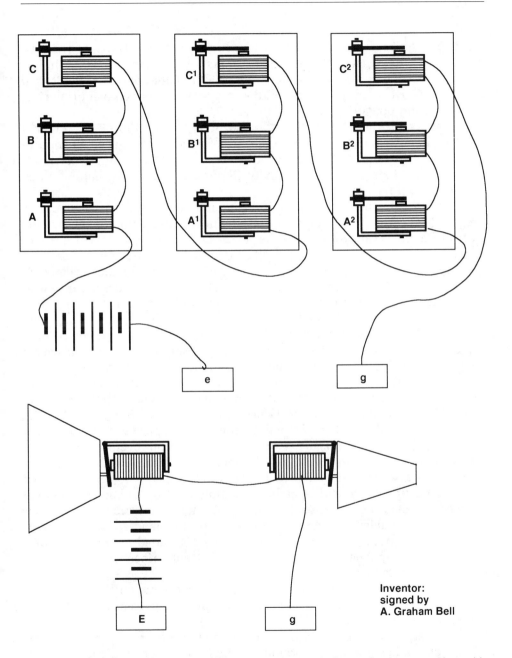

Figure 1.6 Bell's proposal for *vocal or other sounds transmission* as simultaneous telegraphic signals on one line (*After:* U.S. Patent Office Letters Patent 174,465, dated March 7, 1876).

the *Bell Telephone Company* (BTC) together with Mabel Hubbard (who he married two days later), his assistant Thomas A. Watson, his father-in-law G. G. Hubbard, and Thomas Sanders, the wealthy father of one of Bell's deaf pupils. G. G. Hubbard took charge of the company and started manufacturing, installing, and operating telephone lines.

Mabel and Alexander Bell made their honeymoon trip to Europe with a few telephone sets in their luggage that were presented to Queen Victoria in the United Kingdom, Antoine Breguet in France, and the General Postmaster Heinrich von Stephan (1831–1897) of the German Imperial Telegraph Administration in Berlin.

Antoine Breguet (1851–1882)—the grand-grandson of Abraham-Louis Brequet (1747–1832), who made the mechanical construction of the moving parts of Chappe's semaphore, and the son of Louis François-Clément Breguet (1904–1883), who designed and manufactured in 1842 the first electrical telegraph in France—obtained five patents from Bell for the production of telephone sets in France. Antoine Breguet presented his telephone in the same year to the Academie Française des Sciences but produced telephone sets for only a few years until he died in 1882.

Heinrich von Stephan received the two telephone sets in Berlin on October 24, 1877; and two days later the first *local call* was made in Berlin over a 2-km distance. Bell had not patented his telephone in Germany, so German companies, at the request of von Stephan, immediately started the production of partly improved versions of Bell's telephone. Siemens started telephone production in 1877 and Mix and Genest (now in Alcatel-SEL) in 1879. The improvement made by Siemens mainly concerned replacing the cylindrical magnet by a horseshoe magnet. A more significant improvement was made almost simultaneously in 1877–78 by Thomas Alva Edison and Emile Berliner in the United States and by Robert Ludtge in Germany, who replaced Bell's electromagnetic sender with a carbon microphone. The carbon microphone changes its resistance, and thus the battery current, as a function of the speech waves, which enabled telephony over much longer distances.

With Bell back in the United States, the BTC started the operation of a first manual telephone switchboard, serving 21 subscribers in New Haven, Connecticut on January 28, 1878.

The first international telephone line in Europe was put into operation between Bussels and Paris in 1887; regular telephone service between London and Paris started in 1891.

Bell certainly has been the most successful telecommunications inventor. His telephone has given an enormous impetus to telecommunications—over 600 million telephone subscribers worldwide can no longer live without it. The companies founded by Bell have been successfully integrated in AT&T, the world's largest telecommunications operator, and Alcatel NV (International

Bell Telephone Company sold to ITT in 1925; telecommunications part of ITT sold to Alcatel in 1987), which is one of the world's largest telecommunications manufacturers.

After 120 years, telephony has come of age and a successor is evolving: the videophone. Telephony, as useful as it is, still helps communication by sound only. Human communication, however, functions best with a combination of eye and ear contact. Speaking involves the whole body, especially face and hands. The communication sciences teach us that we retain 25% of what we hear, 50% of what we see, and 90% of what we hear and see. Consequently, the videophone can help to improve and personalize communications and could possibly replace the telephone over the course of the next 20 to 40 years. A few years ago the videophone was either prohibitively expensive or of poor quality. The latest videophones, as shown in Figure 1.7, offer excellent quality albeit still at the price of a HD-TV set. Since any narrowband-ISDN line (described in Subsection 1.5.7) can now be used for video-telephony, prices for the videophone will soon follow the rules of the market and be reduced substantially [4–11].

Figure 1.7 Modern videophone (courtesy of Alcatel SEL).

1.5 SWITCHING

1.5.1 Telegraph Switching

Switching started in the 1870s as a concentrating device to connect the various telegraph instruments of a telegraph office with the few lines available for the interconnection of national and international telegraph offices. Switching at these offices was mainly made to enable the telegraph companies to conveniently provide service on a *time-shared* basis between offices and to facilitate the connection of telegraph instruments on lines in working order to a given destination.

1.5.2 Telephone Switching

Switching a calling telephone subscriber to the line of the called subscriber in the early telephone networks was performed manually. Automatic switching was invented by Almon B. Strowger (1839–1902) of Kansas City. Strowger arranged electrical contacts that led to telephone subscribers in rows on the inner surface of a cylinder, called a selector. An arm on a central shaft of the selector activated by individual stepper magnets was to move step-by-step, sweep through the lines, and pause on the particular line of the calling subscriber (where the telephone set had been lifted). The line of the calling subscriber was then connected to another selector (called assignment selector) where likewise the arm on the central shaft, now controlled by impulses coming from the calling subscriber, went up the inner side of the cylinder and then again step-by-step across the cylinder until the contact with the line of the called subscriber was reached.

With such a 100-point assignment selector a maximum of 100 subscribers could be served. For exchanges with more than 100 subscribers 10-point group-selectors had to be inserted before the assignment selector. With one group selector the capacity of an exchange increases to 1,000 subscribers, with two group selectors to 10,000 subscribers, with three group selectors to 100,000 subscribers, and with four group selectors to 1,000,000 subscribers. Figure 1.8 shows how 10,000 subscriber lines can be served with two group selectors. In order to prevent one single calling subscriber from blocking the exchange for further simultaneous calls, a number of group selectors always operated in parallel. In Figure 1.8 this is indicated with "x" and "y", whereby $x > y$ and the values of x and y depend on the traffic load of the exchange. This traffic load is expressed in *Erlang*, the international unit for traffic load. For a single line, 0 Erlang is permanently free, while 1 Erlang means permanently busy.

Strowger manufactured the world's first "girlless and cussless telephone exchange," which was put into operation in La Porte, Indiana on November 3,

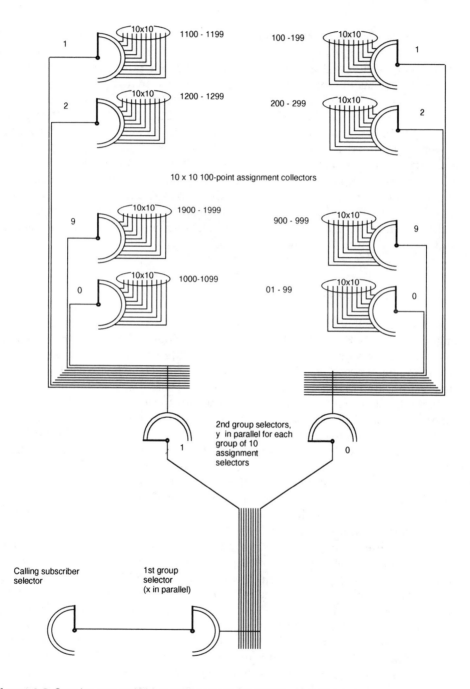

Figure 1.8 Step-by-step switching arrangement for 10,000 subscribers.

1892. His step-by-step two-motion (vertical and horizontal) switch with various improvements remained state of the art until crossbar switching in the early 1930s increased speed and reliability and reduced power consumption and noise. A first major improvement of the Strowger switch was made in 1924 by Siemens; instead of the two-motion (lifting, turning) switch controlled by individual stepper magnets, Siemens constructed a motor-driven purely rotary switch. At the end of a call the selector arm, which returned the same way as it went, continued to the end of the contact bank, thus ensuring equal wear and tear of the contacts and preventing oxidation of seldom-used contacts. In 1954 a further important improvement was made as Siemens replaced the copper contacts by a palladium-silver alloy, thus creating the precious metal rotary switch (in German EMD-Schalter stands for Edelmetal-Motor Dreh-wähler), which substantially improved performance and lifetime, especially under extreme climatic conditions [2,5,6].

1.5.2.1 Crossbar Switching

A patent on crossbar switching had already been issued in 1916 to the American engineers John G. Roberts and John N. Reynolds; the first major crossbar switching exchange was introduced in Sweden in 1923. As the name suggests, a lattice of rectangular crossed bars is involved. Each vertical bar carries a set of contacts that is connected to those carried on a horizontal bar when magnets are activated to move the two bars. Contact is made where the two bars cross, hence the name "crossbar." Its contacts will latch until the communication is over; that is, they will stay attached after the magnets are deactivated so that contacts can be made elsewhere by the other bars. The crossbar switch operates magnetically in such a way that it is free from moving brushes and sliding contacts. There are two main divisions in the switch: the *control* subsystem, which establishes the *talking path* within the application of a *marker,* and the *switching network* subsystem with the crossbars. With crossbar switching the dial pulses are temporarily stored in a *register.*

A modified version of crossbar switching came in the 1960s as the bars were replaced by reed relays. A reed relay (developed in the Bell Laboratories) is a small, glass-encapsulated, electromechanical switching device. A common switching control selects the reed relay to be closed in response to the number dialed. Pulses sent through a coil wound around the relay capsule change the polarity of plates of magnetic material alongside the glass capsules. The contacts open or close in response to the direction of magnetization of the plates, which is controlled by the polarity of the pulses. The contacts latch, so no holding current is required, and are forced to release upon terminating the communication. The latching reed relay, however, was not widely used because the switches easily got out of step with the controller and most reed switches required continuous energy [8,12].

1.5.2.2 Electronic Switching

Electronic switching was first introduced in the United States with a *common control* for crossbar switching. One controller, called a "marker" established calls through idle coordinate switch paths, thus considerably reducing, if not eliminating, the individual switch controls. The advent of integrated circuitry and microcomputers spurred the application of electronics to switching. The common control for electromechanical switching could be extended to *stored program control* (SPC) electronic switching. Just as in the field of computers, special languages have been formulated, such as the CCITT *high-level language* (HLL) for programming, *man/machine languages* for the control from teleprinter keyboard of the supervision and maintenance functions, and a *functional specification and description language.*

The first call through a SPC system was placed in 1958 at the Bell Laboratories and was followed by commercial service via a SPC-switch in Morris, Illinois in June 1960. While SPC and electronics brought improvements in the engineering of switching networks, they also brought in new telephone services such as abbreviated dialing, call forwarding, call waiting, and add-on. To distinguish the thus-far available services from the enhancements that SPC made possible, the phrase *plain old telephone service* (POTS) became an euphemism for electromechanical-switched networks.

At the end of the 1960s came a worldwide transition to electronic switching, and a quick technological evolution started with the following major solutions:

- First-generation equipment (1965–1975): *space division* SPC for switching of analog telephone, telegraph, and telex lines;
- Second-generation equipment (1970–1985): *time division centralized* SPC for switching of digital voice, image, and data lines with all control elements centrally located;
- Third-generation equipment (1985–): *time division distributed* SPC switching with call processing distributed toward the line modules, thus reducing the initial investment and adding control with the increase of capacity.

The term "distributed" also refers to remote switching in *remote switching units* (RSUs) located closer to large groups of subscribers. Whereas large SPC switches may serve 100,000 lines, RSUs can accommodate tens to as many as 5,000 lines. The RSUs may be located up to about 200 km from the central "host" switch. Usually part of the control remains in the host exchange. Some RSUs, called "stand-alone" RSUs, can take over from the host the major switching control for their lines in the event of emergency.

The network of telephone subscribers connected to automatic public telephone exchanges interconnected by transmission circuits is often referred to as the *public-switched telephone network* (PSTN). Until the middle of this century, telecommunications was almost identical with service on the PSTN. Still in 1981, in the United States over 95% of telecommunication charges came from the PSTN. The international telephone network currently has over 600 million telephone lines (compared with some 500 million cars) all over the world and as such is the largest infrastructure connecting mankind. Optimistic forecasters dare to predict that this figure, which is the result of 120 years of telephony, will double by the end of this century to around 1.2 billion!

Automatic switching between those 600 million telephone lines requires that each telephone subscriber is given a specific number, which is the number that must be dialed to move the selectors in the exchanges to the called subscriber. Initially this number could correspond with the order of application of new subscribers of the local exchange. With the introduction of national automatic switching, a code number for each local network had to be added to the local number; and with the introduction of international automatic switching, a country code was also added. To determine the country codes, international cooperation was obviously necessary. The ITU, therefore, divided the worldwide telecommunication network into nine switching regions, as shown in Figure 1.9. Each country code starts with the digit of its region.

To indicate to the exchange that an international connection is required a further combination of usually two digits is added to the telephone number. This combination is 00 in most countries, 01 in the United States, 19 in France (up to October 18, 1996), or 004 in Singapore. A complete telephone number for international direct dialing usually has 12 digits as follows:

00	Prefix for international direct dialing;
9x	Country in region 9, for example, 91 for India;
xxxx	National local network code;
xxxxxx	Telephone number of a subscriber in that local network.

The international telecommunication switching network is divided into three levels as shown in Figure 1.10.

The denomination "CT" is an abbreviation from the French designation *Centre de Transit.*

- CT1: Intercontinental network with intercontinental transit exchanges in New York, London, Moscow, Tokyo, and Sydney. These five transit

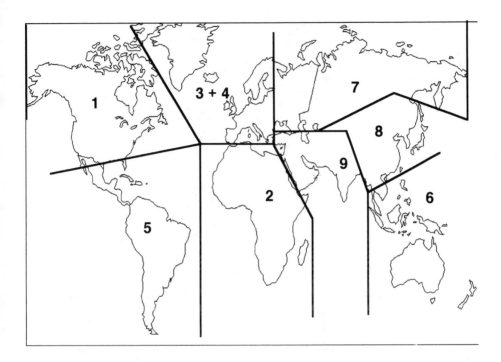

Figure 1.9 World telecommunication network divided by the ITU into nine switching regions. Regional border lines are approximate only; for example, the border between region 7 and the regions 8, 9, and 3 + 4 follows the political borders between the CIS states and the adjacent countries.

exchanges are all directly connected with each other, mostly by satellite and submarine cable.

- CT2: Continental network consisting of the international transit exchanges of all the countries of a continent. All those international transit exchanges are connected to their intercontinental transit exchange. Many of those CT2 exchanges are also directly interconnected.
- CT3: National exchanges with direct lines to national exchanges of other countries [13,14].

1.5.3 Telex Switching

Beyond telegraphy and telephony in the 1910s the teleprinter entered as another form of instantaneous telecommunication. Teleprinters became commonplace in the offices of companies and governmental organizations. With the addition of switching to this new form of telecommunication, teleprinter-switched net-

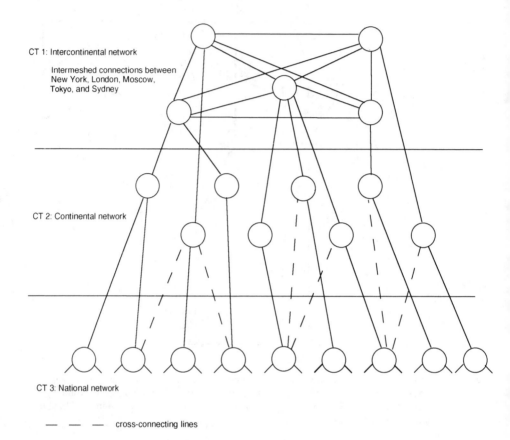

CT 1: Intercontinental network

Intermeshed connections between
New York, London, Moscow,
Tokyo, and Sydney

CT 2: Continental network

CT 3: National network

— — — cross-connecting lines

Figure 1.10 International telecommunication switching network.

works—abbreviated *telex* networks—developed so that company headquarters could communicate with their branch offices and with other companies equipped with teleprinters. Initially modified manual telephone switchboards were used. A unique feature of the telex service was that address information was supplied by the teleprinter keyboard rather than by a dialing disk. Although telex service is being replaced by the much faster data transmission, still some 1.2 million teleprinters worldwide are assisting in the execution of administrative and business transactions.

1.5.4 Data Switching

The advent of computers capable of communicating with each other evoked the introduction of a new type of network: the *public-switched data network*

(PSDN). The first computers were, and many small computers today still are, connected to the PSTN by means of a modem. Modems (an abbreviation of *mo*dulator-*dem*odulator) convert the digital computer signals into a coded combination of voice-frequency signals that then as analog signals can be transmitted via ordinary telephone lines. Modems are available for transmission speeds of 0.3 to 256 Kbps. The word "Bit," derived from "Binary Digit," denominates the unit of data—one bit standing for the smallest information quantity being an "on" or "off" signal. The unit *bits per second* (bps), also called "bit rate," denominates the speed of transmission. A digital telephone speech channel, for example, needs 64 Kbps; whereas the information contained in a color television channel needs 140 Mbps for uncompressed high-quality transmission. For large computers not operating at kilobits per second but megabits per second, the ordinary telephone line and the PSTN were much to slow; so data switching networks had to be created.

In the early 1940s an initial form of data switching evolved from the telex: coupling paper tape perforators and readers so that messages could be stored while awaiting the availability of transmission lines between the corresponding stations. Thus was born the concept of one-way data transmission and data storage at the switching center. This technique allowed delaying delivery of messages on a first-in first-out basis until transmission facilities became available. Fewer transmission circuits are thus required as those circuits can be occupied at an utmost rate. In 1964 this store-and-forward switching assumed a new dimension. To ensure secrecy of military communications, rather than receive-store-and-retransmit each entire message without interruption, the messages became *packetized,* hence enabling the continuous transmission of many partial messages. The message subdivisions were made uniform in size; an address was placed at the start of each subdivision, which became known as a *packet.* At the source, information messages are divided in several equally long packages that are transmitted across the network and reassembled at the point of destination. In the switching equipment the packages can be buffered and recombined with other groups of packages to obtain a more or less constant use of the available transmission links between the switches of a network.

One of the first packet-switched networks was ARPANET, now worldwide popular as INTERNET. This U.S. military *Advanced Research Project Agency Network* (ARPANET) was put into service in 1971, initially connecting UNIX computers at four military locations and soon extended to Harvard, Stanford, and other U.S. universities. The rationale of ARPANET was, and of Internet still is, *resource sharing,* initially on a national basis and one decade later internationally too. Stuttgart University (Germany) was the first foreign university connected by satellite to the ARPANET; many other universities worldwide joined, and to date over 60 million persons all over the globe share resources on the Internet.

Whereas UNIX computers were connected in the ARPANET, other computer companies developed protocols and communication architecture for their computers too. In 1971 IBM developed their *System Network Architecture,* which under the acronym SNA, is still widely used for communication between IBM and IBM-compatible computers.

DEC developed the less-known DEC-Digital Network Architecture used in DECnet.

In 1974 XEROX developed ETHERNET, a low-cost wideband way of sending packets of data between office machines, printers, and computers in so-called *local-area networks* (LANs). The Ethernet protocol is based upon ALOHA-random-access, as described in Subsection 7.2.4.

Rather than staying with a few proprietary data-handling protocols, *comité consultative international téléphonique et télégraphique* (CCITT) in 1976 defined a general applicable protocol for a packet switching data interface called X.25. In the meantime, this protocol has been extended to a family of protocols from X.1 to X.34 for handling data communication at speeds of 4.8 to 64 Kbps. In 1984 CCITT standardized X.400 as the standard for electronic message handling, followed by X.500 in 1988 as a global directory service. In the same year *Frame Relay* was standardized as a protocol mainly for LAN-to-LAN data traffic up to 2 Mbps. *ISDN* (integrated services digital network), described in Subsection 1.5.9, will evolve globally for voice, image, and data transmission and switching, eventually making the aforementioned protocols and separate data switching networks superfluous.

The latest add-on to data switching comes from mobile data. International mobility supported by satellite and cellular radio transmission acquire data to be immediately and globally accessible. Data switching is being implemented on the international cellular radio system *Global System for Mobile radio* (GSM) and will be available on the future *low Earth orbit* (LEO) satellite systems described in Section 10.6 [15].

1.5.5 Signaling

Signaling from a calling subscriber to a central office (subscriber loop signaling) was formerly done using dc current. By lifting the telephone (off-hook status), the calling subscriber closed the dc circuit originating from the central office. Turning the dial disk created a pulse-type interruption of the dc circuit: digit 1 causing one interruption and digit 2 two interruptions, for example. Ringing from the central office to the called subscriber usually is done with a 16-2/3-Hz or 25-Hz signal that may be pulsed in a 5-sec sequence. Signaling between the central offices (trunk signaling) once was *single-frequency* (SF) pulse-tone signaling, employing either in-band or out-band signaling (band standing for the 300- to 3,400-Hz voice spectrum). The in-band signaling was mainly at 2,280, 2,400, 2,600, or 3,000 Hz and out-band signaling at 3,825 Hz or 3,850 Hz above the speech band or in some cases below the speech band with a 50-Hz frequency taken from the mains and thus saving a signal generator. Dual-frequency in-band signaling was later introduced using the frequencies at 2,040 and 2,400 Hz.

The rotary-type dialing with the dc signal interruption process is slow and tiresome. To ease dialing and reduce holding time on registers in the central offices, in the early 1960s push-button dialing was introduced with *dual-tone multifrequency* (DT-MF) signaling. A combination of two continuous tones out of eight were allocated to each digit for the dialing. The excess combinations over the 10 required for dialing are used for special signals. For trunk signaling another group of six frequencies was introduced. Figure 1.11 shows both dual-tone MF signaling codes.

Until the late 1970s speech and associated signaling, which was transmission of dialing only, were kept together in the same channel from subscriber to subscriber throughout the switching network either in-band or out-band. With the advent of the SPC-switching it became advantageous to combine the signaling for groups of calls in a separate so-called *common channel signaling* (CCS) network within the public switching network. This CCS can now combine the signaling information relating to a group of telephone and or data circuits with additional network supervisory information. Both the signaling and supervisory information are not in a sequence of tones but digitally coded as messages. Due to this coding and the higher available bandwidth on the separate channels, much more detailed information about the call, the routing, and the type of service, for example, can be processed across the network. Other advantages of CCS are the faster signaling and the simultaneous application of the signaling in both directions during conversation.

The latest version of CCS is the *CCITT signaling system No. 7* (SS7). It operates in digital networks on 64-Kbps channels and supports ISDN and cellular mobile radio networks.

The evolution from local public automatic telephone switching and *international direct dialing* (IDD) took many years for both technical and political reasons. The first IDD started very modestly in 1958 between two neighboring towns separated by the river Rhine: Basel in the north/west of Switzerland and Lörrach in the south/west of Germany. Belgium and Germany took up IDD in the same year and in 1970 intercontinental IDD started between the United States and Europe [4,12].

1.5.6 Intelligent Networks

Privatization and deregulation of the public network operators result in more competition and the subsequent necessity to use the existing network more

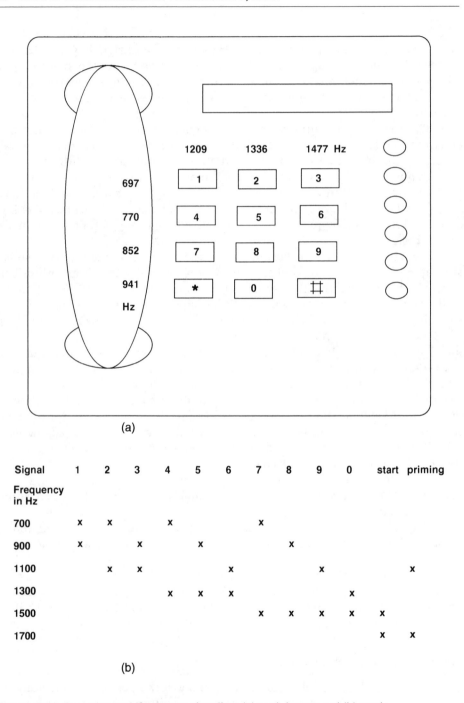

(a)

Signal	1	2	3	4	5	6	7	8	9	0	start	priming
Frequency in Hz												
700	x	x		x			x					
900	x		x		x			x				
1100		x	x			x			x			x
1300				x	x	x				x		
1500							x	x	x	x	x	
1700											x	x

(b)

Figure 1.11 Dual-tone multifrequency signaling: (a) push-button and (b) trunk.

efficiently by adding new services. The integration of computers into switching facilitated the transition of PSTNs to *Intelligent Networks* (INs). The concept of IN is to enable as many service providers as possible to offer their services via the telecommunication network and thus significantly increase the revenue-yielding traffic.

In the United States, almost 30% of all phone calls are already using "freephone" numbers (800 numbers) where the called party picks up the cost for the phone call thanks to close interaction between computers and telephone exchanges. Freephone is one of the best known of a whole list of services offered by INs, which is also called *Advanced Intelligent Network* (AIN) in the United States. There is a close relationship between AIN deployment and the U.S. telecom deregulation. The deregulation efforts that started in 1984 have created a very competitive telecom market in North America, leading to the birth of several independent service providers.

To support a harmonious worldwide deployment of IN, the ITU worked out standards for a set of IN core capabilities in recommendation Q.1211, "Introduction to Intelligent Network Capability Set 1," referred to as the international IN capability set CS-1. CS-1 was published in December 1993 and will be followed by CS-2 in 1997.

Whereas in the past all services had to reside in each local exchange that offered the services, in an IN the traffic handling is separated from the service handling. Therefore, centralized service handling is possible. Less equipment is involved, and new services or service modifications can be implemented more easily and quickly without disturbing traffic handling. Services in the context of IN are divided between conversational services and distribution services.

The category of conversational services includes services whereby a subscriber communicates with a second subscriber or with a remote database or computer in a bidirectional way, that is, information flowing in both transmission directions.

The category of distribution services includes the widescale distribution of television and high-fidelity audio channels (stereo) to residential subscribers. The information flow is in one direction only from the service provider to the subscriber, apart from a small amount of operation and response-information that is possible flowing from the subscriber to the service provider [16].

1.5.7 ISDN

Switching and transmission networks are currently undergoing a worldwide transition from analog to digital. Although installed analog equipment might still be used for another 20 to 30 years, the production of switching and transmission equipment has practically fully changed from analog to digital. One obstacle for a fully integrated digital telecommunication network is now being removed: the telephone itself, still analog, will become digital too—the solution being ISDN.

With ISDN the access network between local switch and the subscribers' telecommunication equipment becomes digitized too. The main purpose of ISDN, however, is not the digitalization of the telephone and the access network but to combine the various existing public telecommunication networks for telephone, data, telex, and video into one single integrated network. Digital signals basically are all the same in that they simply consist of combinations of "on" and "off" octets, so why not a single network to handle them? ISDN, therefore, supports any public telecommunication service on the basis of digital transmission and switching with clusters of various terminals or a multimedia terminal working on a single subscriber line. Preferably, ISDN should be introduced without replacing the existing access network, which worldwide mainly consists of ordinary telephone cable. Therefore, ISDN will be introduced in two stages.

- First stage: *narrowband-ISDN* (N-ISDN);
- Second stage: *broadband-ISDN* (B-ISDN).

The distinction between narrowband, wideband, and broadband signals is made in terms of bit rates. Narrowband is any signal up to 64 Kbps. Signals between 64 Kbps and 2 Mbps are commonly called wideband, while signals above 2 Mbps are called broadband.

The introduction of N-ISDN will be possible on the existing copper-line access network using special ISDN *line terminations* (LT) on the exchange side and *network terminations* (NT) on the subscriber side. For B-ISDN with transmission speeds of 155 or 622 Mbps normally an optical fiber access cable will be required, but broadband radio-relay or coaxial cable systems are possible too.

To introduce ISDN, new allocations had to be made on the data and signaling content of the digital information transmitted in the access network. Therefore, CCITT defined a number of channels.

- "A" Channel: the traditional analog voice channel with a bandwidth of 300 to 3,400 Hz.
- "B" Channel: the fundamental digital information channel operating at 64 Kbps and capable of carrying data and digitized voice and image.
- "D" Channel: primarily defined as a signaling channel to control the route for the subscribers' communication across the network. The D Channel is defined for two transmission rates of 16 and 64 Kbps, depending upon the type of ISDN access for low or high volume traffic. D Channels may be

used in an initial phase for connecting PABXs (see Subsection 1.5.9) via 2-Mbps links to the public ISDN.

- "E" Channel: an alternative type of 64-Kbps signaling channel based upon the SS7 signaling.
- "H" Channels: the higher transmission-rate channels envisaged for use in wideband and broadband ISDN.

H Channels are subdivided as follows.

- H0 channel operating at 384 Kbps;
- H1 channel further divided into the H11 channel operating at 1,536 Kbps and the H12 channel operating at 1,920 Kbps;
- H2 channel operating at approximately 30 Mbps;
- H3 channel operating at approximately 70 Mbps;
- H4 channel operating at approximately 140 Mbps.

For N-ISDN two types of subscriber access are defined by CCITT.

- Basic Access: comprising two B channels for voice and data and one D16 channel for signaling and optionally low-speed packet data. Total capacity is $2 \times 64 + 16 = 144$ Kbps. (For synchronization, timing and control, additional bits are added, resulting in a total transmission rate of 192 Kbps.)
- *Primary Rate Access* (PRA): comprising 30 B channels for voice and data and one D64 channel for signaling and a further 64-Kbps channel for synchronization, timing, and control. Total capacity is $30 \times 64 + 2 \times 64 = 2,048$ Kbps, which is equivalent to the 2-Mbps primary level for 32 PCM channels.

Three different access arrangements are shown in Figure 1.12.

The access arrangement for the basic access mainly used for private or small business subscribers supports point-to-multipoint operation where up to eight subscriber terminals (such as telephone, fax, PC, and video phone) can be connected to a bus and two terminals can be used simultaneously. For small business subscribers a point-to-point access arrangement supports one or a group of *n* ISDN lines connected to a PABX, whereby *n* usually will be smaller than 30 as otherwise the PRA arrangement supporting point-to-point operation for medium to large business subscribers with a digital PABX would be the appropriate solution. As an example, Figure 1.13 shows how ISDN switching can be arranged in a modular distributed control switch architecture. This switch architecture allows adding additional functions or deleting those no longer required. The integrated packet trunk module interfacing with the PSDN,

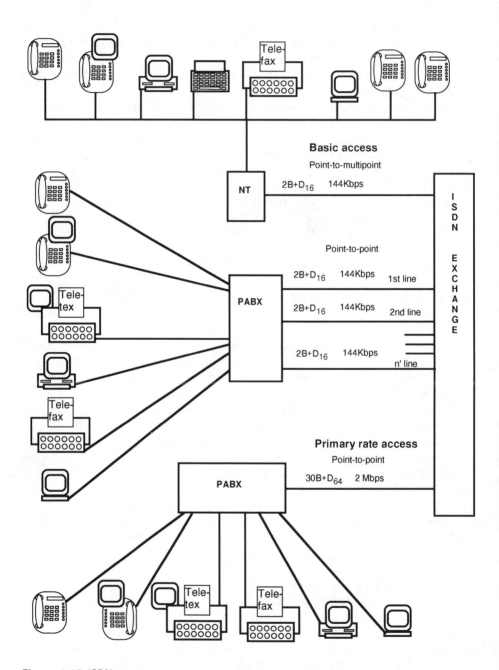

Figure 1.12 ISDN access arrangements.

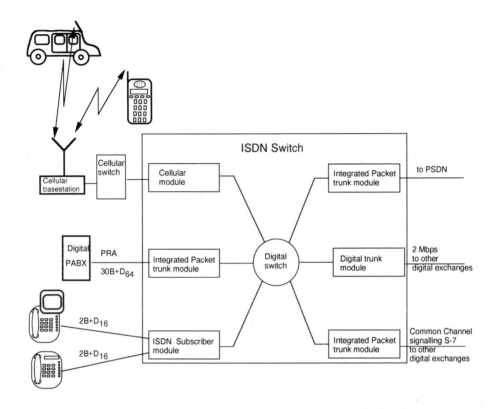

Figure 1.13 Typical ISDN switch with modular distributed control (based upon Alcatel, System 1000 S 12).

for instance, is required only for the transitional stage during which data switching is not yet fully integrated in the ISDN; it can be deleted at a later stage when separate switches for PSDN have disappeared.

The commercial operation of N-ISDN started in Japan on NTT's Tokyo-Nagoya-Osaka network in 1988.

For B-ISDN again two types of subscriber access are defined by CCITT.

- 155-Mbps symmetrical (symmetrical here stands for 155 Mbps in both transmission directions) with an electrical interface for a coaxial cable pair covering a distance of at least 100m and an optical interface for optical fiber cable covering at least 800m;
- 622-Mbps symmetrical and optionally an asymmetrical interface with 622 Mbps toward the subscriber and 155 Mbps from the subscriber.

1.5.8 Broadband Switching

The introduction of various new services within a network require switching equipment operating at higher speeds and thus higher bit rates. Typical new services requiring broadband switching include

- Desktop publishing;
- Medical imaging;
- Video (library) retrieval;
- Color facsimile;
- *Computer-aided development / computer-aided manufacturing* (CAD/CAM);
- Multimedia service (voice, text, graphics, and moving pictures);
- Video conferencing;
- HiFi music and High-Definition TV.

Digital networks used to operate 64-Kbps channels with the switching (and transmission) based on the so-called *synchronous transfer mode* (STM). In this mode a fixed number of bits is periodically available to each connection within a network. That implies that the capacity of the connection is constant even if the information flow is bursty, as is usual with computer communication. STM, therefore, with the increase of data transmission, may lead to a waste of capacity; moreover, in STM (which was developed mainly for transmission) the switching functions are difficult to handle at different bit rates (such as 155 and 622 Mbps).

In the interest of more flexible broadband switching, therefore, a new *asynchronous transfer mode* (ATM) has been developed that can handle traffic relating to services that require widely differing bit rates. In ATM basically the information is put in fixed-length *cells* that are switched and transported through the broadband network and at the point of destination reconstituted in its original synchronous form. Figure 1.14 shows this cell structure as defined by CCITT.

It consists of a 5-octet header field and a 48-octet information (also called "payload") field. The header field contains the data for the routing and control of the payload through the telecommunication network. For this purpose the cell header has a *virtual path identifier* (VPI) and a *virtual channel identifier* (VCI). The combination of VPI and VCI identifies the cell within a network. Figure 1.15 demonstrates what is meant by those two virtual connections. A *virtual channel* (VC) is a usually semipermanent, software-routed communication channel between end points of an ATM link. A *virtual path* (VP) carries a group of VCs, all sharing the same VPI. Virtual path connections enable the creation of virtual trunks between two users in virtual private networks and

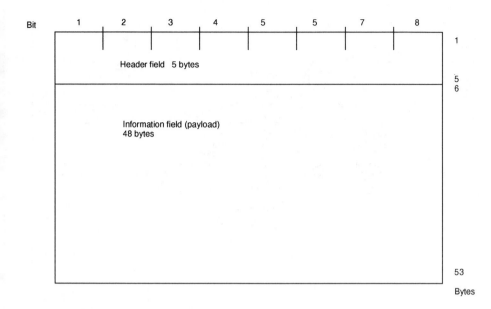

Figure 1.14 ATM cell structure.

enable the operator to transport aggregate traffic within network nodes. Both the VC and the VP have a variable bit rate capacity. The desired VP routing is obtained by *cross-connecting* (CC) VPs within an ATM node. When a VC is switched from one VP to another, its VPI is changed. Figure 1.16 demonstrates the principle of the (self-)routing of cells in an ATM broadband switch. The incoming ATM cell contains a header with a specific VPI/VCI pair related to the incoming link. Under instruction of the *call control*, the *header processor* translates the incoming VPI/VCI pair into an outgoing VPI/VCI pair and adds a pair of internal routing tags T1 and T2. With T1 the cell will be guided through the first switching stage and with T2 through the second.

For most services the stream of ATM cells in either direction is nonperiodic. Where a source produces continuous output, the ATM cell stream will be quasi-periodic, while for other types of communication the flow of cells in both directions may occur in bursts with idle periods of unpredictable duration in between. Due to this nonperiodic nature of ATM, the cells originating from different sources with highly different information rates may easily be combined and switched, thereby avoiding the inflexibility of STM. When an information source is temporarily not producing any output, no ATM cells are produced; consequently the occasional waste of capacity associated with STM is avoided. Furthermore, if temporary peaks occur in the output of an information source, a properly dimensioned broadband network will be able to cope with the

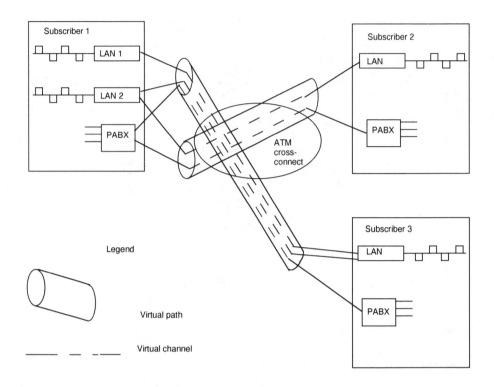

Figure 1.15 ATM virtual path and virtual channel.

resulting increase in ATM cells. This flexibility enables an operator to offer *bandwidth-on-demand* (BOD). ATM thus has some similarity with conventional packet switching, however, operating at a much higher speed and capable of handling voice as well as data. Figure 1.17 shows a simplified schematic of a typical broadband switch, still incorporating narrowband services.

At Telecom 95 NEC unveiled an ATM switching product program for public switching at speeds ranging from 10 to 160 Gbps!

1.5.9 Private Switching

Private switching refers to switching effected on the premises of a public network subscriber to support

- The internal communication within the subscriber's own organization;
- The external communication with the PSTN in a time-sharing mode among the users of the private switch.

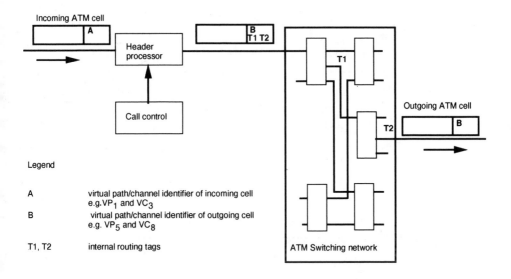

Figure 1.16 Routing of cells in an ATM switch.

Private switching is usually referred to as *private branch exchange* (PBX), respectively, *private automatic branch exchange* (PABX), as compared with *central office* (CO) for a public switch.

PBXs/PABXs are similar to COs except that they include only a few of the operational and network management functions. In fact, a PABX used to be a scaled down CO switch. On the other hand, a PABX usually provides many functions and features not available in the PSTN, such as

- Office automation facilities;
- Data processing applications;
- Restricted dialing into the PSTN;
- Automatic route selection on the PSTN;
- Radio connectivity (cordless service);
- Generation of billing and traffic reports.

Large corporate users that have office facilities in several buildings within a city or at several distant locations can operate for intercompany use a private network of interconnected PABXs located at the various LANs. Interconnection between the PABXs of a LAN is normally via leased lines from the public network. With increasing deregulation and privatization of the public operators, the use of proprietary links between the PABXs may become possible and advantageous.

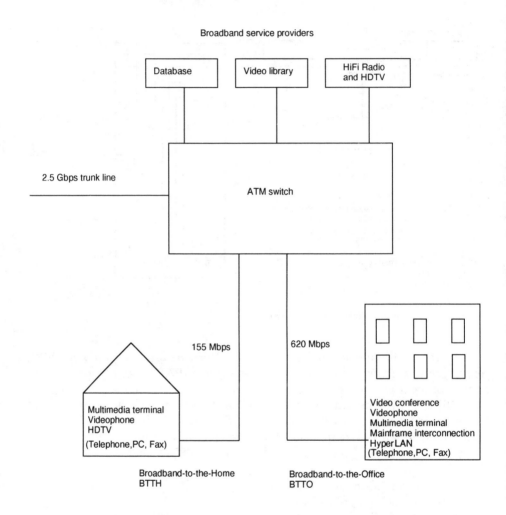

Figure 1.17 Typical broadband switch.

The present convergence of telephony and computer technology is thoroughly changing the features of the PABX. Computer application is certainly taking place, especially at the premises of large companies, organizations, and institutions—the domain of private switching; so data switching in parallel to telephone switching is rapidly expanding. The intelligence and processor power of PABXs is therefore moving from telephone operation to data handling. Data switching, however, means packet switching; whereas telephone switching is a constant bit rate switching. Both data and telephone switching can be supported by ATM, so PABXs are very likely to evolve into ATM switching. On the other hand, ATM offers an advantage mainly to the telephone traffic that is routed on an outside transmission line to the public PSTN or the corporate LAN or *wide-area network* (WAN) but not for the high volume of telephone traffic that remains within the local area of the PABX. Similarly for the public networks, where quick bandwidth increases and new facilities are introduced by means of high-speed overlay networks, special corporate ATM switches are appearing for private switching, as shown in Figure 1.18. In such corporate ATM switching, an "overlay" ATM switch takes care of the data switching and the outside telephone switching as a gateway to a LAN, WAN, MAN, or ISDN network but keeps the handling of the local telephone switching with the existing PABX. Eventually with new investments and once appropriate standards for voice on ATM have matured, the local voice switching function will evolve into the corporate ATM switch too [17].

1.5.10 CENTREX

CENTREX (an abbreviation of "Central exchange") is an alternative solution for a PABX or even a LAN that provides intercompany communication with a

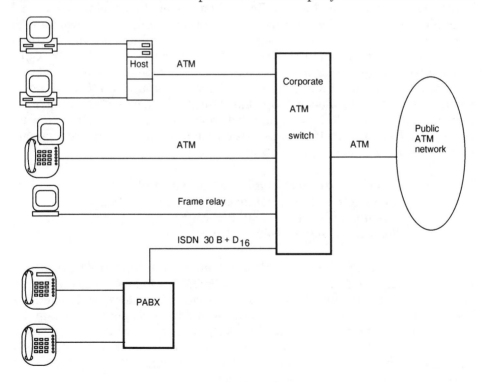

Figure 1.18 Corporate ATM switching.

minimum of investment. Instead of investing in one or more PABXs, a public network subscriber can permanently hire switching capacity at a CO switch. This service can be economical for medium or large business customers that have a number of business locations in a single urban area, such as multibranch banks, scattered municipality offices, and retail chains with many stores in one city. CENTREX users economize on space requirements and operating and maintenance staff and material and can enjoy updates without investment. Wide-area CENTREX offers CENTREX service for customers with multilocations using various CO switches.

1.6 TELECOMMUNICATION MANAGEMENT NETWORK

The *telecommunication management network* (TMN) is not another public switching network like PSTN, PSDN, or ISDN but an additional network—on top of the previously mentioned public networks—to manage those networks. The public networks historically have their own independent (mainly noncompatible operator and vendor-specific) network operation, maintenance, and management systems. The liberalization of telecommunication networks with the rapid introduction of competing new networks and new services such as cellular radio, mobile satellite services, video-on-demand, and multimedia, together with an increasing network complexity due to the introduction of new transmission and switching technologies have encouraged the ITU to define common standards for network operation under the name TMN. A *network management forum* (NMF) was founded to develop a basic architecture for managing telecommunication networks consistently and efficiently using common protocols. The major protocols are

- *Common management information protocol* (CMIP);
- *Common management information service element* (CMISE);
- *Remote operation service element* (ROSE);
- *Association control service element* (ACSE).

A TMN basically consists of a *data communication network* (DCN); *mediation devices* (MD); an *operation system* (OS); and the to-be-managed elements of the PSTN, PSDN, and ISDN. The mediation devices store, route, and link CMIP management traffic between the various network elements such as multiplexers, cross-connects, and cellular base stations. A specific hierarchical "manager-to-agent" model with four management layers facilitates an orderly allocation of management competence. Figure 1.19 shows the four layers and illustrates an allocation of management competence.

TMN basically covers following functions:

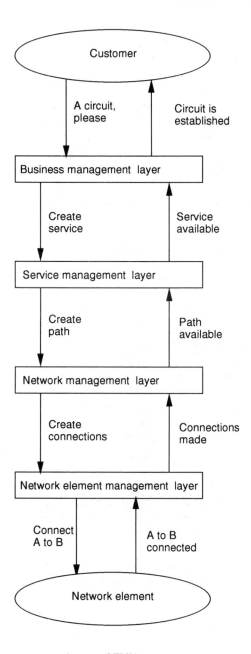

Figure 1.19 The four management layers of TMN.

- *Network configuration:* provisioning of trunks, groups of trunks, trunk testing and routing, and remote cross-connecting of links;
- *Traffic management:* traffic control, traffic evaluation, planning, and internal administration accounting;
- *Network surveillance:* fault detection and diagnosis, organization of fault clearance and maintenance, and security management;
- *Subscriber management:* provisioning of services, provisioning of equipment and access lines/ports, subscriber fault clearance, and administration and billing.

In an "electronic bonding" procedure, administrative traffic between operators can be carried over the TMN, too, instead of through the public networks. A further target for management integration into the TMN are the customer facilities ranging from LANs to set-top boxes for video-on-demand services [18,19].

1.7 TRANSMISSION

The word *transmission,* from the Latin "trans mettere" for transfer or transport in the figurative sense, quite confusingly, is used for many purposes. Probably first in the Industrial Revolution as *transmission system* for the transmission of power from a central steam engine to the various production machines in a factory. In electrical power technology, *high-tension* (HT) *transmission line* and *HT-transmission grid* are well-known names for high-voltage power overhead electricity distribution. In the book *Transmission Systems for Communications* published by members of the technical staff of the Bell Telephone Laboratories in 1954, which used to be the "bible of transmission," transmission is described as: "The primary function of a transmission system is to provide circuits having the capability of accepting information-bearing electrical signals at a point and delivering related signals bearing the same information to a distant point."

In this book transmission within the context of telecommunications is concisely defined as: *the technology of information transport.* In the context of telecommunications, a transmission system transports information between a source of a signal and a recipient. Transmission thus executes the "tele" part of the word "telecommunications" and as such is the basis of all telecommunications systems.

Transmission systems are applicationwise divided into three groups as follows:

- *Corporate transmission systems:* connecting office telecommunication equipment usually via a PABX with the access network;

- *Access transmission systems:* connecting the terminal equipment of individual subscribers or PABXs of corporate users through local loop systems with the nearest local public exchange;
- *Transport transmission systems:* providing the interconnection between local, regional, national, and international exchanges.

Figure 1.20 presents a basic telecommunication network with indication of the three different applications.

Technologywise transmission equipment is divided into line transmission, for transmission via open wire lines and cable, and radio transmission, for wireless transmission.

Line transmission is the technology of sending electrical signals via copper wire, and nowadays increasingly via optical fiber pairs, on overhead lines, in underground, and in submarine cables by means of line transmission equipment. Line transmission equipment serves to combine, send, amplify, receive, and separate the electrical signals in such a way that the long-distance transmission of information is possible. The line transmission technology is described in Chapters 3 and 4.

Radio, in the context of telecommunications, stands for the technology of information transmission by means of electromagnetic waves. Short-wave radio propagation is of limited interest and therefore not further covered in this book, whereas the other versions of radio transmission are described in detail in Chapters 5 to 7. Figure 1.21 indicates the frequency bands in which the radio transmission systems operate [20].

1.8 TWO HUNDRED YEARS OF TELECOMMUNICATIONS

Telecommunications, starting with optical telegraphy during the French Revolution on August 15, 1794, got its first major growth with the introduction of electrical telegraphy in the 1840s and obtained the interest of the general public with the invention and implementation of the telephone by A. G. Bell in 1876. For more than 100 years telecommunications was almost identical to telephony. At its centenary telephony provided over 75% of worldwide total telecommunication services, and over 90% of telecommunications services were provided via fixed line networks.

The advent of transistorization, very large scale integrated electronic engineering, computerization, and optical transmission provided the basis for an almost revolutionary development of telecommunications in the last quarter of this century. Digital optical transmission bandwidth, and thus transmission capacity, is no longer an issue; wireless technology ensured that people can get their information wherever they are moving on the globe or however isolated they might be living far away from civilization's infrastructure.

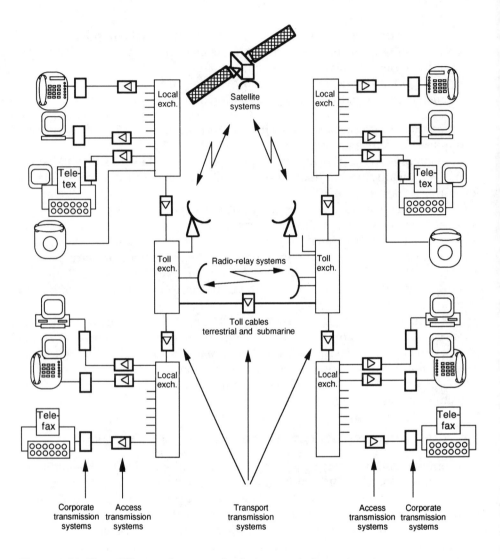

Figure 1.20 Three different telecommunications transmission systems.

The major steps in this evolution of telecommunications from semaphore to *broadband-to-the-person* (BTTP) are shown in Figure 1.22.

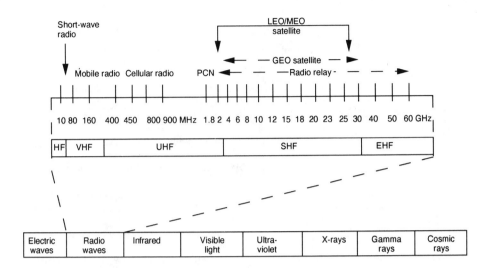

Figure 1.21 Radio transmission systems.

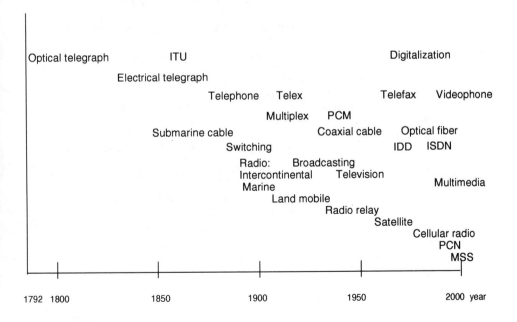

Figure 1.22 Major steps in the 200-year evolution of telecommunications.

References

[1] Bertho, Catherine, *Histoire des Télécommunications en France,* Toulouse, édition Érès, 1984.

[2] Michaelis, Anthony R., *From Semaphore to Satellite,* Geneva: International Telecommunication Union, 1965.

[3] Aschoff, Volker, *Geschichte der Nachrichtentechnik,* Band 2, Berlin, Heidelberg, and New York: Springer-Verlag, 1995.

[4] Reuter, Michael, *Telekommuniation Aus der Geschichte in die Zukunft,* Heidelberg: R.v. Deckers's Verlag, G. Schenk, 1990.

[5] Gööck, Roland, *Die großen Erfindungen Nachrichtentechnik Elektronik,* Künzelsau, Sigloch Ed., 1988.

[6] Young, Peter, *Person to Person: The International Impact on the Telephone,* Cambridge: Granta Editions, 1991.

[7] Bell, Alexander Graham, *Improvement in Telegraphy,* United States Patent Office, Letters Patent No.174,465, dated March 7, 1876; application filed February 14, 1867, Salem, Massachusetts.

[8] Joel, Amos E., Jr., "The Past 100 Years in Telecommunications Switching," New York, *IEEE Communications Magazine,* Vol. 22, No. 5, 1984, pp. 64–70.

[9] Breguet, Claude A. J., "The Breguet Dynasty Two Centuries of Interdisciplinary Scientists and Engineers," *Interdisciplinary Science Reviews,* Vol. 5, No. 2, 1980, pp. 149–164.

[10] Brodbeck, Didier, *Journal Imaginaire d'Abraham-Louis Breguet,* La Conversion/Lausanne (CH): Editions Scriptar S.A., 1990.

[11] Libois, Louis-Joseph, *Genése et Croissance des Télécommunications,* Paris: Masson S.A., CNET-ENST, 1983.

[12] Minoli, Daniel, *Telecommunications Technology Handbook,* Norwood, MA: Artech House, 1991.

[13] Chapius, Robert J., "Present status and trends in digital switching," *Telecommunication Journal,* Vol. 60, IV/1993, pp. 161–167.

[14] Siegmund Gerd, *Grundlagen der Vermittlungstechnik,* Heidelberg: R.v. Deckers's Verlag, G. Schenk, 1993.

[15] Green, P. E., Jr., "Computer Communications: Milestones and Prophecies," *IEEE Communications Magazine,* Vol. 22, No. 5, 1984, pp. 49–63.

[16] Petterson, Gunnar, "Intelligent Networks—The Key to Advanced Telephony Services," *Telecommunications,* Vol. 29, No. 12, 1995, pp. 55–61.

[17] McGarvey, Brendan, "Where PBX Meets ATM," *Telecommunications,* Vol. 29, No. 9, 1995.

[18] Taschenbuch der Nachrichtentechnik, Ingenieurwissen für die Praxis, Alcatel SEL AG, Berlin: Schiele & Schön, 1994.

[19] Graham, Moore, "TMN: Network Management's Golden Thread," *Telecommunications,* Vol. 30, No. 5, 1996, pp. 55–58.

[20] Members of the Technical Staff, *Transmission Systems for Communications,* Winston-Salem: Bell Telephone Laboratories, Inc., Revised 4th Edition, 1971.

Multiplex 2

2.1 INTRODUCTION

The basic function of a transmission system is to get information from one place to another with reasonable quality and cost. Rather than using individual pairs of wire for each channel, specific numbers of *voice-frequency* (VF) channels for telephony and data are grouped together in multiplex equipment (much like wagons in a train or parcels in a container) before they are sent over the transmission media. The transmission media are open wire, cable, radio relay, and satellite. Multiplex equipment is not a transmission medium but a prerequisite for transmission via a transmission medium. In principle, the type of multiplex equipment used is independent of the chosen transmission medium apart from the capacity, which needs to be low for transmission via open wire and can be extremely high for transmission via optical fiber cable.

Multiplex equipment is also called "carrier equipment" as individual channels and groups of channels are "carried" on specific frequencies. In digital multiplex equipment individual channels or groups of channels are allocated in specific time slots. Instead of the names analog and digital multiplex the more explanatory names *frequency division multiplex* (FDM) and *time division multiplex* (TDM) are now commonly used.

The superior quality of digital transmission has technically outdated the analog transmission systems. The production of analog multiplex equipment has practically stopped, it can be expected, however, that installed FDM equipment will still remain in operation for many years. For the sake of completeness and to make it easier to understand the multiplexing technology, FDM is still briefly described at the beginning of this chapter.

Digital multiplexing is then described in detail, followed by high-capacity optical fiber *wavelength division multiplexing* (WDM) systems. This chapter further features the description of transmission equipment that is neither multiplex equipment in the classical sense nor transmission media equipment but is similar to multiplex equipment interfaces between telecommunications (switching, terminal, or even multiplex) equipment and the transmission media, such as

- Circuit multiplication equipment for more economical use of long-line channels;
- Inverse multiplexing equipment for the transmission of a digital broadband signal over a number of narrowband digital channels;
- Data transmission interface equipment: modems for analog transmission of digital data signals and *data circuit-terminating equipment* (DCE) interfacing digital data terminals with digital data networks;
- Codecs for transmission of video and sound signals.

Figure 2.1 shows typical applications of multiplex and transmission media interface equipment (the numerals in the relevant equipment boxes refer to the respective subsections of this chapter). For the sake of simplicity, the special analog-to-digital transmultiplexers, which are required to connect analog multiplex groups with a digital exchange (or digital transmission medium), have not been shown in this drawing and should here be considered as part of the analog multiplex boxes.

2.2 ANALOG MULTIPLEX

In analog multiplexing, several VF channels, which are transmitted via common transmission media and come from an exchange or in the access network from the PABX of large subscribers, are connected parallel to multiplex channel modulation equipment. This channel modulation equipment first limits the bandwidth of each VF channel to 4 kHz for the accommodation of a 300- to 3,400-Hz speech band and a signaling channel. The signaling channel—either inband and thus at a frequency within the speech band; or outband, for example, 3,850 or 3,825 Hz—is added in the exchange to each speech channel for the transmission of the relevant dialing, calling, and switching criteria, thus creating the VF channel. With this VF channel as the starting point, historically two different multiplexing technologies emerged: pregroup translation and single-channel translation.

The pregroup translation technology combines three VF channels by modulating each VF channel on a HF carrier that is spaced 4 kHz from the next carrier in the same pregroup, for example, on 12, 16, and 20 kHz. For the VF channel translation, single-sideband modulation with a surpressed carrier is applied so that each carried VF channel occupies only 4 kHz instead of 8 kHz. Apart from reducing crosstalk, the carrier surpressing has the advantage that correct channel demodulation is independent of the exact phase of the demodulating carrier generated at the receive side. The carriers and the lower sidebands

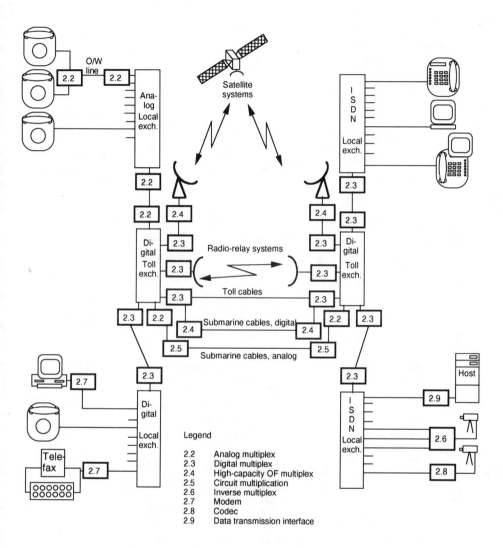

Figure 2.1 Typical multiplex applications.

are filtered out, resulting in a 12- to 24-kHz pregroup band. Four such pregroups are then modulated on four HF carriers spaced 12 kHz apart, for example, on 84, 96, 108, and 120 kHz, resulting in a 60- to 108-kHz basic group. An example of pregroup modulation is shown in Figure 2.2(a). Pregroup translation has been dictated by the engineering that prevailed at the time when the original equipment was designed, especially in view of the industrial production of crystals for exact frequency-generating and effective filters for specific fre-

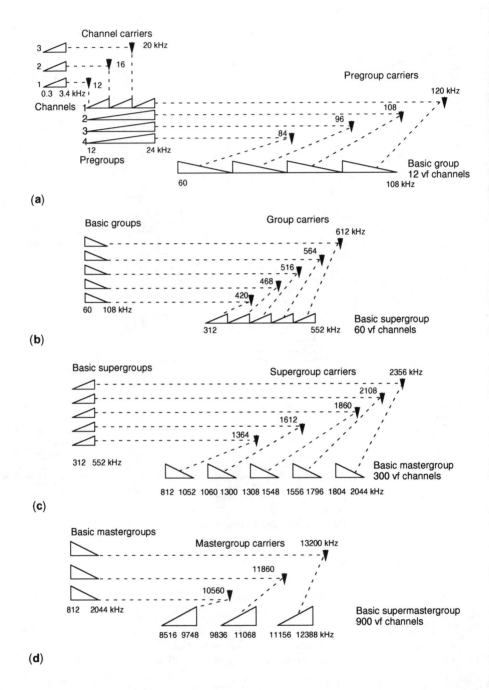

Figure 2.2 CCITT FDM basic group compositions: (a) group, (b) supergroup, (c) mastergroup, and (d) supermastergroup.

quency bands. To accommodate 12 VF channels within one group, by means of the intermediate modulation of three VF channels in one pregroup, only three channel carriers and four pregroup carriers are required instead of twelve if each VF channel has been modulated by a separate carrier.

The single-channel translation applies an individual carrier for each of the twelve channels. An example is the Bell system applied in North America and some parts of Asia, in which channel 1 is modulated with a 108-kHz carrier. The carrier and the upper sideband (108 to 112 kHz) are filtered out, and the lower sideband (108 to 104 kHz) is placed into the basic group. Similarly channel 2 is modulated with a 104-kHz carrier; channel 3 with 100 kHz; and so on down to channel 12, which is modulated with a 64-kHz carrier and placed at the beginning of the likewise 60- to 108-kHz basic group. Single-channel translation saves one translation stage and thus creates less modulation noise and signal distortion.

Starting from the aforementioned 12-channel basic group, two higher translation schemes emerged: CCITT-standardized analog multiplex and Bell analog multiplex system.

2.2.1 CCITT-Standardized Analog Multiplex

Beyond the 12-channel basic groups, CCITT has defined basic supergroups for 60 channels, basic mastergroups for 300 channels, and basic supermastergroups for 900 channels, as summarized in Table 2.1.

The multiplexing according to CCITT recommendations of five basic groups to a basic supergroup is shown in Figure 2.2(b). The subsequent multiplexing in accordance with CCITT recommendations of five basic supergroups to a basic mastergroup is shown in Figure 2.2(c), whereas Figure 2.2(d) shows the multiplexing of three basic mastergroups to a CCITT basic supermastergroup.

The translation of 300 VF channels in one basic mastergroup thus requires three channel carriers, four pregroup carriers, five group carriers, and five supergroup carriers—thus, in total only 17 carriers instead of 300. Consequently,

Table 2.1
CCITT Basic FDM Groups

CCITT Denomination	Number of VF Channels	Frequency Band (kHz)
Basic group	12	60–108
Basic supergroup	60	312–552
Basic mastergroup	300	512–2,044
Basic supermastergroup	900	8,516–12,388

the limited variety of carrier generators, translators, and band filters enabled a more economical production, fewer spare parts, and lower maintenance costs.

Figure 2.3 shows the principle arrangement of the translation of a basic group into a basic supergroup.

Because the carriers are not transmitted with the single sidebands, and thus a new carrier with practically the same frequency has to be generated at the receiving side of a transmission system, a very stringent frequency accuracy is required. Fortunately, the human ear notices a frequency shift of 20 Hz as a slight change in timber only. In fact, the carrier frequency accuracy was dictated by the more stringent requirements for VF-telegraph and broadcast program transmission on multiplexed VF channels, so a 2-Hz frequency accuracy became the quality standard. Usually a 4-kHz master frequency is generated in a crystal-controlled carrier oscillator and stabilized the generated frequency within ±1 part in 10^7 per month. All other carriers are then derived as harmonics of the 4-kHz master frequency. The channel carriers, for example, are three, four, and five times 4 kHz; the pregroup carriers 21, 24, 27, and 30 times 4 kHz. All the master oscillators of the various stations of an administration are usually synchronized to a reference signal derived from an incoming pilot, which in a national synchronization network is connected with a national (or international) reference frequency standard with an accuracy in the order of ±1 part in 10^{10}

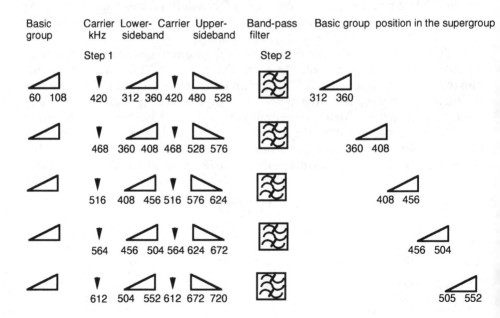

Figure 2.3 FDM basic group translation to supergroup.

per month or even better if a Caesium-atom-standard frequency generator is used.

Pilot frequencies are added in the multiplex equipment to each basic group (respectively, supergroup, mastergroup, and supermastergroup) for system monitoring, supervision, level adjustment, group blocking, automatic standby switching, and carrier synchronization. These pilot frequencies are standardized, too, by CCITT. Table 2.2 summarizes the various pilot frequencies. Moreover, for carrier synchronization frequency pilots at 60 and 308 kHz are recommended [1].

2.2.1.1 Multiplex Equipment for Submarine Operation

For transmission via submarine cable it was justified to spend special efforts on more complex filtering and higher frequency accuracy so that 16 instead of 12 VF channels could be accommodated in a basic group with 3 kHz per VF channel at the expense of a slight bandwidth reduction of the speech from the range 300 to 3,400 Hz to the range 200 to 3,050 Hz. A basic supergroup for submarine cable thus accommodates $5 \times 16 = 80$ 3-kHz VF channels [2–4].

2.2.2 The Bell Analog Multiplex System

In the Bell analog multiplex system the translation from five basic groups into a basic supergroup is the same as shown in Figure 2.2 for the CCITT multiplexing scheme. The multiplexing scheme for the basic master- and supermastergroups, however, differs from the CCITT scheme. Instead of a 300-channel mastergroup, a lower and an upper mastergroup for each 600 VF channels and a supermastergroup for 3,600 VF channels are applied. The composition of these master- and supermastergroups is shown in Figure 2.4.

Table 2.3 summarizes the Bell scheme and indicates the pilot frequencies as well.

Table 2.2
CCITT Recommended Group Pilot frequencies

CCITT Denomination	Pilot Frequency (kHz)	Frequency Accuracy (±kHz)
Basic group	84.08	1
	84.14	3
Basic supergroup	411.92	1
	411.86	3
Basic mastergroup	1,552	2
Basic supermastergroup	11,096	10

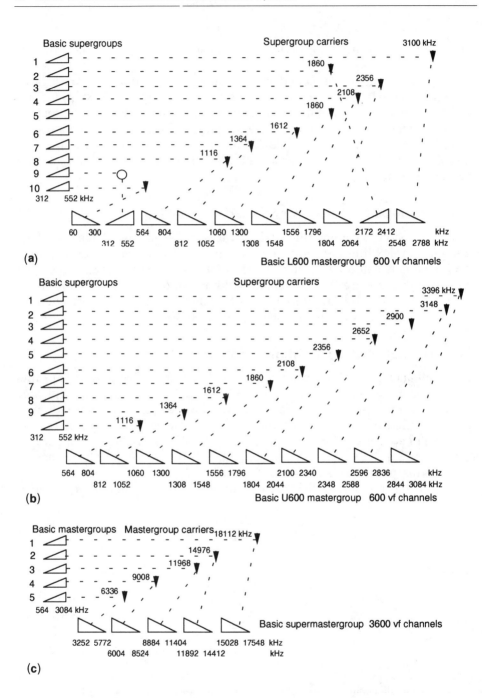

Figure 2.4 Bell system basic mastergroup compositions: (a) L600 mastergroup, (b) U600 mastergroup, and (c) supermastergroup.

Table 2.3
Bell Telephone System Basic FDM Groups

Bell Denomination	Number of VF Channels	Pilot Frequency (kHz)	Frequency Band (kHz)
Basic group	12	84.08/84.14	60–108
Basic supergroup	60	411.92/411.86	312–552
Basic mastergroup L	600	2,840	60–2,788
Basic mastergroup U	600	2,840	564–3,084
Basic supermastergroup (also called basic jumbogroup)	3,600	*	564–17,548

*The basic supermastergroup has six pilots, one in each carried mastergroup, as follows: 2,840 kHz, 3,496 kHz, 6,248 kHz, 9,128 kHz, 12,136 kHz, and 15,272 kHz [2,3].

2.2.3 Analog Transmission Systems

The basic groups/supergroups/mastergroups are the building blocks for the various systems for the transmission of large numbers of multiplexed VF channels via the transmission media. A system for transmission of, for instance, 24 VF channels is composed of two basic groups, one of which is down translated with a 114 kHz carrier to 6 to 54 kHz so that the 24-channel multiplex system occupies a band from 6 to 108 kHz. Figure 2.5 shows systems that are derived from supergroups, whereas Figure 2.6 shows systems derived from mastergroups.

FDM systems have been developed for the transmission of a maximum of 2,700 VF channels via radio-relay systems (in Japan up to 3,600 and in the United States a maximum of 6,000 VF channels) and a maximum of 13,200 VF channels via coaxial cable. Table 2.4 summarizes the major analog transmission systems and indicates the respective transmission media.

For the Bell Telephone Network two further systems with still higher transmission capacities have been developed, namely, an AR6A system for the transmission of 6,000 VF channels via one RF-channel of a single-sideband AM-modulated radio-relay system and an L5E system for the transmission of 13,200 VF channels via coaxial cable [4,5].

2.3 DIGITAL MULTIPLEX

2.3.1 General

Digital transmission started with the introduction of *pulse code modulation* (PCM) in the United States on trunk cables in 1962 and in the access network

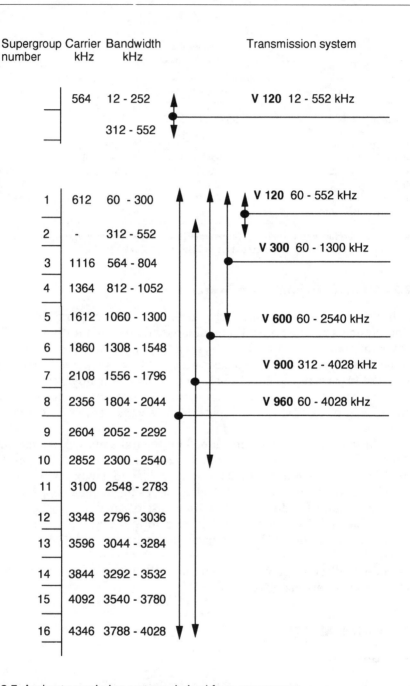

Figure 2.5 Analog transmission systems derived from supergroups.

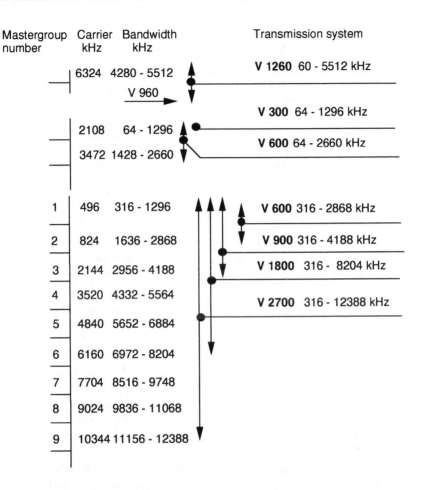

Figure 2.6 Analog transmission systems derived from mastergroups.

in 1973 mainly because with PCM it became possible to increase the number of telephone circuits on existing symmetrical pair cable. These cables had been designed and laid for the transmission of one telephone channel per pair. The crosstalk on those cables was too high to accommodate a capacity extension with analog multiplex.

With analog transmission systems, each time the signal is amplified at a repeater station or a back-to-back terminal, the inevitable signal distortions and unwanted (noise) signals generated in the cascaded repeater sections are simultaneously amplified. Consequently, the signal quality, measured as *signal-to-noise ratio* (SNR), of analog signals deteriorates as a function of distance. Digital signals, however, essentially consist of coded *on* and *off* signals. As

Table 2.4
Major Analog Transmission Systems

Number of VF Channels	Frequency Band (kHz)	Transmission Medium
3	4–16/18–31	Open wire lines
12	36–84/92–143	Open wire lines
12	6–54/60–108	Symmetric cable pairs
12	60–108	Radio relay
24	(6) 12–108	Radio relay
60	12–252	Symmetric cable pairs
	60–300	Radio relay
120	12–552	Symmetric cable pairs
	60–552	Radio relay
300	60–1,300 (64–1,296)	Coaxial cable, radio relay
600	60–2,540 (64–2,660)	Coaxial cable, radio relay
960	60–4,028	Coaxial cable, radio relay
1,260	60–5,512 (60–5,680)	Coaxial cable, radio relay
1,800	316–8,204	Radio relay
2,700	316–12,388	Coaxial cable, radio relay
10,800	4,322–59,684	Coaxial cable

long as a repeater can recognize when an *on* signal has been received, it can then regenerate (not amplify the weak signal, as is the case with analog signals) a new undistorted *on* signal and transmit this over the next section of the transmission link. Consequently, in contrast to analog transmission, the quality of digital signals [measured by the *bit error rate* (BER)] is, up to a certain limit, practically independent of the transmission distance. Moreover, digital transmission suffers not from crosstalk, and the quality is independent of the kind of service applied (telephony or data or video)—digitized all signals are equal! Digital transmission thus offers substantially higher quality than analog transmission. For technical and economical reasons, practically all new transmission equipment is now digital.

2.3.2 Pulse Code Modulation

Transmission digitalization starts with the conversion of analog telephone signals into a digital format. An analog signal can be converted into a digital signal of equal quality if the analog signal is sampled at a rate that corresponds to at least twice the signal's maximum frequency. Analog VF channels, which as mentioned are limited to the 300- to 3,400-Hz band, are therefore sampled at an internationally agreed rate of 8 kHz. Each time the analog signal is sampled, the result (the measured value of the signal at the sampling moment) is then encoded using an 8-bit (= one octet or one byte) code. Because the sampling happens at a sampling rate of 8 kHz (8,000 samples per second), and each sample is coded with 8 bits, the resulting transmission speed of a digitized VF speech channel is 8,000 samples/s × 8 bits = 64,000 bits/s or 64 Kbps.

This *analog-to-digital* (a/d) conversion is carried out by PCM equipment, which furthermore time-multiplexes a number of digitized VF channels into a standard digital frame, similarly to the analog multiplex where 12 VF channels are frequency-multiplexed to form a basic group. Unfortunately, as with the analog multiplex, two different PCM systems and, consequently, two (in fact three) different digital multiplex hierarchies have been developed: a 30-channel European system and 24-channel North American and Japanese systems.

The 30-channel PCM system is standardized by CCITT and has found worldwide application. The 24-channel system was developed in the Bell Laboratories a few years before the development of the 30-channel system in Europe and has found application in North America, Japan, and Korea.

2.3.2.1 30-Channel PCM

After the a/d conversion, in line with CCITT Recommendation G.701, 30 VF channels plus one channel for the signaling of the 30 VF channels and one channel for frame synchronization, maintenance, and performance monitoring—thus a total of 32 channels—are time-multiplexed in a standard PCM digital frame called the primary digital frame. The transmission speed of this primary digital frame thus is 32 × 64 Kbps = 2,048 Kbps. This transmission speed is usually referred to as the 2-Mbps primary level, or first-order PCM, as well as CEPT-1 (CEPT standing for *conférence Européenne des postes et télécommunications*) or E1 (E standing for *European* standard). Figure 2.7, in a simplified way, shows the pulse coding of a VF channel and the subsequent multiplexing of 32 channels into this 2-Mbps primary level. The duration of the standard primary digital frame thus is 1s/8,000 Hz = 125 μs; it contains 8 bits/sample × 32 channels = 256 bits, and per second 8,000 such frames with each 256 bits produce the aforementioned transmission speed of 2,048 Kbps.

The a/d conversion as described changes the continuous analog signal to a pulse-type signal that at the receive end is converted again to a continuous analog signal. This a/d and d/a conversion inevitably introduces a slight distortion of the original signal that is determined in a *signal-to-distortion ratio* (SDR). The sampling specifically leads to an approximation of the original signal that will be better the shorter the sampling intervals are until with an (uneconomical) infinite number of samplings the digital signal remains totally identical with the analog signal. To optimize this encoding and reduce the SDR historically two different logarithmic signal companding methods are used and were included in

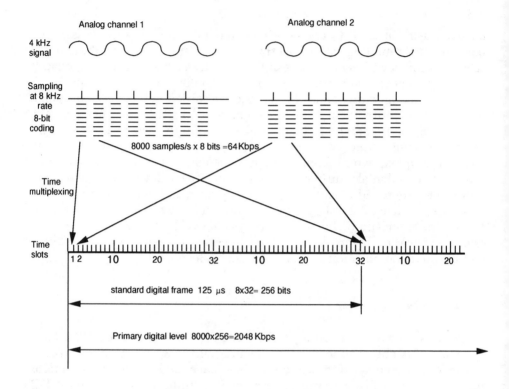

Figure 2.7 Conversion of the analog channels into the 2-Mbps primary level.

CCITT recommendation G.711: the A-law method, used in the 30-channel PCM, and μ-law method, used in the 24-channel PCM.

2.3.2.2 24-Channel PCM

In North America and Japan a 1.5-Mbps primary level, also called T1, has been standardized corresponding to 24 VF channels, in which each channel includes its signaling in the 64-Kbps rate. The two systems further differ in the applied signal valuation at sampling. For the 24-channel system the μ-law, instead of the A-law, is used to assign the PCM encoding values.

Furthermore, the 24-channel PCM applies the basic bipolar *alternate mark inversion* (AMI) line code, whereas the 30-channel system applies the more sophisticated HDB3 that allows clock information to be embedded in the data stream. To facilitate a better understanding of digital multiplexing and the significance of the line code difference, a short explanation of digital line codes follows.

2.3.2.3 *Digital Line Codes*

Digital signals are coded in multiplex equipment with binary 0/1 values represented by zero value and positive value signals in a so-called *non-return-to-zero* (NRZ) coding. As shown in Figure 2.8, two or more "1" values are thereby grouped together into one positive signal without *return-to-zero* (RZ) between the two or more separate "1" values. Similarly, two or more "0" values are grouped together to one zero-level signal.

For transmission and subsequent synchronizing and decoding on the receive side, the NRZ coding has, in principle, the following disadvantages:

- The signal obtains a positive dc potential, which conflicts with the basic requirement that equipment interfaces shall be potentialfree.
- Synchronizing at the receive side becomes difficult when the signal starts with a large number of equal binary states (all "0" or all "1"), thus the signal clock frequency cannot be recognized soon enough.
- Although a NRZ signal requires less bandwidth than an RZ signal, a relatively high bandwidth is still required for the transmission of the resulting square wave signal.

To overcome those disadvantages, a number of line codes have been created. Three such improved line codes are included in Figure 2.8 and together with other major line codes summarized in Table 2.5 [4,5].

2.3.2.4 *The Future of PCM*

Digital switching equipment, introduced in the early 1970s, includes already a/d conversion on all incoming subscriber lines and provides a 2-Mbps interface to the trunk transmission line, thus eliminating the major application of PCM equipment. With ISDN the a/d conversion of the speech channel is even further moved from the digital exchange directly to the subscriber terminal equipment so that subscribers can be connected to digital transmission and switching equipment without additional a/d converters. PCM equipment, therefore, which was a major product of the transmission equipment industry, will only survive in niche applications, for example, where analog telephone channels of subscriber clusters need to be concentrated for transmission as E1/T1 on twisted pair cable.

2.3.3 Plesiochronous Digital Multiplex Hierarchy

2.3.3.1 *30-Channel Hierarchy*

Similar to the analog multiplex where basic groups of VF channels are combined into basic supergroups and mastergroups, CCITT has defined a *plesiochronous*

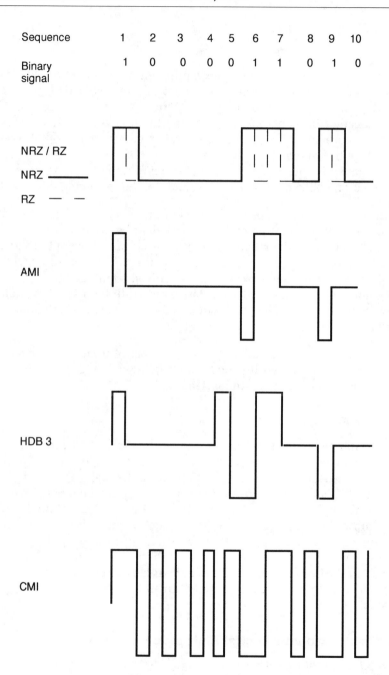

Figure 2.8 Digital line codes.

Table 2.5
Summary of Major Digital Line Codes

Line Code	Abbreviation	Characteristics
Return-to-Zero	RZ	Value "1" part time positive only (Bipolar).
Alternate Mark Inversion	AMI	The "1" values are alternately coded as +1 and −1. Disadvantage: The "0" values are still grouped, thus clock retracting is still difficult.
High-Density Bipolar	HDB-n	The "1" values as for AMI. After n "0" values follows a signal at the same polarization as the preceding "1" value; thus the signal remains potentialfree, and the clock frequency is exactly recognizable. Violations are forced to alternate in polarity by including an extra "1" at the beginning of a sequence of zeros when needed. HDB3-coded line signals are widely used below 140 Mbps.
Coded Mark Inversion	CMI	The "1" values as per AMI, the "0" values are separated in two equal parts—the first part with a negative value, the second part with an equal positive value. The resulting signal is thus potentialfree, and the clock frequency is easy recognizable. Above 140 Mbps, CCITT recommends application of the CMI code.
2 Binary, 1 Quarternary	2B1Q	This code converts blocks of two consecutive bits into a single four-level pulse. The information rate is thus double the baud rate. 2B1Q is applied for ISDN.
3 Binary, 2 Ternary	3B2T	This code converts blocks of three consecutive bits into a two-symbol three-level pulse.
4 Binary, 3 Ternary	4B3T	Converting blocks of four consecutive bits into a three-symbol four-level pulse.

digital hierarchy (PDH). Each higher level is set at "almost but not exactly" four times the bit rate of the previous order, as shown in Table 2.6.

Table 2.6 indicates that for each multiplex level a frame structure has been standardized with obviously no relation between them. This implies that at each higher order level the lower level frames are "rotating" with regard to the higher order frame. The practical result is that dropping a lower level signal from a higher order bit stream, similarly as with analog multiplex, can be realized only through step-by-step demultiplexing. Thus, in order to drop, say, 10 channels from a 140-Mbps data stream, demultiplexing is required from 140 to 34 Mbps, from 34 to 8 Mbps, and from 8 to 2 Mbps.

Different from analog multiplex, where the basic groups/supergroups/mastergroups are the building blocks for the various systems for the transmis-

Table 2.6
Digital Multiplex Hierarchy Based Upon 30-Channel PCM

Multiplex Level	Bit Rate (Mbps)	Bits per Frame	Frame Frequency (kHz)	Frame Duration (μs)	Number of VF Channels
1	2.048	256	8.0000	125.0000	30
2	8.448	848	9.9622	100.3788	120
3	34.368	1,536	22.3750	44.6927	480
4	139.264	2,928	47.5628	21.0248	1,920
5	564.992	2,688	210.1904	4.7576	7,680

sion of large numbers of multiplexed VF channels via the transmission media, PDH transmission systems are available for each of the above digital hierarchy levels. The PDH transmission network, thus, is designed to function in a hierarchical fashion. In the interest of multivendor equipment compatibility, the signals that pass between various network equipment have been standardized by CCITT in two distinct ways, specifically, according to the *form of the signals* with their characteristics specified in CCITT Recommendation G.703 and the *structure of the signals* and the way they are combined in the multiplex as defined in the G.750-Series of CCITT Recommendations.

Figure 2.9 illustrates this grouping of digital channels into the digital multiplex hierarchy together with the relevant transmission systems.

2.3.3.2 24-Channel Hierarchy

The higher level multiplexing hierarchies used in North America, Japan and Korea, based upon the 24 VF channel PCM, are shown in Table 2.7.

2.3.3.3 Intelligent Multiplexers

In addition to the above fixed hierarchical PDH multiplex equipment and systems, with the introduction of computer technology in the multiplex equipment a new generation of intelligent multiplex equipment came on the market in the early 1990s. This intelligent multiplex with software-controlled flexibility enables a more economical use of available transmission lines in both the access and transport networks. Four typical intelligent multiplexing features are as follows.

Subscriber multiplexing. With customer premises multiplex equipment a customer (or a group of customers) can use additional telecommunication services without needing additional cable lines. Both analog and digital (including

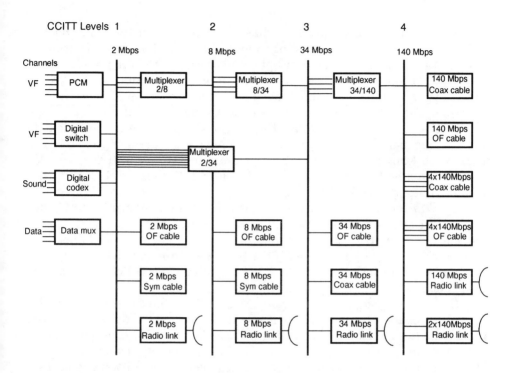

Figure 2.9 Hierarchical PDH multiplex environment.

Table 2.7
Digital Multiplex Hierarchy Based Upon 24-Channel PCM

Multiplex Level	Bit Rate in Mbps	Number of VF Channels
In North America		
D1	1.544	24
D2	6.312	96
D3	44.736	672
D4	139.264	2,016
D5	564.992	8,064
In Japan and Korea		
1	1.544	24
2	6.312	96
3	32.064	480
4	97.728	1,440

ISDN) terminal equipment can be connected. The access network interface is 2 Mbps for transmission over existing copper wire (typically 3 to 5 km unrepeatered on 0.4- or 0.6-mm wire), over optical fiber cable, or via radio relay.

Flexible multiplexing, beyond the facilities of subscriber multiplexing, allows an easy adaptation to variable traffic load from a (group of) subscriber(s). A flexible multiplexer can connect $n \times 64$ Kbps (n_{max} typically 480) terminals or subscriber lines to $m \times 2$ Mbps (m_{max} typically 32) access lines. Flexible multiplexing includes additional features like voice compression from 64 Kbps to 32 or 16 Kbps (thus enabling the operation of two, respectively, four VF channels on a 64-Kbps channel), supervision and diagnosis of incoming and outgoing lines, and control by PC.

Drop/Insert (D/I) multiplexing can be used in linear networks (LAN; railways, utilities) in a ring configuration to distribute traffic within the ring independent from the PSTN. It can operate in a 1 + 1 mode on two 2-Mbps lines automatically selecting the better of the two lines. Dropped channel capacity can be re-used for insertion. $n \times 64$ Kbps (n_{max} typically 60) channels can be dropped and inserted per equipment location.

Cross-connect multiplexing. Rerouting traffic at transmission nodes (for example, to change the allocation of leased lines) was once and is still achieved at many nodes, by manually changing color-coded cable pairs at the stations' *main distribution frame* (MDF). With cross-connect multiplexing equipment, 64-Kbps channels can be electronically regrouped under software control on different 2-Mbps transmission paths by applying service "grooming" and circuit "compacting" techniques. With service grooming, common services such as data channels on different incoming 2-Mbps lines are automatically grouped together in a common outgoing 2-Mbps line, whereas, say, incoming leased lines are grouped in another outgoing 2-Mbps line. With circuit compacting, channels from partially loaded 2-Mbps incoming lines can be compacted into one outgoing 2-Mbps line.

A cross-connect multiplexer can typically handle up to 1,024 2-Mbps ports. The various characteristics of cross-connect multiplexing are summarized in the CCITT Recommendation G.797 "Characteristics of a flexible multiplexer in a PDH environment."

An application of these intelligent multiplexing features in addition to standard digital multiplexing is shown in Figure 2.10.

2.3.3.4 Explanation of the Term Plesiochronous

Before going on to the next section on synchronous digital multiplex, the term *plesiochronous* in the title of this section still needs to be explained. The prefix "plesio," which is of Greek origin, means "almost equal but not exactly." As

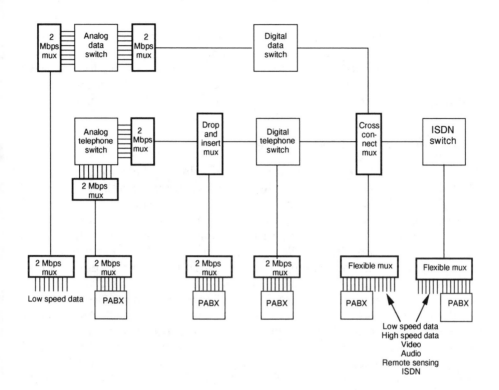

Figure 2.10 Multiplexing applications.

already indicated, each higher level in the CCITT hierarchy has been defined as "almost but not exactly" four times the bit rate of the previous level. For example, level 2 has a bit rate of 8.448 Mbps compared with the primary level rate of 2.048 Mbps. However, $4 \times 2.048 = 8.192$ Mbps, whereas $8.448/4 = 2.112$ Mbps. The difference of $2.112 - 2.048 = 64$ Kbps. The reason for this "little difference" of 64 Kbps is partly because the so-called four tributaries that are combined to a common higher level can, in principle, come from different transmission stations that are "almost but not exactly" synchronous with one another. To compensate for this "plesiochronousness" at each higher level a number of bits are added (bit stuffing) to each tributary, depending on the rate of plesiochronousness (that is, divergence from the exact frame bit rate) of the tributary.

Level 2 tributaries are synchronized at $8.448/4 = 2.112$ Mbps. Thus, a 2-Mbps tributary with a bit rate of 2.040 Mbps, say, instead of 2.048 Mbps, will get $2,112 - 2,040$ Kbps $= 72$ Kbps added to its transmission rate; whereas a tributary coming from another station with a bit rate of 2.050 Mbps will get

2,112 − 2,050 Kbps = 62 Kbps added to its transmission speed. The bit rate of 2.0112 Mbps has been chosen to be high enough to cater for the maximum expected plesiochronousness of a tributary. If this short explanation of what is a complex matter was only *plesio*understandable, it might be worthwhile to read it again before continuing with synchronous digital multiplex.

2.3.4 Synchronous Digital Multiplex

With increasing network digitalization, the inflexibility of plesiochronous digital multiplexing became obvious. In particular, interconnections via leased lines often required cross-connecting and D/I functions. This is time consuming and costly because it is necessary to demultiplex to and remultiplex from the level to be accessed. In addition, rerouting may involve manual changing of cabling at distribution frames. To eliminate these problems and to make use of 20 years of significant advanced technology with computerization and high bite rate transmission via optical fiber, a new practically exact synchronous digital multiplex hierarchy known as SONET (from Synchronous Optical Network) was proposed in the United States by Bellcore in 1985. As the name indicates, SONET was originally conceived to provide a standard optical interface signal specification. The SONET concept was then extended to a new multiplexing hierarchy, especially because the PDH in the United States had no standard beyond 45 Mbps, and adopted worldwide by CCITT under the name *synchronous digital hierarchy* (SDH) at a CCITT conference in Seoul in February 1988. The major SDH standards for physical interface, optical line rates, frame structure, network node interface, multiplexing structure, and transmission protocols enabling interconnection of equipment from different manufacturers were defined in the following three recommendations:

- G.707: SDH bit rates;
- G.708: Network node interface for the synchronous digital hierarchy;
- G.709: Synchronous multiplexing structure.

2.3.4.1 SDH Frame Structure

The SDH concept is designed so that all digital signals at any synchronous multiplex level have a frame repetition frequency of exactly 8 kHz and thus a frame duration of 125 μs instead of varying at five values between 4.7576 and 125.0000 μs. Moreover, the SDH signals consist of 64-Kbps (one VF channel) interleaved bytes. As a consequence, the digital exchanges in an all-digital network must also be synchronized to this 8-kHz frame frequency in order to prevent bit slips, which cause loss of information.

The SDH concept makes a clear distinction between multiplex functions and transport functions. It defines *synchronous transport modules* (STM-N) as the synchronous signal blocks that have to be transported from node to node in an SDH network. The affix "N" stands for the hierarchical level that thus far has been defined for N equal to 1, 4, and 16. STM signals have a frame structure that again have a frame frequency of exactly 8 kHz. The STM frame contains $9 \times 270 \times N = 2{,}430 \times N$ bytes, each of 64 Kbps. The STM-1 bit rate is thus $2{,}430 \times 64$ Kbps = 155.520 Mbps; the STM-4 bit rate is four times higher, thus 622.080 Mbps; and STM-16 is sixteen times higher than STM-1, thus 2,488.320 Mbps. The STM-4 frame is obtained by phase-locking and byte inter-leaved multiplexing four STM-1 frames. Likewise, the STM-16 frame is made of four STM-4 frames. Consequently, the STM-1, -4, and -16 frames have the same frame structure with the exception that each higher level contains four times more bytes. It is customary to show the STM frame graphically as a field of nine lines each with $N \times 270$ bytes (columns). Figure 2.11 shows the graphical frame configurations for both SONET and SDH. In SONET the basic building block is not called STM but *synchronous transport signal-level* (STS), where STS-1 has a bit rate of $90 \times 9 \times 64$ Kbps = 51.840 Mbps.

The actual information being carried, which, in the case of STM-1, consists of 261 bytes less 1 byte for POH equals 260 bytes (of 64 Kbps) $\times$ 9 rows = 2,340 VF channels, is defined as the *payload*. As will be explained, however, the corresponding number of VF channels accommodated in STM-1 is only 1,890 instead of 2,340.

An integral *section overhead* (SOH)—respectively, in the case of SONET a *transport overhead* (TOH)—contains all the required information for operation and maintenance, order wire, performance monitoring, framing, alarm, and standby switching, for centralized network management, and remote control of section and line, including substantial spare capacity for future applications. These SOH/TOH are processed, evaluated, and reconstructed section by section in the network. The SOH of a STM-1 frame has a transmission capacity of $9 \times 9 \times 64$ Kbps = 5,184 Kbps, thus a generous 5-Mbps capacity for network operation; STM-4 with 20 Mbps and STM-16 with 80 Mbps thus have plenty of spare capacity for future applications.

A *path overhead* (POH) is given to each tributary lower level signal in the payload and accompanies the payload on its entire path through the network. The POH of a STM-1 frame has a transmission capacity of $1 \times 9 \times 64$ Kbps = 576 Kbps for path identification and quality control.

Figure 2.12 clarifies the meaning of path, line, and section.

Administrative unit (AU) pointers are another key element in the SDH concept. In general, when a tributary signal arrives at a network node, its frame will not be in phase with the STM frame of that node. However, in the SDH concept it is not the phase of the signal that is aligned but the phase relationship

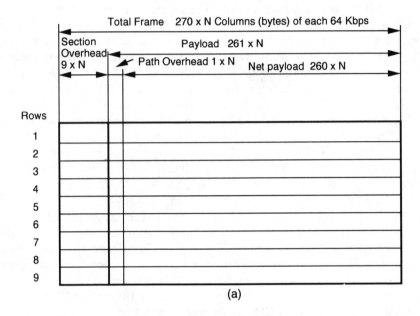

(a)

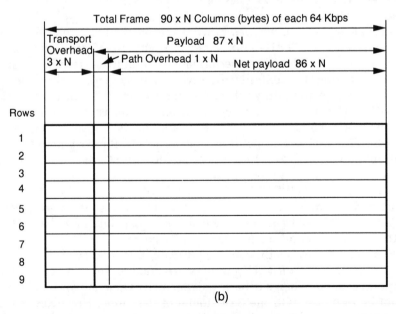

(b)

Figure 2.11 Synchronous frame structure: (a) SDH STM frame and (b) SONET STS frame.

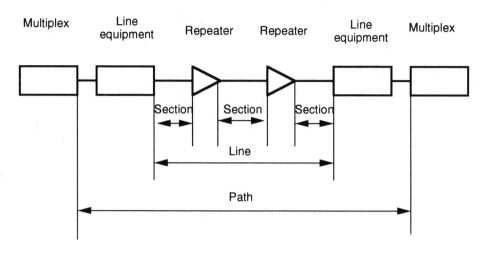

Figure 2.12 Link division into path, line, and sections.

itself that is maintained and written into a pointer. The pointer provides information about where the first byte of a lower level signal starts within the higher level frame, thus making it possible to find any tributary signal in the STM frame. This AU pointer and the byte interleaved multiplexing enable a direct access to lower level signals without need for cascaded demultiplexing—which is the biggest advantage of SDH over PDH multiplexing! Figure 2.13 illustrates this advantage in a link consisting of two terminals and a back-to-back terminal.

The STM-N signal is synchronized to a 8-kHz (or 2,048-kHz or 155,520-kHz) national master clock. The SDH signals are transmitted via optical fiber cables or radio-relay systems the same way as with PDH signals because the line system always synchronizes itself to its input signal. Phase fluctuations or wanderings, which may happen because of transmission path variations and regulation impairments, can easily be corrected thanks to the pointers.

2.3.4.2 SDH Virtual Containers

To enable the initial parallel operation of PDH and SDH followed by an easy integration of PDH in SDH and finally replacing PDH by SDH worldwide, the concept has been made such that SDH can transport the plesiochronous CCITT-recommended bit rates of 2, 34, and 140 Mbps as well as the 1.5-, 6-, and 45-Mbps rates of the American and Japanese standards. To this purpose, SDH uses a concept of *virtual containers* (VCs). These containers are made wide enough to accommodate the plesiochronous signal, including bit rate

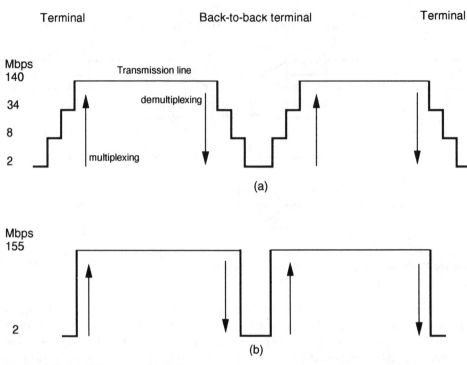

Figure 2.13 (a) PDH versus (b) SDH.

justification signals to match the PDH to the SDH transmission speed as well as its own POH. These containers are designated as listed in Table 2.8.

2.3.4.3 SDH Multiplexing Structure

The complex SDH multiplexing structure conceived in order to accommodate the various PDH bit rates is shown in Figure 2.14.

Figure 2.14 is presented in two versions to facilitate a better understanding of the complex structure. Figure 2.14(a) in a principle diagram showing the multiplication stages for the hierarchical integration of the PDH tributaries into the STM-N signal. Figure 2.14(b) shows in a functional diagram the successive stages of signal mapping, containerization, section and path overhead insertion, frame alignment, pointer insertion, and multiplication from PDH tributary signal to SDH line signal.

As the figure shows, SDH multiplexing starts at the 1.5-Mbps, respectively, 2-Mbps primary rate level; thus the 1.5- and 2-Mbps primary levels themselves and lower bit levels will not be converted from PDH to SDH. The seldom

Table 2.8
SDH Virtual Containers

Container Designation	PDH Bit Rate (Mbps)
VC-11	1.544
VC-12	2.048
VC-2	6.312
VC-3	34.368 and 44.736
VC-4	139.264

used PDH 8-Mbps level is not included because 2 Mbps is very often directly multiplexed to 34 Mbps.

ETSI recommends that the European operators use a slightly simplified structure that eliminates the 6.3-Mbps level too. Moreover, in the interest of equipment compatibility of various manufacturers, ETSI has specified an equipment practice for the SDH equipment. This standard, called European Telecommunications Standard S-9, defines matters like mechanical dimensions (up to two times six equipment units back-to-back in a standard 2,200- by 600- by 600-mm rack, similarly as standardized for switching), power supply, environmental conditions, and *electromagnetic compatibility* (EMC).

SDH covers high to very high bit rates; however, it slightly reduces the traffic transmission capacity versus comparable PDH levels due to the incorporation of significant software in the VCs. The 2-Mbps 30-channel primary rate multiplexing, according to this structure, follows in steps of 3 × 7 × 3 instead of 4 × 4 × 4 with PDH; consequently the STM-1 level accommodates 3 × 7 × 3 × 30 VF channels = 1,890 VF channels compared with 4 × 4 × 4 × 30 = 1,920 VF channels in the 140-Mbps PDH hierarchy.

The next STM levels are always in exact steps of four. Table 2.9 summarizes the SDH hierarchy.

SDH optical fiber transmission systems exist for STM-1, -4, and -16 (and prototype systems for STM-64); radio-relay transmission equipment is available for STM-1 and -4; whereas satellite transmission at the STM-1 rate is being introduced by INTELSAT.

2.3.4.4 SONET

SONET, as mentioned, has been the pacemaker for SDH. SDH has become the worldwide standard for synchronous transmission, and SONET can be considered a subset of the SDH standard. The most significant difference between SONET and SDH occurs in the frame as already shown in Figure 2.11; SONET was conceived to serve the existing 24-channel systems, and SDH

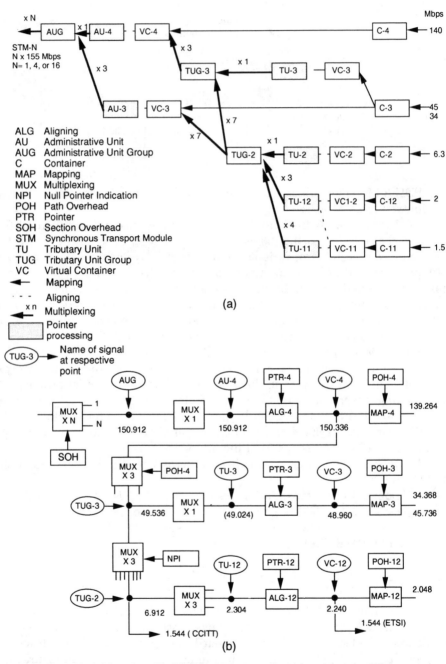

Figure 2.14 SDH multiplexing structure: (a) principle and (b) functional diagrams.

Table 2.9
SDH Hierarchy

Multiplex Level	Bit Rate (Mbps)	Referred to as	Equivalent Number of VF Channels
STM-1	155.520	155 Mbps	1,890
STM-4	622.080	620 Mbps	7,560
STM-16	2,488.320	2.5 Gbps	30,240
STM-64	9,953.280	10 Gbps	120,960
STM-256	39,813.120	40 Gbps	483,840

serves both the 24-channel and the 30-channel systems. The SDH multiplexing structure was then made such that SONET and SDH meet again at STM-1 and STS-3 levels so that beyond STS-3 SONET and SDH are "almost but still not exactly" equal. Pointer and overhead processing are slightly different in SONET from SDH; and whereas with SDH each higher level is exactly four times the preceding level, SONET makes smaller steps as shown in Table 2.10.

Figure 2.15 summarizes the SONET and SDH hierarchies and shows the major differences between the two standards.

For SONET the STS-N denomination is used for electrical signals. A parallel denomination OC-N (OC standing for *optical carrier* level) is used for optical signals (which directly interface with optical fiber lines). OC-48 thus denominates a synchronous 2,488.320-Mbps signal for transmission via optical fiber cable.

2.3.4.5 SDH Multiplexing Equipment

PDH networks are built up from separate multiplex and line terminal equipment for each level of the PDH hierarchy, for example, 2 + 8 + 34 + 140-Mbps multiplex

Table 2.10
SONET Hierarchy

Synchronous Transport Signal	Bit Rate (Mbps)	Comparable SDH Level
STS-1	51.840	
STS-3	155.520	STM-1
STS-9	466.560	
STS-12	622.080	STM-4
STS-18	933.120	
STS-24	1,244.160	
STS-36	1,866.240	
STS-48	2,488.320	STM-16

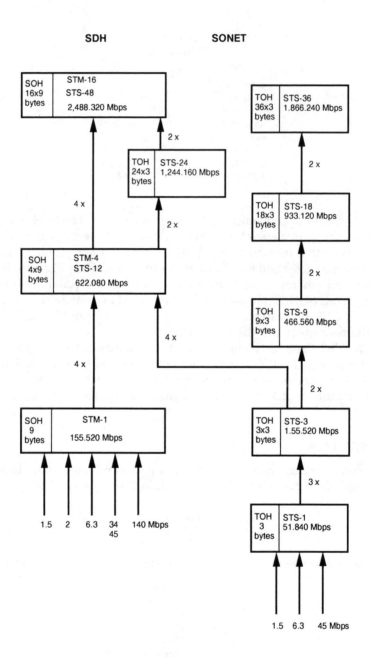

Figure 2.15 Summary of SDH and SONET hierarchies.

equipment and 140-Mbps line equipment for a 140-Mbps transmission link via optical fiber cable. The SDH multiplexers for the STM-N levels perform both multiplexing (from 2 Mbps up to $N \times 155$ Mbps) and line terminating functions. For a 155-Mbps SDH transmission link via optical fiber cable one equipment type only, the STM-1 multiplexer, is required instead of five equipment types for the comparable 140-Mbps PDH transmission link.

In addition to this new type of multiplexer with integrated line termination, SDH features two additional types of equipment: the add/drop multiplexer and the cross-connect multiplexer. Likewise, as for the previously described intelligent PDH multiplex equipment, the add/drop and cross-connect functions can be combined in common equipment units.

2.3.4.6 Add/Drop Multiplexer

The add/drop multiplexer provides the functions of a back-to-back terminal along a route where part of the traffic terminates and new traffic will be added, but the main stream continues on the route without demultiplexing/remultiplexing. A combination of SDH multiplexer and add/drop multiplexer can be used for another feature of SDH transmission: in "self-healing ring configurations". The SDH ring configuration automatically reconfigures the signal in another direction in case of an interruption in the line. Figure 2.16 shows such a ring configuration for a 155-Mbps route between two terminal stations A and B with two smaller stations C and D. Terminals A and B interface with 140-Mbps PDH equipment. On the stations C and D the add/drop multiplex interfaces with a 34-Mbps PDH multiplex to drop traffic coming from station A and with 2-Mbps PDH multiplex to insert or add traffic to terminal B. The traffic routing has been indicated in a basic schematic too. Traffic from A to B normally will flow on the section A-B; in case of an interruption on section A-B the traffic will flow from A via C and D to B.

Add/drop multiplexers are likewise available for STM-4 (622-Mbps) and STM-16 (2.5-Gbps) routes. Thus, on a 2.5-Gbps SDH route with add/drop multiplexer along the route, traffic can be dropped and added at a choice of the following bit rates: PDH 1.5, 2, 6.3, 34, 45, and 140 Mbps; and SDH 155 and 622 Mbps.

2.3.4.7 Cross-Connect Multiplexer

A cross-connect multiplexer, in essence, is an electronic line distribution frame incorporating an SDH multiplexer. Electronic line distribution is also possible with the aforementioned intelligent PDH cross-connect multiplexer; however, the additional advantage of the SDH cross-connect multiplexer is that the cross-

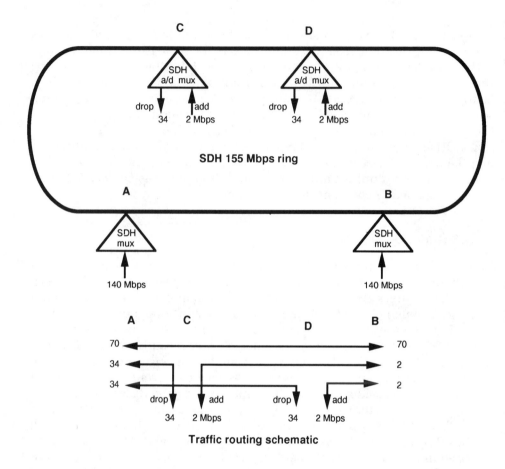

Figure 2.16 SDH ring configuration.

connecting happens without any demultiplexing. The SDH frame structure with the help of pointers and virtual containers allows any signal to be located within the SDH frame. Consequently, in addition to adding and dropping signals directly from and to the SDH signal, combining extracted and reinserting combined signals enables traffic to be rerouted without demultiplexing.

Thus, SDH cross-connecting is the rerouting of tributary signals under computer control. Cross-connecting multiplexers can be used for rapid rerouting in the event of link failures in order to adapt routing to time-of-day and seasonal traffic load variations such as carrying cable TV programs during off-peak hours for telephone traffic.

Besides the advantages of rapid computer-controlled cross-connecting at network nodes, the SDH structure enables the volume of the equipment, station cabling, and power supply to be greatly reduced compared with even the most modern PDM solutions as it replaces higher order multiplex and distribution frames in medium and large stations. Figure 2.17 clearly depicts this advantage. The PDH version shown in this figure relates to "nonintelligent" PDH. Intelligent PDH can replace the distribution frame but still needs multihierarchy demultiplexing. This figure clearly demonstrates the significant equipment and labor-reducing effect of the SDH technology.

Figure 2.18 shows a high-capacity state-of-the-art cross-connect system for a mixed synchronous and plesiochronous network. Figure 2.18(a) shows the system in a test field; a simplified functional diagram of the system in given in Figure 2.18(b). This high-capacity cross-connect can switch up to 16,384 equivalent 2-Mbps ports (equal to 256 STM-1 signals) between STM-1 SDH lines and 2-, 34-, and 140-Mbps PDH lines. Switching in the matrix takes place at AU-4, TU-2, TU-3, and TU-12 levels (see Figure 2.14) in such a way that each incoming 2-Mbps port can be routed to any outgoing 2-Mbps port. The clock unit can be synchronized with a 10-MHz clock frequency derived from a central national Caesium-atom-standard frequency generator, an external 2,048 kHz standard signal frequency, or a digitized 2,048-kHz signal extracted from the SOH of an incoming STM-1 signal. The node control unit with two memory boards, each with a capacity of 105-Mbyte and a 32-Mbyte RAM (remote access memory) processor, can be operated from a local workstation or remotely, via the standardized Q-interface, by a TMN. The node control unit as well as the clock unit and the matrix are duplicated with automatic standby.

2.3.4.8 SDH Networks

The introduction of SDH technology can have an additional cost-saving effect on the network configuration. The number of main nodes with switching equipment can be reduced and the capacity of the lines between the remaining main nodes increased. Moreover, the network can be structured on two administrative levels, 2 Mbps and 140 (and 155) Mbps, thereby simplifying overall network management. Figure 2.19 illustrates such a typical architecture for a SDH network. The main trunk network along a STM-16 backbone link consists of 155- (140-)Mbps cross-connects on a 2.5-Gbps optical fiber cable. The regional and urban networks typically consist of 2- to 155-Mbps cross-connects in a 622-Mbps ring via optical fiber cable and or radio-relay systems. The access network may be based on a 155-Mbps ring with 2- and 34-Mbps add/drop and cross-connect multiplexers connecting local exchanges, RSUs (see Subsection 1.5.2.2), and large customers.

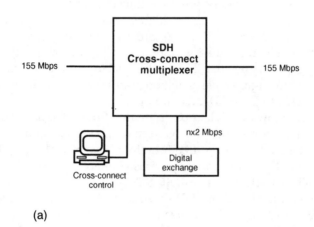

(a)

(b)

Figure 2.17 (a) SDH versus (b) PDH cross-connecting.

(a)

Figure 2.18 Cross-connect (a) in test field and (b) functional diagram (courtesy of Siemens).

2.3.4.9 SDH Implementation

Implementing an SDH network does not mean that all the installed PDH equipment becomes obsolete. Just as SDH systems can accept PDH signals at the levels 1.5, 2.0, 6.3, 34, 45, and 140 Mbps, a PDH network can progressively be upgraded to an SDH network whereby part of the higher order PDH multiplex and the distribution frames can be taken out of operation thus regaining floor space. Installing SDH equipment provides the immediate advantage of increased flexibility, easier network monitoring and supervision, and increased network capacity with significantly less additional equipment compared with an extension with PDH equipment. New services at higher transmission rates can be cost-effectively integrated and easily reconfigured. The major differences between PDH and SDH multiplex are summarized in Table 2.11.

2.3.4.10 SDH Standardization

The core of SDH standards, as mentioned at the beginning of this subsection, are given in the CCITT Recommendations G.707, G.708, and G.709 as published

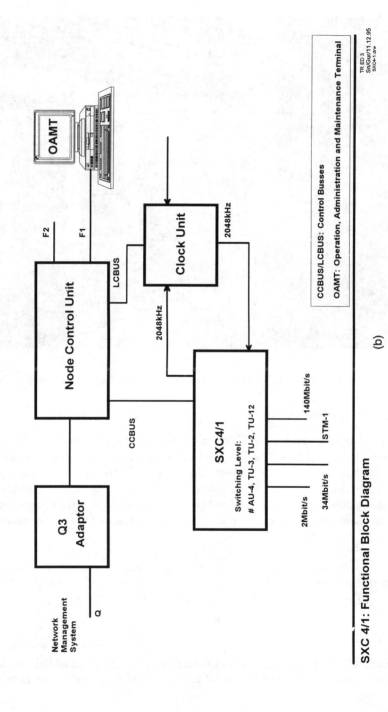

SXC 4/1: Functional Block Diagram

(b)

Figure 2.18 (continued)

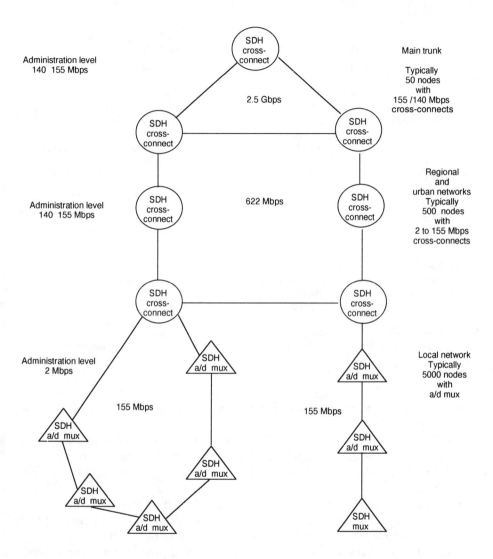

Figure 2.19 Typical SDH network architecture.

in 1988. Standardization of SDH is still ongoing and further important SDH standards were subsequently published in the following recommendations.

- G.774: SDH management information model for the network element view.
- G.781: General aspects of digital transmission systems. Terminal equipment. Structure of recommendations on multiplexing equipment for the SDH.

Table 2.11
Comparison Between SDH and PDH Multiplex

Characteristic	PDH	SDH
Multiplexing mode	bit interleaved	byte interleaved
Frames at different levels	uncorrelated	fixed relationship
Frame duration	all different	all 125 μs
Drop and insert	with hierarchical de/ remultiplexing	direct access to VF channel level
Extra capacity	none, unless pre-installed	plenty for various applications
System design	gradually extended to higher bit rates as traffic increased and technology progressed	top-down design, complexity not cost-sensitive because of use of VLSI
Technology	hardware based	software based
Cross-connection	manual	computer controlled
Network management	possible with additional equipment	system incorporated, highly flexible, efficient and versatile
Required floor space	large	small
North American/CCITT interface	difficult	system incorporated

- G.782: General aspects and general characteristics of SDH multiplexing equipment.
- G.783: General aspects of digital transmission systems terminal equipment. Characteristics of SDH multiplexing equipment functional blocks.
- G.784: General aspects of digital transmission systems terminal equipment. Characteristics of SDH management.
- G.803: Architecture of transport networks based on the SDH.
- G.957: Digital networks, digital sections, and digital line systems. Optical interfaces for equipment and systems relating to the SDH.
- G.958: Digital networks, digital sections, and digital line systems. Digital line systems based on the SDH for use on optical fiber cables.
- G.sdxc1/2/3: Cross connect equipment.
- G.tna/1/2: Network architecture.
- G.81s: Equipment clocks.

2.3.4.11 SDH and ATM

With synchronous multiplexing being presented as a significant improvement compared with plesiochronous multiplexing, the question arises: Why introduce asynchronous multiplexing? The answer is that the ATM (briefly described in Subsection 1.5.8) is not a multiplexing technology. The word "asynchronous" indicates that cells of an information unit may pass over network links at irregular (asynchrony) intervals. Whereas SDH has been developed to improve the efficiency and capacity of transmission networks, ATM has been developed as a switching technology allocating bandwidth on demand to all forms of traffic as voice, data, image, and multimedia. ATM has been standardized by CCITT in 1990 as a particular technique and architecture for the switching, multiplexing, and transmission aspects of B-ISDN networks. In fact, ATM will reduce the distinction between transmission and switching. ATM, in essence, is a packet switching and multiplexing technique that uses short, fixed-length cells to carry information in a high-speed network. Whereas SDH concerns the physical network that the operator creates and manages for himself, ATM concerns virtual networks for the user.

ATM was conceived primarily for broadband communications. ATM *user-network-interfaces* (UNIs) were initially defined for 45, 100, and 155 Mbps and subsequently followed by UNIs for 1.5, 2.0, 34, 51, and 140 Mbps. At an even lower level of 56/64 Kbps a so-called *Frame*-UNI (FUNI) has been defined to allow low-speed data to be carried in ATM mode on standard TDM in the access network and then to be integrated in broadband ATM on the transport network.

ATM, as shown in Figure 1.14, is made up of cells, each consisting of a 5-byte header and 48-byte payload, each byte constituting 64 Kbps. The 5-byte header (thus $5 \times 64 = 320$ Kbps) identifies the "virtual path" to be followed by the cell through the network and contains information on cell loss priority and error control. The ATM header thus is comparable with the function of the POH in an SDH signal. ATM has no SOH, whereas the SOH in an STM-1 signal has over 5-Mbps transmission capacity for network operation; retrospectively, this shows that ATM was not conceived as a transport vehicle. ATM does not incorporate an error correction mechanism to compensate for poor line quality. SDH, with its comprehensive performance monitoring and error correction facilities, is better than PDH. For the transmission of ATM signals via networks with PDH and SDH multiplex, special *network node interfaces* (NNIs) have been defined. As first, and major, NNI between ATM and SDH, the SDH VC-4 container with an interface of 139.620 Mbps was already defined in G.709, in the meantime followed by NNIs at 25.6, 51, and 622 Mbps.

Whereas SDH uses a synchronous frame with byte interleaved multiplexing, ATM uses a form of TDM whereby nonsynchronous cells are added as a function of the required bandwidth. Channel association is by means of a label (in the header). A sequence of cells with the same label constitutes a "virtual channel." The transfer mode is asynchronous in the sense that the rate of transmission of cells within a particular virtual channel can be variable depending on the source rather than on any clock reference within the network. Consequently an ATM-based network is bandwidth transparent, allowing the handling of a dynamically variable mixture of services at different bandwidths.

Figure 2.20 shows the major difference between SDH, with a regular frame repetition, and the multiplexing function of ATM with a sequence of irregularly loaded cells.

SDH will be the solution for new transmission networks based on optical fiber cable as well as for overlay networks to copper-based networks that require bandwidth extensions. The deployment of ATM will be driven by the requirement for effective data switching in the B-ISDN network. ATM, thereby, will provide the flexible interface to variable-bandwidth users with multimedia terminals, LANs, and computer hosts and workstations. ATM, providing all communications needs via a single interface, already appears as the next generation for LAN technology and may provide a seamless LAN/WAN integration.

The major advantage of ATM is that a single technique can be used to switch and transmit any type of digital information from a low to very high bit rate on the same switching and transmission medium. The variable bit rate offered by ATM facilitates the introduction of new services.

Whatever broadband services are offered in the future, they will use ATM for cost-effective network interface and switching as well as SDH for highly reliable transmission. Therefore, it can be expected, at least for the next ten years, that ATM and SDH deployment will be parallel. ATM, rather than superseding SDH, more likely will be superseded together with SDH by another common probably optical switching and transmission technology [6,7].

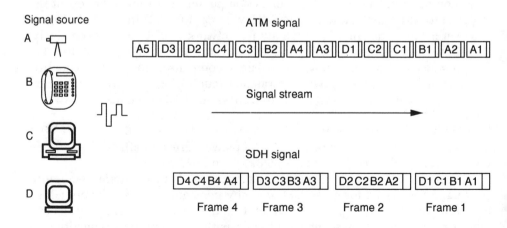

Figure 2.20 Comparison of SDH and ATM multiplexing functions.

2.4 WAVELENGTH DIVISION MULTIPLEXING

The high cost involved for the implementation of submarine cables has always been an incentive to operate submarine links at their highest possible transmission capacity. Special efforts were justified in using systems at higher transmission capacities than existed for landline systems. In the analog days special multiplexing systems with 3-kHz channel spacing instead of the conventional 4 kHz were used. The first circuit multiplication equipment described in the next section increased the capacity of an analog 12-channel group by a factor of 2.5, thereby recouping the investment cost of the multiplication equipment in a few months only. In the digital era 280-Mbps systems were used on submarine cable (for example, on TAT-8) while 140 Mbps was the state of the art for terrestrial systems. Whereas 2.5 Gbps is the highest bit rate for terrestrial systems currently in use, TAT-11 has operated on 5 Gbps since 1995 and TAT-12/13, which is currently being laid, will operate 2×5 Gbps.

Here the question arises why and how on a single fiber pair 2×5 Gbps and not 1×10 Gbps? The answer lies in the physical limits of optical fiber and optical amplifiers as explained in Chapter 4 and in the fact that, with the present technology, generating transmission speeds becomes increasingly difficult near 20 Gbps.

A solution for achieving a higher transmission capacity of optical fiber cable without increasing the transmission speed beyond practical limits is, apart from adding optical fiber pairs, to apply WDM. Similar to analog multiplex, with WDM a number of digital TDM signals are modulated on individual optical carriers at slightly different frequencies. The modulated optical signals operating parallel on adjacent frequencies are applied to one single fiber pair. The simplest form of WDM, which has already been applied for many years, is to transmit one modulated signal through the 1.3-μm window and a second modulated signal through the 1.5-μm window of the same fiber. The optical signals applied on fiber have a bandwidth of typically only a few nanometers, whereas the transmission window at 1,550 nm has a bandwidth of approximately 50 nm. State-of-the-art *erbium-doped fiber amplifiers* (EDFAs) have a flat gain bandwidth of typically 10 nm, allowing two 5-Gbps signals to be transmitted in parallel through the 1.5-μm window on one single optical fiber resulting in the aggregate 2×5 Gbps = 10 Gbps transmission capacity. To distinguish the operation of multiple signals within one window from parallel operation through the two windows, the operation of multiple signals within one window is also called *Dense*-WDM (DWDM).

Availability of EDFAs with a flat gain bandwidth similar to the bandwidth of the fiber transmission window combined with laser emitting extremely narrow pulses thus would allow several optical signals to be transmitted in parallel on the same fiber. The feasibility of 40-Gbps transmission with 16 wavelength-multiplexed 2.5-Gbps signals within a 25-nm bandwidth has recently been demonstrated. The aggregate 40-Gbps signal could be transmitted over 440 km of standard optical fiber cable using four cascaded experimental fluoride-based EDFAs.

An example of eight parallel 2.5-Gbps signals covering a bandwidth of 40 nm and producing an aggregate 20-Gbps transmission capacity on one single optical fiber pair is shown in Figure 2.21 [8].

Further progress in (D)WDM, for example, with 8 × 10 Gbps = 80 Gbps or even 16 × 10 Gbps = 160 Gbps will pave the way for the super high-speed data highways mentioned in Chapter 10.

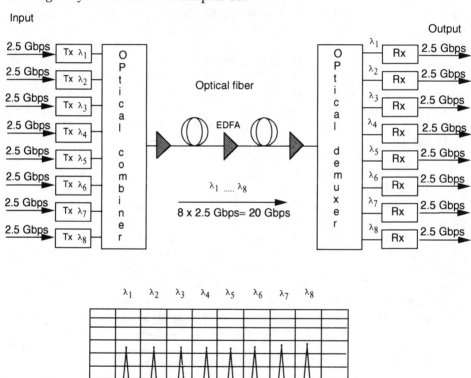

Figure 2.21 WDM 20-Gbps optical transmission (*After:* Alcatel).

2.5 CIRCUIT MULTIPLICATION

Circuit multiplication was originally developed in the Bell Laboratories for submarine application. It was first applied in 1960 to increase the capacity from 48 telephone circuits to 74 on both TAT-1 (Transatlantic Telephone cable number 1) and TAT-2 by means of a technique called *time assignment speech interpolation* (TASI). Both TASI and the presently used *digital speech interpolation* (DSI) are based on the fact that telephone conversations are essentially a one-way activity—when one subscriber speaks, the other usually listens; thus, the two-way circuit is used only 50% of the time. In practice the utilization is even less because of pauses in the conversation and the fact that lines are seized by the telephone exchange several seconds before speech can start, so the average utilization of each transmission path is around 40%.

A *circuit multiplication system* (CMS) consists of two terminals, one at each end of the transmission route whose capacity is to be multiplied. The CMS terminals are inserted between the trunk side of an exchange (public or even PABX in private networks) and the first stage multiplexers. The near-end CMS terminal scans the trunk lines connected to it; and as soon as a subscriber starts to speak, it routes this speech down to any transmission path that is free at that instant. The far-end CMS terminal reroutes the "speech-burst" (which continues for as long as the subscriber continues to speak without an appreciable pause) back to the correct subscriber line. As soon as a subscriber pauses for more than a few milliseconds, that transmission path becomes available for any other speaking subscriber.

In this way CMS "fills up" the unused transmission capacity of each transmission path—which had previously consisted of 60% silent periods—thus increasing the installed capacity typically by a factor of 2.5.

It should be noted that this capacity increase is obtained by adding the two CMS terminal stations only, without any modifications to the repeaters along the existing route.

The presently used *Digital Circuit Multiplication System* (DCMS) beyond applying DSI incorporates *variable bit rate adaptive differential* PCM (VBR-ADPCM), which replaces the standard primary rate PCM equipment and adapts the bite rate to the traffic. The phrases, words, or syllables of speech that are recognized and separated from the pauses by the DSI are applied to the VBR-ADPCM, in which the standard 8-bit digital coding (of primary rate PCM) is reduced as a function of the traffic load and type of traffic to 5-, 4-, 3-, and even 2-bit coding [also called *very low rate encoding* (VLRE)]. With VBR-ADPCM typical speech circuits during off-peak hours are 4-bit (32-Kbps) encoded; during busy hours 3-bit (24-Kbps) and even 2-bit (16-Kbps) encoding is applied to ensure a trunk-to-bearer connection for a maximum of active speech channels. Analog voice channels loaded with data up to 4,800 bps are

4-bit encoded and 5-bit (40-Kbps) encoded for data between 4,800 and 9,600 bps. Standard 8-bit encoding is applied for 64-Kbps data channels. To enable deployment on transatlantic submarine or satellite routes (the major application of DCMS) the DCMS can include transcoding from VBR-ADPCM to and from 8-bit μ-law 1.5-Mbps primary rate as well as to and from 8-bit A-law 2.0-Mbps primary rate, which is important for the transatlantic telephone traffic.

The combination of DSI and VBR-ADPCM allows a multiplication factor from 6 to 8, depending on speech/data mix, time of the day, and time-zones involved; so 180 to 240 VF channels can be connected via DCMS to one single 30-channel primary group.

Figure 2.22 shows a simplified schematic of a DCMS terminal serving typically 180 and maximum 240 trunk lines on a 2- or 1.5-Mbps bearer. The trunk access and adaptation units include line code/binary conversion and synchronization of the digital streams. The delay line inserted in the trunk lines, as shown in the upper part of the figure, introduces a typical 20-ms delay to offset speech detection time and interterminal signaling processing time. The noise generator in the middle of the figure is included for listening comfort. On the receive side the natural speech pauses not only have to be inserted

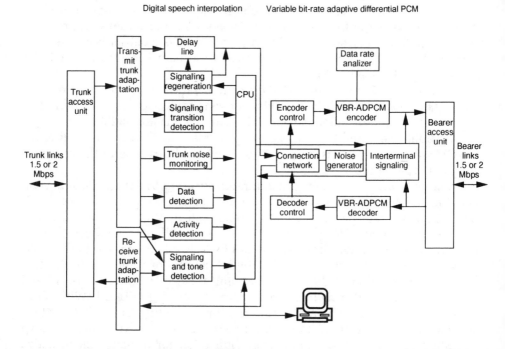

Figure 2.22 DCMS terminal (*After:* Alcatel).

again but also filled with artificial noise to prevent the listener from getting the impression of a line interruption. For the same reason noise has to be inserted on the trunk side of the voice channel during pause elimination periods. The noise generator, therefore, provides a typically 16-level adaptive noise matching signal. The VBR-ADPCM encoder and decoder ensure transcoding from 8-bit μ-law or A-law to the variable-length ADPCM. The operator station can control the operational functions of a terminal, collect status reports and statistics, and on larger stations, control automatic protection switching for a maximum configuration of seven such terminals and one for standby.

The various characteristics of DCMS are summarized in CCITT Recommendation G.763.

DCMS is mainly used on transoceanic submarine and satellite routes, but it can be used on any four-wire transmission including open wire or troposcatter and is often applied on long terrestrial cable and radio-relay routes, which are used to their maximum capacity and otherwise would need to be replaced. Even provisional use of DCMS can be economically justified to delay unavoidable extensions or to solve temporary congestions. DCMS can be implemented from the time of initial commissioning of a transmission link to reduce the cost per channel or as a retrofit to add extra capacity to keep pace with increasing traffic. Both the implementation of WDM on optical submarine cable and the dramatic shift from telephony to data transmission may significantly reduce future DCMS deployment [1,9,10].

2.6 INVERSE MULTIPLEXING

Contrary to classical multiplexing where multiple channels are grouped together for transmission on a single line, *inverse multiplexing* (IM) separates a wideband signal into x narrowband signals to enable transmission over x parallel narrowband lines. Typically a 384-Kbps video signal from a video conference codec can be split into six separate ISDN B-channels with each 64 Kbps for transmission over six N-ISDN lines. These six B-channels may then follow different paths through the ISDN switched network, resulting in a slight difference when each channel arrives at the point of destination. At the point of destination this difference is eliminated and the original wideband signal recreated by means of a special synchronizing algorithm.

IM is a new technique based upon a so-called "Bonding" (an abbreviation of: Bandwidth ON Demand INteroperability Group) standard that is likely to be incorporated into the ITU-H series recommendations for videoconferencing, but the IM technique can be employed in applications other than videoconferencing that require an aggregation of 64-Kbps channels.

IM exists likewise with a bandwidth-on-demand feature and data splitting onto two primary rate ISDN channels thus providing a 4-Mbps public network

access that can be shared, with dynamic allocation depending on actual requirement and time of the day, by such services and applications as high-definition
video conferencing, electronic imaging, traffic overflow from LAN, PABX or
leased lines at peak times, and high-speed data.

A further version of IM is presently (early 1996) being prepared by the
ATM Forum (a worldwide assembly of over 750 companies, ATM producers,
and users set up to accelerate the global acceptance and deployment of ATM
switching and transmission) as ATM-IM (AIM). Whereas IM is conceived for
rates up to 2 or 4 Mbps, AIM will combine multiple 2-, 8-, and 34-Mbps PDH
and eventually even 155-Mbps SDH lines and convert them to a single high-
speed ATM interface.

The AIM interface is being conceived to send voice video and data in the
form of parallel streams of cells across multiple lines, dynamically balancing
the cells over these lines. At the receiving side the cell streams are reconstructed,
placing the cells in their original sequence. Depending on the path of each line,
the ATM cells may experience different delays in reaching their destination,
which means that cells may be received in the wrong order and so it is necessary
to correct this by detecting and compensating the relative path delays. The AIM
control protocol in the ATM header will handle this function taking less than
1% overhead. Initially an AIM interface standard is planned to be available
in 1996 for up to eight E1/T1 circuits and thus up to 16 Mbps, respectively,
12 Mbps [11,12].

2.7 DATA TRANSMISSION INTERFACE EQUIPMENT

2.7.1 General

In the early days of electronic data processing when switched data networks did
not yet exist, a device called "modem" was introduced as a data transmission
interface to enable the transmission of data, converted from digital to analog,
via the subscriber line and the analog PSTN. When PSDNs eventually appeared,
the necessity arose to define interfaces and provide *data circuit-terminating
equipment* (DCE) between the PSDN and the *data terminal equipment* (DTE).
Data transmission interface equipment can be subdivided, therefore, in two
categories (shown in Figure 2.23):

- *Modems*: for analog transmission of digital data signals via a PSTN;
- *DCEs*: interfacing digital data equipment with digital data networks via a
  PSDN.

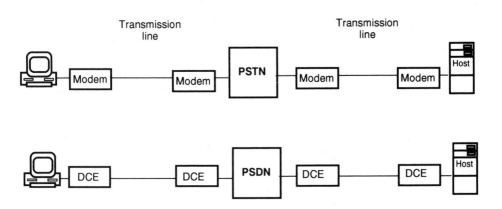

Figure 2.23 Basic data communication links.

2.7.2 Modems

A modem, as the word implies, consists of a modulator and a demodulator. A modulator changes certain characteristics of an electrical signal—the carrier—in accordance with the characteristics of a second electrical signal. A demodulator derives the characteristics of the second signal again from the modulated carrier. In the field of transmission there are two categories of modems: those that interface digital data terminal equipment with analog telephone networks and those used in the radio-relay and satellite technique to modulate/demodulate a baseband signal onto/from an *intermediate frequency* (IF) in heterodyne radio transmission systems (see Chapter 5).

A data modem has the function to convert digital signals to such analog ones that efficient transmission is possible over the telephone network. A VF telephone channel in the telephone network has a bandwidth of 300 to 3,400 Hz. The digital data signals thus need to be converted to analog signals within that frequency band. Figure 2.24 shows the principle of the following basic modulation modes:

- ASK: amplitude-shift keying with amplitude modulation;
- FSK: frequency-shift keying with frequency modulation;
- PSK: phase-shift keying with phase modulation.

ASK is the first applied and simplest technique that, apart from exceptionally on optical fiber, is seldom used anymore because of its sensitivity to line errors.

FSK is still widely used for low-speed modems (up to 1,200 bps) or with multiple carriers, instead of two, for higher speeds or in the case of duplex

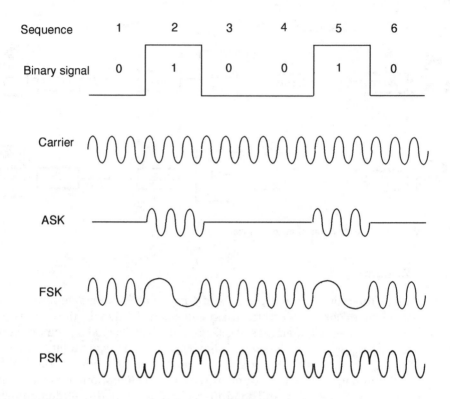

Figure 2.24 Basic modulation modes.

operation (simultaneous transmitting and receiving). A typical duplex two-wire FSK modem transmits 1,070-Hz and 1,270-Hz signals to represent a binary 0 (space) and binary 1 (mark), respectively, and receives 2,025-Hz and 2,225-Hz signals as binary 0 and binary 1.

PSK is increasingly being used for higher speeds, especially in a variation called "differential-PSK" whereby each bit in the data is coded as a phase change relative to the previous change of the carrier. For example, a "0" bit can be coded as a 90-degree phase change and a "1" bit as a 270-degree phase change.

For further explanation about modulation/demodulation please refer to Chapter 5.

Modems are furthermore divided in two groups:

- VF-modems: for operation on telephone circuits within the PSTN;
- Baseband modems: for connection via a subscriber's analog access lines to the PSDN.

The transmission speed on baseband modems, contrary to the VF-modems, is not limited to the 300- to 3,400-Hz bandwidth of the PSTN but is limited to the bandwidth of the subscriber line in the access network, which is typically around 2 Mbps. With baseband modems, therefore, much higher speeds such as 19.2, 48, 56, 64, 72, 96, 128, 144, 256, 288, 512, and 576 Kbps can be used.

Modems exist for a large variety of applications, and modems at both ends of a circuit need to be compatible. CCITT has therefore published in the *1988 Blue Book VIII.1* "Data communication over the telephone network. Series V" a comprehensive "V-series" of recommendations, frequently updated, for transmission speeds from 300 bps to 64 Kbps.

As mentioned at the beginning of this section, modems were introduced (in 1958) to enable the transmission of data, converted from digital to analog, via the subscriber line and the analog PSTN. Now, after almost 40 years, a new type of modem appears on the market for the transmission of voice simultaneously with data! This *simultaneous voice and data-modem* (SVD-modem) exists in two versions:

- ASVD-modem: analog simultaneous voice and data modem;
- DSVD-modem: digital simultaneous voice and data modem.

The ASVD-modem compresses the voice, similar to TASI, thus providing space within the 4-kHz voice band for data a modern version of narrowband voice plus VF-telegraphy within a single VF-channel.

The DSVD-modem first applies digital speech interpolation and then this reduced voice signal is packetized and multiplexed together with data and the DTMF-signaling into the full 4-kHz speech channel. DSVD is implemented via a mode of operation that allows the modem to initiate the SVD mode without suspending an online connection. DSVD modems thus can transfer office functions to individual homes or remote locations, thus facilitating telecommuting without requiring an ISDN line. A telecommuter with their home telephone, their PC, and in the near future their fax connected to a DSVD modem can call into their office where a further DSVD modem will provide a connection to the office PABX and the office data processing equipment, or to a LAN, as if they were in the office. The telecommuter thus can make use of the IN facilities of the office PABX.

Finally, modems are also used to connect laptops to digital cellular radio mobile units. These modems are the size of a credit card, also called *data service adapter* (DSA) or *PCM common interface adapter* (PCMCIA), and support fax and data operation via cellular networks.

Modems share a reputation with radio-relay transmission—both technologies are repeatedly predicted to disappear but still continue to find wide application. Whereas radio-relay transmission will increasingly be used in the access network, modems with applications like the above DSVD and DSA will continue to be used for many years to come [4,5,13,14].

2.7.3 Digital Data Interface Equipment

The large variety of data processing equipment such as computer, PC, plotter, CAD/CAM-terminal, word processor, facsimile, telex, teletype, teletex, videotex, email, and directory supplied by numerous manufacturers, which should communicate in an orderly and structured manner via a data network, require detailed agreements about the interface to and the way of communication via the PSDN. Such detailed agreements called "protocols," which are to be implemented in the DCE, have been standardized by the *International Organization for Standardization* (abbreviated ISO from its former name, the International Standards Organization) and by CCITT.

The ISO has developed a comprehensive standard for "open systems" (systems that, instead of a proprietary interface with access limited to equipment of the same or of a licensed supplier, interfaces with equipment from any supplier that adheres to the common "open" standards), that was published in 1984 as the *open systems interconnection reference model* (OSIRM) in their specification ISO 7498. CCITT has incorporated this ISO model in Recommendation X.200 Open Systems Interconnection Model and Notation.

In OSIRM the many communication functions are defined and grouped in a hierarchical system of seven clusters of functions called "layers" as summarized in Table 2.12 and shown in Figure 2.25.

Table 2.12
Open Systems Interconnection Reference Model

Layer Number	Layer Name	Layer Functions
1	Physical	Control of data circuits, transfer of bits
2	Data link	Transfer of blocks of data between nodes including error detection and recovery
3	Network	Routing blocks of data through a network
4	Transport	End-to-end control of transportation of data; optimization of usage of network resources
5	Session	Communication initiation, dialogue control, synchronization and release
6	Presentation	Data formats, coding, transfer syntax, and representation
7	Application	System and application control and identification of corresponding user

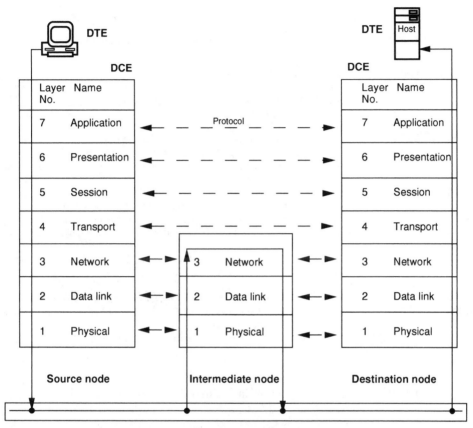

Figure 2.25 ISO 7-layer reference model.

The layers 1 to 4, also called the lower layers, concern the transport functions; whereas the layers 5 to 7, the upper layers, define the protocols for the data application. The function clusters of the seven layers are briefly described in Subsections 2.7.3.1 to 2.7.3.7.

The conditions for the actual transmission of the data through the transmission media are not covered by the seven OSI layers. The transmission interface of the DCEs with the nodes shown in Figure 2.25 normally will be either a direct or a multiplexed subscriber access line to the next public or (for example, in a LAN) private, data or ISDN exchange. The data transmission interface conditions (among other data transmission conditions) of the DCEs with the indicated nodes and with the DTEs are defined by CCITT and ITU-T in the X- and I-series of recommendations as summarized in Table 2.13.

Table 2.13
CCITT X- and I-Series of Recommendations for Data Transmission Interface

Recommendation	Description
X.1	International user classes of services in public data networks and ISDNs
X.2	International data transmission services and optional user facilities in public data networks and ISDNs
X.7	Technical characteristics of data transmission services
X.10	Categories of access for DTE to public data transmission services
X.21	Interface between DTE and DCE for synchronous operation on public data networks
X.25	Interface between DTE and DCE for terminals operating in the packet mode and connected to public data networks by dedicated circuit
X.28	DTE/DCE interface for a start-stop mode DTE accessing the packet assembly/disassembly facility PAD in a public data network situated in the same country
X.30	Public data networks: Interfaces-Support of X.21, X21bis, and X20bis-based DTEs by an ISDN X.36, Frame relay UNI specifications
X.76	Frame relay NNI specifications
I.122	Framework for providing additional packet mode bearer services
I.233	Frame mode bearer services: ISDN frame relaying bearer service, ISDN frame switching bearer service
I.372	Frame relaying bearer service network-to-network interface requirements

A brief description of the major data transmission protocols and interfaces is given in Subsections 2.7.3.8 to 2.7.3.22. Reference [4], from which significant information has been drawn for those brief descriptions, can be recommended for further study of this complex matter [4,15].

2.7.3.1 Layer 1 Physical

Layer 1 is concerned with the physical, mechanical, electro-optical, functional, and procedural characteristics necessary to establish, maintain, and disconnect the physical link of a communication network. It defines attributes of a data communication system such as

- Wiring connections;
- Electrical, electromagnetic, or optical characteristics of the data signal;
- Details of the mechanical connectors such as dimensions, number and utilization of pins;
- Signal synchronizing;
- Grounding.

2.7.3.2 Layer 2 Data Link

This layer covers the data segmenting, protection, error correction, and data flow functions. The data segmenting provides easily manageable frames (blocks) to be moved in and out of storage buffers of the DTE.

2.7.3.3 Layer 3 Network

This layer provides the protocols for the routing of the data via the network from DTE to DTE, establishing and supervising the route from source to destination.

2.7.3.4 Layer 4 Transport

The transport layer provides the protocols for the correct assembling of the data blocks (from layer 2) and transparent transfer of data between the DTEs. This layer includes such functions as multiplexing a number of independent message streams over a single connection and segmenting data into appropriately sized units that can be handled by the network layer.

2.7.3.5 Layer 5 Session

The session layer establishes and maintains and terminates the communication connection, including the negotiation of transmission rules and dialogue coordination and recovery from prematurely terminated conversations.

2.7.3.6 Layer 6 Presentation

This layer is responsible for the task of ensuring a consistent, mutually acceptable representation of data while in transit. It established the syntax of the information including a translation to the local system if required.

2.7.3.7 Layer 7 Application

The application layer is concerned with communication activities on behalf of the ultimate *application service elements* (ASEs) of the DTE. This layer consists functionally of two parts: *common application service elements* (CASE) and *specific application service elements* (SASE).

CASE includes association control with other applications within the OSI environment, identification of the corresponding DTE, establishment of an association so that communication can take place between corresponding DTEs, and provision of coordination between multiple application entities. Under SASE a still growing number of service standards are defined, of which the major service standards are

- ISO 10021 CCITT X.400 and F.400 families. *Message handling system* (MHS). MHS is a store-and-forward electronic messaging architecture as a platform for services as email, mail boxes, interpersonal messaging service, public message transfer service, and electronic data interchange.
- ISO 8571 1-5. *File transfer, access, and management* (FTAM) providing the ability to transfer and access files remotely across multiple computer systems.
- ISO 9594 1-7 CCITT X.500 family. *Directory services* (DS). This family of standards specifies common protocols to retrieve and update information stored in globally distributed data banks, for example, the access to a worldwide telephone directory.
- ISO 8831 and 8832 CCITT X.216 and X.226. *Job transfer and manipulation* (JTM) enables users to define, submit, and receive results of processing on remote-end systems and to suspend, cancel, or resume processing.
- ISO 10026 1 and 2. *Transaction processing* (TP) handling a wide range of computer-to-computer transactions.
- ISO 9040 and 9041. *Virtual terminal* (VT) allowing a device-independent way of representing data and control actions supporting communication between, respectively, two human terminal users, two computer systems, or a human terminal user and a computer.

2.7.3.8 X.25

Recommendation X.25—drafted in the 1960s; published by CCITT in 1976; and updated thrice in 1980, 1984, and 1988—is applied worldwide as the access protocol for public packet switched data transmission at speeds of 4.8 to 64 Kbps and high-speed X.25 up to 2 Mbps.

X.25 essentially acts as a cost-effective means of multiplexing data from a number of sources to a number of destinations. The packets are routed through a data network interleaved with a variety of packets from other sources. After passing through sophisticated error correction circuitry at each link and destination node, upon arrival in the destination DCE each packet is stripped of its control and address information and passed to the end user. The control information that is sent at the outset of the packet transmission ensures that the packets are reassembled in the correct order, regardless of the order in which they were received.

2.7.3.9 Transmission Control Protocol/Internet Protocol (TCP/IP)

ITU-T standardized *transmission control protocol/Internet protocol* (TCP/IP) is a ARPA-based pre-OSIRM protocol already defined in the early 1970s. TCP/IP basically specifies the details of how computers communicate and provides a set of conventions for interconnecting networks and routing information. In essence, TCP takes care of the integrity and IP moves the data. TCP/IP architecture and protocols acquired their current form in the early 1970s as ARPANET, now Internet, protocols were developed for communication between UNIX computers. In the early 1980s ARPA made TCP/IP available to the academic community at a low cost, which paved the way for worldwide Internet access. In the meantime, over 150 vendors support TCP/IP for both LAN and WAN operation. TCP/IP, either incorporated in a PC or attached to a PC by means of a low cost adapter, enables worldwide access to hosts on remote networks and file and mail transfer via Internet [4].

2.7.3.10 SNA

SNA, or *system network architecture*, is one of the oldest proprietary protocols introduced by IBM in 1971. SNA is still widely used for communication between IBM and IBM compatible computers. SNA interface to public networks are possible via *synchronous data link control-lines* (SDLC-lines).

2.7.3.11 Ethernet

Ethernet was developed by Xerox in their Research Center at Palo Alto in 1973 as a low-cost wideband method of sending packets of data between Xerox, and Xerox-compatible, office machines like computers, printers, and copying machines in the emerging LANs. The name "Ethernet" (obviously inspired by "ARPANET," which was in service since 1971; see Subsection 1.5.4) was derived from a long-ago discarded notion that electromagnetic energy is transmitted through a fluid substance that permeates the Universe, called "ether." The Ethernet protocol is based upon ALOHA random access and is so-named because the access to a common data line is based upon a principle developed (for satellite transmission) at the University of Hawaii. With the Aloha protocol, now called *carrier sense multiple access/collision detection* (CSMA/CD) any user can send packets of data without checking whether the line is free. If the line is free, the user obtains an acknowledgment that the information has been sent. Should such an acknowledgment not be received within a specified time, a collision of the sender's data with other data will have occurred and a second attempt has to be made.

Ethernet initially was defined for operation on coaxial cable at speeds up to 10 Mbps. Based upon this proprietary Ethernet protocol, the standard IEEE 802.3 was published in 1985. The IEEE 802.3, respectively, the original Ethernet, has a frame with a maximum of 1,518 octets of information. Network management, source routing, and transparent bridging within LANs are defined in IEEE 802.1; whereas IEEE 802.2 under the title LLC (*Local Link Control*) defines peer-to-peer (equal level) protocol procedures for information transfer.

With *full duplex Ethernet* (FDE) simultaneous transmittal and receipt is possible. A latest addition to Ethernet duplex operation is given with standard IEEE 802.9, which defines *integrated voice, video, and data interfaces* to LAN (IVDLAN). This standard for interface via *unshielded twisted-pair* cable (UTP; see Chapter 3) includes the use of a special *terminal access* (TA) allowing PCs and telephones to be adapted to the interface and line *access units* (AUs) permitting access to ISDN via an *integrated services* PBX (ISPBX). This interface supports isochronous (voice) and asynchronous traffic (computer data). Isochronous channels provide a guaranteed bandwidth and minimal delay. Asynchronous channels require an arbitration scheme to resolve contention among multiple users. The transfer rate is defined at multiples of 64 Kbps up to initially 4.096 Mbps.

Important security matters for LAN operation are defined in security standard IEEE 802.10, which includes provisions for authentication, access control, data integrity, and confidentiality. Authentication prevents stations from reading packets destined for other stations access control limits the use of resources (such as files, servers, and gateways) by unauthorized users data integrity guarantees that the data are not modified before they reach destination and confidentiality uses encryption to mask the information to users without the appropriate key.

Ethernet with various modifications is now available for network speeds up to 100 Mbps [4].

2.7.3.12 X/Base Systems

The aforementioned IEEE 802.3 standard, which was defined for coaxial cable, has been modified by IEEE for operation on twisted pair cable [the usual access cable for private and *small office home office* (SOHO) subscribers] and for coaxial cable into the following xBaseY (in which x indicates the bit rate, Base stands for Baseband, and Y is the serial number or service indicator) standards.

- 10Base-T: standardized in 1990 for baseband LAN operating at 10 Mbps over two twisted pairs, allowing a distance of 100m between station and hub and the connection of a maximum of 2 units of equipment;

- 1Base5: for baseband LAN operating at 1 Mbps over one twisted pair, allowing a distance of 250m between station and hub;
- 10Base5: for baseband LAN operating at 10 Mbps over thick coaxial cable ("yellow cable"), allowing a distance of 500m between station and hub and the connection of a maximum of 100 units of equipment;
- 10Base2: for baseband LAN operating at 10 Mbps over thin coaxial cable (for example, type RG-58), allowing a distance of 185m between station and hub and the connection of a maximum of 30 units of equipment.

In early 1996 two competing extensions of 10Base-T for operation on 100 Mbps are being considered: (1) 100BaseVG—where VG stands for "voice grade," indicating that operation on normal telephone cable is envisaged; and (2) 100BaseVG-Anylan—which constitutes a versatile compromise between Ethernet and Token Ring and covers operation on two twisted pairs of Category 3 UTP cable, or two screened pairs, or optical fiber.

For operation on optical fiber cable IEEE is developing special 10Base-F standards in the following versions:

- 10Base-FB, for fiber backbone systems;
- 10Base-FL, for link segments;
- 10Base-FP, for passive star configurations [4].

2.7.3.13 Token Passing

The "token passing" concept—obviously inspired by an old railway collision prevention procedure on single-track sections, whereby the engine driver personally had to take a token before entering the track—uses a special control packet that circulates around the network from node to node. The token thus avoids the collision inherent in Ethernet by requiring each node to defer transmission until it receives a "clear-to-send" message—the token. Each node constantly monitors the network to detect any packet that is either addressed to it or which (without address) is the token. When the token is received by a node and the node has nothing to send, the token is passed along to the next node. If the token is accepted, it is passed on after the node has completed transmitting their data. The token must be surrendered within a specific time so that no node can monopolize the network resources. The token-passing concept was developed by IBM in the early 1980s, and as such it is mainly known as the "token ring" method, although bus operation is possible too. The IBM proprietary token ring method supports speeds up to 16 Mbps operating on twisted pairs. The token-passing method has been standardized for speeds up to 4 Mbps in 1985 under IEEE 802.4 (since 1989 ISO 8802-4) for bus access and in IEEE 802.5 (ISO 8802-5) for ring access [4].

2.7.3.14 Fiber Distributed Data Interface (FDDI)

FDDI stands for a series of standards being developed by the Task Group X3T9.5 of ANSI since 1982. FDDI is mainly for private (rather than public) application of high-performance high-speed packet switched data for asynchronous and synchronous transmission over various types of optical fiber cable. FDDI grew out of the need for high-speed interconnections among mainframes, minicomputers, and associated peripherals in general-purpose multiple-station networks. FDDI basically encompasses a token-passing network employing two pairs of fiber in a counterrotating ring. The *medium access control* (MAC) briefly operates as follows:

- A token (a small control packet) is circulating around the ring.
- A station must wait for and remove the token before transmitting (access delay is typically 0.3 ms at 25% network loading and 0.5 ms for 50% loading).
- Each station repeats the frame it receives to the downstream neighbor.
- If the destination address matches the stations address, the frame is copied into the stations buffer.
- The receiving station sets indicator symbols in the frame status field to confirm reception.
- The transmitting station receives the data frame with the reception confirmation and then is responsible to remove its data frame from the ring.

The major characteristics of FDDI are summarized in Table 2.14 [4].

The success of FDDI and the emerging technologies supporting the use of ordinary twisted pair cable in the access network for data transmission have lead to the following modifications of FDDI for operation on copper lines:

Table 2.14
Major FDDI Characteristics

Standardized by Topology	ANSI Task Group X3T9.5 dual counterrotating token-passing ring
Capacity	100 Mbps on each ring
Bandwidth	125 MHz
Coding	4B5B NRZI
Frame length	4,500 octets (bytes) including header and trailer
Ring access	multimode fiber up to 2 km; single-mode fiber up to 50 km
Path-length	100 km (per fiber)
Network capacity	500 stations
Station-station spacing	maximum 2 km
Isochronous traffic	defined in FDDI-II
Overall BER	$< 1 \times 10^{-9}$

- FDDI-over-UTP: Fiber distributed data interface over UTP;
- CDDI: Copper distributed data interface;
- TPDDI: Twisted-pair distributed data interface;
- TP-PMD: Twisted pair-physical medium dependent.

2.7.3.15 Frame Relay

Frame relay was standardized in 1988 as a variable-length packet technology using statistical multiplexing to allow users to share WAN bandwidth. The term "relay" refers to the procedure that the data frame, which is initiated in layer 2 at the source DCE, is relayed to its final destination without termination and processing at the in-between nodes of links in the entire network. In the source DCE the data is encapsulated in a *link access procedure*-D (LAP-D, taken from ISDN and one of the reasons why frame relay can easily be handled in ISDN) into packets with a frame, as shown in Figure 2.26.

The boundary of each frame is determined by a one-byte "flag." The routing information is in a 2-byte header. The payload is contained in a variable-length maximum 262-byte user data part followed by a two-byte frame sequence checking part. As this frame passes through the frame relay network it is switched to the proper destination by means of the *data link connection identifier* (DLCI). The *forward explicit congestion notification* (FECN) and the *backward explicit congestion notification* (BECN) are used to warn the receiving or sending devices that the frames have encountered or will encounter congestion. The C/R (*command/response*) indication is application-specific. The EA (*address field extension*) defines the destination. The DE (*discard eligibility*) indicator is set in the source node on frames that are bursting above their

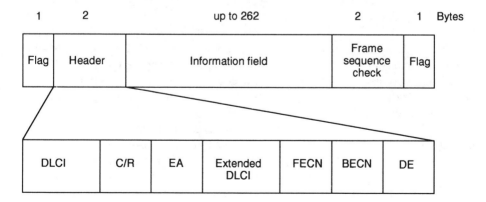

Figure 2.26 The frame of frame relay.

associated CIR (*committed information rate*) to inform the frame relay switches which frames can be discarded in case of acute congestion problems. This lean frame header occupies typically 1% only of the bandwidth compared with 10% for cell-based ATM.

Whereas X.25 was still conceived within a mainly analog environment, frame relay has been designed to take advantage of the rapidly expanding digital transmission and switching infrastructure with high bandwidths and extremely low error rates. Frame relay provides an effective flexible solution for bursty data traffic with bandwidth sharing in an operation range between 9.6 and 45 Mbps.

The attractiveness of frame relay for corporate networks has even led to the development of procedures for the integration of voice over frame relay. The Frame Relay Forum—an international group of service providers, equipment vendors, users, and other interested parties promoting the acceptance and implementation of frame relay based on international standards (formed in 1991)—in early 1996 worked out the *Implementation Agreement* (IA, a foundation of practical frame relay solution) for packetizing voice for transport over frame relay networks in compliance with ITU-T Recommendation G.764 Packet Voice Protocol.

Before packetizing a 64-Kbps digital voice channel, the data stream is compressed by a factor of four, using a form of DSI and VBR-ADPCM similar to that explained in Section 2.5. The resulting digital information together with a voice-packet-header is packaged into the standard frame relay data frame. An appropriate CIR level eliminates the danger of discarded packets disrupting voice conversation [4,16,17].

2.7.3.16 X.400

The MHS standard series X.400 was published by CCITT in the *1988–1989 Blue Book* together with the X.500 standard for Directory Services. X.400 closes the bridge between private and public network domains, offering an end-to-end audit trail that enables the sender to check that a message reaches its destination regardless of actual routing.

Basically a message submitted by a user, the originator, is conveyed by a *message transfer system* (MTS) and delivered to one or more message recipients. *Message transfer agents* (MTA) cooperate to perform the store-and-forward MTS function. Messages stores provide storage for messages and enable their submission, retrieval, and management. *User agents* (UAs) help users access the MHS. *Access units* (AUs) provide links to other communication systems and services (for example, other telematic or postal services) [4].

2.7.3.17 X.500

This family of standards specifies common protocols or retrieving and updating information stored in globally distributed data banks, for example, access to a worldwide telephone directory.

Basically a user accesses a directory via its *directory user agent* (DUA), which interacts with *directory system agents* (DSAs). A DSA provides access to other DUAs or to the *directory information base* (DIB) in which all information entries are arranged in the form of a tree called the *directory information tree* (DIT). All directory services are provided in response to requests from DUAs that may concern interrogations or modifications of the directory.

2.7.3.18 Distributed Queue Dual Bus (DQDB)/IEEE 802.6

The DQDB system, unlike FDDI, is for public high-performance high-speed packet-switched and circuit-switched application in *metropolitan-area networks* (MANs) operating at speeds of 34, 45, 140, or 155 Mbps. MANs are conceptually equivalent to LANs but cover a much larger area and operate as public networks typically covering citywide areas.

Interestingly, DQDB was invented by students of the University of Western Australia in 1987. The university together with Telecom Australia founded a company called QPSX Communications Ltd. for the further development and production of the technology, whereas the IEEE published the technology as standard IEEE 802.6. This standard is closely related with the CCITT cell-based ATM standards to ensure worldwide interworking; in fact, DQDB can be considered an interim solution for broadband services until it will be technically outdated by, but still might continue to co-exist with, ATM-based B-ISDN networks.

DQBD, like ATM, defines a 125-μs frame with cells with a 5-byte header (slightly different from ATM) and 48-byte payload. The DQDB architecture uses a pair of contradirectional buses allowing full-duplex communication between any two attached nodes and automatic rerouting in case of an interruption on one of the buses. DQDB, unlike FDDI, is not exclusively specified for transmission on optical fiber and can be used on radio-relay and satellite links too.

To achieve equitable access among the nodes to the shared bus, the variable-length packets originating from the connected LANs are split up into fixed-length segments for transport in the queue arbitrated slots of the selected bus. At the destination node, the segments are assembled into the original packet.

Disciplined democratic access to the bus is obtained by a special (queuing) reservation method. Each station maintains a counter for each direction of transmission. The counter is incremented for each slot request (from other stations) arriving from the direction of the destination and is decremented for each vacant slot toward destination. If the counter is zero when a station wishes to transmit, it can use the next vacant slot. If, however, requests from other stations are still pending, then the counter will have a nonzero value, say 10; then the next 10 arriving vacant slots cannot be used because those have to serve the requests already pending at the moment that the station decided to send a message [4,18].

2.7.3.19 Switched Multimegabit Data Service (SMDS) and Connectionless Broadband Data Service (CBDS)

With the availability of MANs, multimegabit data streams can be switched, as opposed to point-to-point operation on leased lines; thus, in anticipation of the ATM-based B-ISDN networks, an interim solution is required for access to MANs. As such, a "technical advisory" has been published by Bellcore in 1988 on SMDS. In this advisory a *subscriber-network interface* (SNI) is defined that complies with IEEE 802.6. According to this advisory SMDS is offered via a SNI that is dedicated to an individual subscriber. The network validates that the source address associated with every data unit transferred using SMDS is an address that is legitimately assigned to the SNI from which the data unit originated. SMDS uses addresses similar to telephone numbers with extensions. The SDMS address protocol is currently being replaced by the ITU-T address protocol E.164, which initially was recommended for routing in the ATM network, thus ensuring efficient international routing. This address protocol provides a high degree of data security; it includes billing and enables application-specific security measurements.

SDMS basically is a LAN interconnect service that may run over DQDB or ATM platform. SDMS provides a datagram service in which each information packet is switched separately without the prior establishment of a virtual circuit connection. Up to 9,188 octets of user information can be delivered in one single transmission. SMDS, with bit rates from 64 Kbps to 155 Mbps, is defined for fiber-based transmission only.

Connectionless broadband data service (CBDS) is the ETSI-defined European counterpart of SMDS [4].

2.7.3.20 High-Speed Serial Interface (HSSI)

To meet the requirements of increasing bandwidth on copper lines, a still nonstandardized HSSI was created in the early 1990s. HSSI comprises the electrical and functional specifications first written in EIA Standard proposals SP 2795 and SP 2796. Comparable to other interfaces such as X.21/V.11 or V.36, HSSI is defined as the reference point between DTE and DCE, however, for a maximum speed of 52 Mbps (SONET ST-1) providing simple interface control in terms of handshake. The interface cable (25 twisted copper pairs) between DTE and DCE may have a maximum length of 15m. For subsequent long-distance transmission, a HSSI-to-fiber converter converts the HSSI signals to light pulses and adds an overhead with synchronizing and monitoring data into an optical fiber transmit frame [19].

2.7.3.21 Multilink Protocol (MP)

The combination of a recently proposed Internet MP that supports bandwidth-on-demand allocation to simultaneous or successive telephony, videoconferencing, fax, and data communication, and the *integrated access devices* (IADs) will enable small SOHO subscribers to connect their video, voice, fax, and data terminals to a single ISDN basic access line, thus enabling those subscribers to make more economical use of their existing telecommunication infrastructure. The IAD can be connected externally to a single ISDN BRI (two B-channels) line and internally to one Ethernet and two analog ports. The Ethernet port connects to the users workstation, or the SOHO-LAN. The two analog ports allow users to attach telephones, videophone, fax, modems, and answering machines and place and receive calls for these devices via the ISDN line. With compression, bandwidths of up to 512 Kbps can be internally distributed [20].

2.7.3.22 Application Overview

For a better understanding of the various protocols and interfaces described above, see Figure 2.27, which gives an indication of their major applications. ATM, N-ISDN, and B-ISDN have been included in that figure as a demonstration of their all-encompassing technology, which eventually will outdate the numerous data interfaces.

2.8 CODECS

2.8.1 General

To enable the transmission in a telecommunication network, voice signals are adapted in analog carrier frequency equipment or digital PCM, analog data signals are adapted in modems, digital data signals are adapted in DCEs, and audio and video signals finally are adapted for transmission in so-called codecs. A codec, as the name implies, consists of a digital COder and a DECoder.

Video and audio transmission qualitywise can be divided into the following four categories.

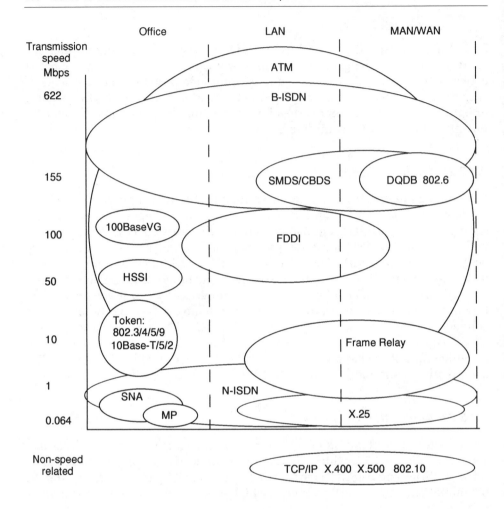

Figure 2.27 Data communication protocols and interfaces.

- *Contribution*: For professional high-quality one-way transmission to carry video and audio signals to production centers where post-production processing may take place, for example, on feeder links between TV studios. The essential feature of contribution is that the signal is not yet intended for the TV-viewer and does not necessarily represent a complete program because it may be subject to further editing and processing before release.
- *Distribution*: For professional one-way transmission to distribute television programs when no further post-production processing is expected, for example, within (inter)national TV distribution networks, to TV transmitters, CATV head stations, or Earth station for direct broadcasting services.

The term "distribution" refers to primary distribution (for example, to the input of a broadcast transmitter) and not to secondary distribution (for example, delivery to the TV viewer).

- *Videoconferencing*: For commercial interactive, thus both-way, transmission between conference studios or offices (CCITT G.722).
- *Videophony*: For transmission (in the access network) of personal interactive videoconferencing per PC or ultimately via the videophone (CCITT H.320).

2.8.2 Audio Codecs

For audio the coding and decoding is similar to the voice coding as explained in Subsection 2.3.2. As stated in that subsection "an analog signal can be converted into a digital signal of equal quality if the analog signal is sampled at a rate that corresponds to at least twice the signal's maximum frequency." A 15-kHz audio channel for HiFi sound broadcasting, therefore, as per ITU-T Recommendation J.41, is sampled at the rate of 32 kHz and then encoded using a 12-bit code to obtain a digital transmission speed of 32×12 Kbps = 384 Kbps, which happens to correspond with six 64-Kbps voice channels in the PCM primary rate. By means of *time-slot interchange* (TSI) up to five HiFi sound (for example, two stereo- plus one mono-) broadcasting channels can be accommodated in the 2-Mbps primary level. Similarly, up to ten 7-kHz audio broadcast channels, sampled at the rate of 16 kHz and 12-bit coded, can be transmitted within the 2-Mbps primary level. Figure 2.28 shows the accommodation of the coded audio channels in the 2-Mbps primary level frame. Music encoding frequently is done at 128 or 192 Kbps using MPEG (see Subsection 2.8.3) according to ISO standard 11172.

For commentary and videoconferencing audio channels, which do not require the full dynamic and bandwidth of music channels, CCITT in Recommendation G.722 has reduced the coding for 7-kHz audio signals from 12 to 4 bit resulting in a transmission speed of 16-kHz sampling $\times$ 4-bit coding = 64 Kbps. To enable the transmission of audio in a standard digitized voice channel, G.711 specifies the coding for both the North-American μ-law and the otherwise worldwide applied A-law.

Audio coding is, moreover, defined in the following recent ITU-T recommendations:

- J.21: Performance characteristics of 15 kHz-type sound-program circuits— circuits for high-quality monophonic and stereophonic transmissions.
- J.51: General principles and user requirements for the digital transmission of high-quality sound programs.

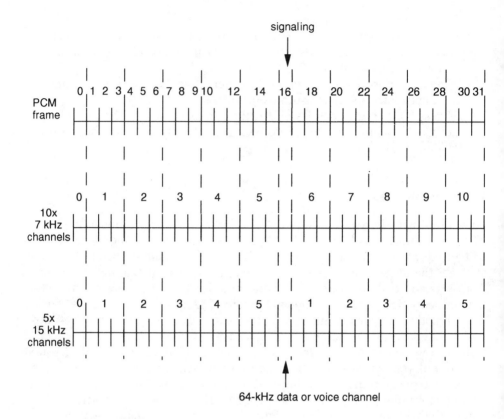

Figure 2.28 Accommodation of audio channels in the 2-Mbps primary level frame.

- J.52: Digital transmission of high-quality sound-program signals using one, two, or three 64-Kbps channels per mono signal (and up to six per stereo signal).

2.8.3 Video Codecs

2.8.3.1 TV Signals

TV signals for transmission exist in three different forms:

- Composite TV signal;
- Component TV signal;
- Studio standard TV signal.

A color TV camera basically has three separate pick-up tubes that generate the three primary colors: red, green, and blue. The luminance signal Y contains the three primary colors in a very specific mixture such that it creates a black-and-white picture, thus a compatability is given between color-TV and black-and-white TV. This mixture is derived from the chromatic sensitivity of the human eye and is internationally agreed upon as

$$Y = 0.3U_r + 0.59U_g + 0.11U_b \tag{2.1}$$

The chrominance signal logically—if color TV could have been developed independently from black-and-white TV—would have contained the direct information about the primary red, green, and blue colors. For historical reasons, however, in order to ensure compatibility with the existing black-and-white TV sets, the color information is squeezed within gaps of the initial black-and-white signal. To reduce the information content, therefore, a complicated procedure is applied whereby the color is defined in two color-difference components. These two color-difference components, C_r and C_b—which define the deviation from white—are obtained by subtracting the luminance value of (2.1) from the values of the red and the blue color signals and multiplying the resulting values, again in line with the chromatic eye sensitivity and in order to prevent overmodulation, with specific internationally agreed upon factors as

$$C_r = 0.877(U_r - Y)$$
$$C_b = 0.493(U_b - Y)$$

The generation of the Y, C_r, and C_b signals is in a much simplified form illustrated in Figure 2.29.

In black-and-white TV sets only the Y signal can be detected, whereas in color TV sets the C_r and C_b signals can also be detected and used for primary color recomposition in accordance with the formulas

$$U_r = C_r/0.877 + Y \quad \{= 0.877(U_r - Y)/0.877 + Y = U_r - Y + Y = U_r\}$$

$$U_b = C_b/0.49 + Y \quad \{= 0.49(U_b - Y)/0.49 + Y = U_b - Y + Y = U_b\}$$

$$U_g = -0.51C_r - 0.19C_b + Y \quad \{= -0.447U_r + 0.4477\ Y - 0.094C_b + 0.094Y + Y$$
$$= -0.447U_r - 0.094C_b + 1.541Y$$
$$= -0.447U_r - 0.094C_b + 0.462U_r$$
$$+ 0.909U_g + 0.17C_b\# U_g\}$$

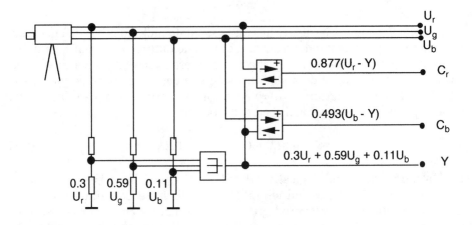

Figure 2.29 Generation of luminance and chrominance signals.

2.8.3.2 Composite TV Signal

A TV signal is composite when the color-difference components signals C_r and C_b are interleaved into gaps of the luminance signal. The composite TV signal thus consists of

- Luminance signal Y;
- Chrominance color-difference signals C_r and C_b;
- Synchronization pulses to mark the beginning of the image;
- Blanking pulses for the intervals that the CRT beam returns to the beginning of the next scanning line.

The internationally used PAL, SECAM, and NTSC TV signals in their analog form are always composite signals.

2.8.3.3 Component TV Signal

For the analog picture processing in TV studios, instead of the composite signal a component signal is used that consists of the luminance signal Y and the color-difference signals C_r and C_b.

2.8.3.4 Studio Standard TV Signal

With the advance of digital image processing a new standard for digital TV signals has been created and standardized in CCIR Recommendation 656, the studio standard "4:2:2", so-called because of the applied sampling ratio between luminance and the two chrominance components C_r and C_b.

2.8.3.5 Video Coding

Contribution and Distribution Quality

For the digitized transmission of video in contribution quality, for example, a PAL, SECAM or NTSC color television signal, with a bandwidth of approximately 7 MHz, in accordance with CCIR Recommendation 601, the video signal is sampled at the rate of 17 MHz and encoded using a 8-bit code resulting in a transmission rate of 17 MHz × 8 bit = 136 Mbps. The composite TV signal plus two high-quality sound channels (for example, one stereo channel) plus service channels can thus be accommodated and transparently transmitted in a standard PDH 140-Mbps or SDH 155-Mbps signal. Before continuing with coding, two terms in these sentences need clarification, namely, CCIR recommendation and transparent transmission.

Whereas CCITT has defined the recommendations concerning audio—which was first distributed via cable, which was CCITT domain—the recommendations concerning video contribution and distribution transmission, for historical reasons, are defined by CCIR because the first transmission of TV signals (apart from early applications on coaxial cable) took place via radio-relay links, which were CCIR domain. The recommendations for videoconferencing and videophony, however, are CCITT domain.

A (TV) signal is transmitted transparently if the output signal corresponds to the input signal, that is, if no information is changed or manipulated, apart from the transmission impairments. If the signal processing is controlled by the image content itself, as is the case in "compressed" signals, the transmission of the signal is not transparent. A transparent channel transmits all signals without modifications as long as these fit in the bandwidth available and remain within the permitted level ranges and tolerances. Transparent transmission, which maintains the full content of the signal, is of importance in the contribution network especially in the case of reprocessing TV signals in interconnected studios.

The application of linear digitizing of TV video signals for transparent transmission requires large line bit rates that, for example, for HDTV, easily reach 1 Gbps. This has paved the way toward two different approaches, specifically, uncompressed fully transparent composite video signal transmission via optical fiber and compressed video transmission.

Uncompressed Video Transmission on Optical Fiber

The high transmission bandwidth capacity of optical fiber makes it an ideal medium for the transparent uncompressed transmission of TV signals. CCITT, nor CCIR, have issued a recommendation for this transmission mode, but a few proprietary systems are on the market.

Modern HDTV systems like PAL-Plus and HD-MAC easily occupy a 12-MHz bandwidth. In a state-of-the-art optical video transmission system this 12-MHz band is digitized with a sampling rate of 31.5 MHz and a 10-bit coding resulting in a nonstandard 315-Mbps transmission speed. Simultaneously, a 2-Mbps data signal and four audio signals are transmitted over one single-mode fiber. Each audio channel, with a bandwidth of 20 kHz, is digitized with a sampling rate of 48 kHz and a 16-bit code, thus at a 768 Kbps transmission speed. The overall transmission speed is 315 + 2 + 4 × 0.768 + 26.428 (spacing between the signals) Mbps = 346.5 Mbps.

A multichannel optical video transmission system can typically transmit up to four composite studio quality (6-MHz bandwidth) video signals plus 24 audio channels over one single-mode fiber unrepeated over distances up to 100 km. In this multichannel system each video signal is digitized with a sampling rate of 15.4 MHz and a 10-bit code, thus with a 154-Mbps transmission speed. The audio signals are digitized as described, so that the overall transmission speed is 4 × 154 + 24 × 0.768 + 43.568 (signal spacing) Mbps = 678 Mbps.

Such optical video transmission systems are particularly suitable for point-to-point contribution network feeder links as well as for traffic and safety surveillance applications over long distances and requiring a high video quality. The transmission capacity can be even doubled if WDM is applied, using both the 1.3- and 1.5-μm optical fiber windows [21].

Compressed Video Transmission

An important characteristic of video signals is the redundancy of information content. The characteristic of a pixel (spot of light, pixel, abbreviated from picture element) is that there is a high probability of correlation with neighboring spots; for example, if a pixel is black, it is very likely that the neighboring pixels will be black or dark gray but probably not be white or red. This correlation can exist in space (thus in the same frame) or in time (thus in successive frames). To reduce the high speed required for the transmission of uncompressed video signals to a more economical range, special signal compression redundancy elimination techniques are applied. The major compression techniques are

- *Discrete cosine transform* (DCT);
- *Differential pulse code modulation* (DPCM);

- *Variable-length coding* (VLC);
- Prediction;
- *Conditional replenishment* (CR).

Discrete Cosine Transform (DCT). DCT is an algorithm for two-dimensional signal transformation applied on the redundancy of video signals in space (thus within a frame). Due to the nature of the human eye, it is permissible, as applied with DCT, to quantize the higher frequencies more coarsely than the lower frequencies without causing a subjectively perceptible degradation of the image quality. This effect can be exploited even more with color difference signals since the eye is less adept at differentiating color structures than the equivalent brightness structures. Coarse quantization produces large compression factors, however, with increasing compression detectable coding errors arise. With DCT these errors are used to obtain a balance between quantity of generated data and the resulting image quality. DCT compression is even more effective if in addition to evaluating the spatial correlation, the temporal correlation of samples in successive frames will also be evaluated. This type of spatial and temporal implementation of DCT is known as *hybrid DCT* (HDCT).

Differential Pulse Code Modulation (DPCM). DPCM, similar to that in Section 2.5 as VBR-ADPCM, exploits the correlation between successive images and codes the deviation of the actual values of the picture elements instead of the actual value themselves, thus allowing typically 4-bit coding instead of 8-bit coding and thereby reducing the transmission bit rate by 50%.

Variable-Length Coding (VLC). VLC, also called "Huffman coding," evaluates the statistical occurrence of sampled values. Short code words are used for frequently occurring values and longer code words for rarely occurring values. VLC reduces the transmission bit rate typically by 25%.

Prediction. Prediction is based on the statistical probability of the behavior of pixels depending on the behavior of the surrounding pixels; for example, the development of a shade of a pixel can be predicted with high accuracy from its surrounding pixels. Basically, after the predictor has "understood" the behavior of the pixels, ideally, deviations only of a predicted behavior need to be coded.

Conditional Replenishment (CR). CR is the name for a procedure taking advantage of a special structure of the moving pictures. For an image showing people against a static background, the changes between two successive frames are limited, usually not more than 25%. With CR the codec, therefore, codes only those picture elements that have changed since the previous frame and, on the receive end, only those parts that reflect actual changes in the frame are updated, thus reducing the required bit rate in the order of 75%.

Compression, of course, will always be a compromise between transmission speed, image quality, and cost. To obtain multivendor compatible solutions with a minimum impairment of the image quality for a given application, essentially three different compression standards are applied, namely, JPEG, MPEG, and H.261.

JPEG stands for *Joint Photographic Expert Group*, the name of Working Group 10 (WG10) of Subcommittee SC 29 of the *Joint Technical Committee* (JTC) created by ISO and the *International Electrotechnical Commission* (IEC). JPEG is for still pictures; it uses 16 × 16 DCT and does not provide interframe coding.

MPEG stands for *Moving Pictures Expert Group*, which is the name of WG11 of SC 29. MPEG uses 16 × 16 DCT and encodes differences between frames. This compression technique is widely used for video-CDs and interactive multimedia recordings.

H.261 is a recommendation of ITU-T for videoconferencing and videophony titled, "Line transmission of non-telephone signals. Video codec for audio-visual services at $p \times 64$ Kbps." This recommendation defines a set of algorithms for video compressing as well as two standard picture formats: CIF and QCIF. CIF stands for *common intermediate format* and is a standard that defines the number of scan lines, the number of pixels, and the frame frequency as shown in Table 2.15.

H.261 uses 8 × 8 DCT and encodes differences between frames on QCIF (Quarter CIF).

Explaining the merits of these standards would go beyond the scope of this book. To demonstrate, however, the compression effectiveness of the standards, in Table 2.16 a very approximate comparison of the achievable compression is given.

Figure 2.30, as an example, shows a highly simplified block diagram for coding a distribution quality video signal with its stereo channel onto a 34-Mbps standard PDH bit rate. The video interface samples the video signal at the rate of twice the video bandwidth, for example, at 17 MHz for a PAL signal. The digital interface performs a conversion of the digitized composite signal into the components luminance Y and chrominance C_r and C_b. The DCT processor calculates the redundancy reduction to match the coding rate in the

Table 2.15
ITU-Recommended Standard Picture Formats

Standard	Scan Lines	Pixels	Frames per Second
CIF	228	352	30
QCIF	144	176	30

Table 2.16
Comparison of Compression Performance

Video Format	*Video Resolution*		*Approx. Uncompr. (Mbps)*	*IPEG Bit Rate (Mbps)*	*MPEG Bit Rate (Mbps)*	*H.261 Bit Rate (Mbps)*
	Pixels	*Lines*				
QCIF	176	144	10	–	–	0.064–2
CIF	352	288	40	1.2–3	3–8	0.064–2
Standard TV	720	486	140	5–10	15–25	–
HDTV	1,920	1,080	1,000	20–40	60–100	–

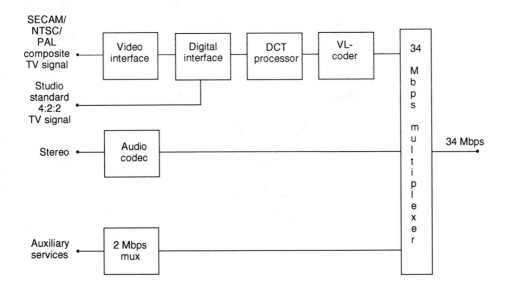

Figure 2.30 34-Mbps codec for video and sound.

VLC (typically between 2 and 18 bits) to the line transmission speed. The applied video signal compression still leaves space for two mono 15-kHz audio channels, or one stereo channel, as well as a 2-Mbps primary rate capacity for auxiliary services [21–24].

Videoconferencing Quality

As an example of the compression of videoconferencing signals a procedure is briefly described that reduces the transmitted data by a factor of 70 from 140 to 2 Mbps. Figure 2.31 shows the basic steps of this procedure.

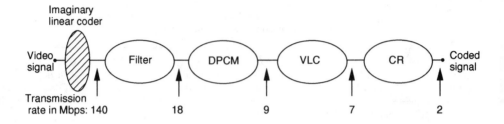

Figure 2.31 Video compression procedure.

A video signal, taken by a camera, with 625 lines per frame, 50 fields per second, and a bandwidth of typically 7 MHz, if connected to a linear coder would result in a theoretical bit rate of 140 Mbps. The signal, however, is not connected with a linear coder but to a filter unit where it is first converted into the components Y, C_r, and C_b and then each component is limited in bandwidth: the Y signal to 2.5 MHz and the C signals each to 0.5 MHz. This bandwidth limitation causes a reduction of the bit rate after the a/d conversion of 50%, with the resulting picture quality still acceptable for video conferencing. The a/d conversion uses sampling frequencies of 5 MHz for the Y and 1 MHz for the C signals and an 8-bit coding, which would result in a theoretical bit rate of 48 Mbps. The Y and C signals, however, are sent through digital filters, which perform a vertical, low-pass filtering of the picture, halving the number of lines and thereby reducing the required bit rate to approximately 18 Mbps. Using DPCM halves the transmission bit rate to 9 Mbps. A further 25% reduction of the bit rate from 9 to 7 Mbps is obtained in the VLC. Finally conditional replenishment reduces the required bit rate by about 75% to well below the 2-Mbps primary rate, leaving enough capacity for an audio channel and signaling [25].

Videophone Quality

The telephone of the future, as mentioned in Section 1.4, will be a videophone; ahead of, or parallel with, the development of videophony goes the development of the market for audio/video on PCs in the emerging evolution toward multimedia. In this consumer domain the state of the art is MPEG-1 coding/decoding with the following features:

- Full-screen, full-motion video at 30 frames per second;
- Video coding between 1.2 and 9 Mbps, covering the range of applications from single-speed to six-speed PC CD-ROM;

- Selectable 44.1- or 48-kHz CD-quality audio;
- Video interface with monitor or TV set in YUV or RGB format.

To pave the way to a smooth worldwide introduction of affordable videophony, effective international standards are required to ensure industrial production of fully compatible equipment from various manufacturers. ITU-T, therefore, has issued an H-series of recommendations for audio-visual services of which the major ones are listed in Table 2.17.

Video codecs are based upon the H.261 interface with CIF and QCIF video inputs. A videophone, for example, as shown in Figure 1.7, includes a high-performance motion estimation and compression codec using DCT and DPCM in one of the following combinations with audio codec and operational data transfer (including signaling between terminals):

- Video at 68.8 Kbps, data at 3.2 Kbps, and audio as per G.711 from 3.1 kHz bandwidth to 56 Kbps;
- Video at 76.8 Kbps, data at 3.2 Kbps, and audio as per G.722 from 7 kHz bandwidth to 48 Kbps;
- Video at 108.8 Kbps, data at 3.2 Kbps, and audio as per G.728 from 3.1 kHz bandwidth to 16 Kbps.

The videophone communicates on N-ISDN basic access using both 64-Kbps B-channels for video, speech, and operational data transfer. The audio

Table 2.17
ITU-T H-Series of Recommendations for Audio-Visual Services

Recommendation	Description
H.221	Line transmission of nontelephone signals; frame structure for a 64- to 1,920-Kbps channel in audio-visual teleservices
H.230	Line transmission of nontelephone signals; frame synchronous control and indication signals for audio-visual systems
H.231	Multipoint control unit for audio-visual services
H.233	Confidentiality system for audio-visual services
H.242	Line transmission of nontelephone signals; system for establishing communication between audio-visual terminals using digital channels up to 2 Mbps
H.243	Procedures for establishing communication between three or more audio-visual terminals using digital channels up to 2 Mbps
H.261	Line transmission of nontelephone signals; video codec for audio-visual services at $p \times 64$ Kbps
H.320	Line transmission of nontelephone signals; narrowband visual telephone systems and terminal equipment

codec in compliance with the ITU-T Recommendation G.711 includes the μ-law/A-law coding for unimpaired conversation between North America and "the-rest-of-the-world" [23,26].

References

[1] Huurdeman, Anton A., *Transmission: A Choice of Options*, Paris, France: Alcatel Trade International, June 1991.

[2] ITT, *Reference Data for Radio Engineers*, 5th ed., Indianapolis, Kansas City, and New York: Howard W. Sams & Co, Inc., 1986.

[3] Members of the Technical Staff, *Transmission Systems for Communications*, Winston-Salem: Bell Telephone Laboratories, Inc., Revised 4th ed., Dec. 1971.

[4] Minoli, Daniel, *Telecommunications Technology Handbook*, Norwood, MA: Artech House, 1991.

[5] Huurdeman, Anton A., *Radio-Relay Systems*, Norwood, MA: Artech House, 1995.

[6] Newall, Christopher, et al., *Synchronous Transmission Systems*, London: Northern Telecom Europe Ltd., 1992.

[7] Eneborg, Mats, and Bengt Lagerstedt, "Foundations for Broadband: The Transport Network," *Telecommunications*, Vol.. 27, No. 3, 1993, pp. S9-S15.

[8] Chesnoy, J., et al., " Ultrahigh Bit Rate Transmission for the Years 2000," *Electrical Communication*, 3rd quarter 1994, pp. 241–250.

[9] Shalit, Mike, "Circuit Multiplication," in *Developing World Communications*, London: Grosvernor Press International, 1986, pp. 259–264.

[10] Alcatel 1611 QX Celtic 3G Digital circuit multiplication system, data sheet of Alcatel Network Systems 9882-0991, 1991.

[11] "Inverse Multiplexing," Technology Note in *Telecommunications*, Vol. 29, No. 11, 1995, p. 58.

[12] Dean, Phil, "AIMing at ATM," *Telecommunications*, Vol.. 30, No. 2, 1996, pp. 28–30.

[13] Edwards, David, "Modems: Getting the Message Across," in *Developing World Communications*, London: Grosvernor Press International, 1986, pp. 386–395.

[14] Mistry, Dilip, "Modems: Still Gathering Speed," *Telecommunications*, Vol.. 30, No. 2, 1996, pp. 56–59.

[15] Siegmund, Gerd, *Grundlagen der Vermittlungstechnik*, Heidelberg: R.v. Deckers's Verlag, G. Schenk, 1993.

[16] Llana, Andres, "Realigning the Corporate Network," *Telecommunications*, Vol.. 29, No. 12, 1995, pp. 36–41.

[17] Hopkins, Harman, "Frame Relay: Setting Standards for the Market," *Telecommunications*, Vol. 30, No. 2, 1996, pp. 31–34.

[18] "Broadband Communications," booklet within *Managing Network Evolution*, Paris: Alcatel Network Systems, 1991.

[19] Buchholz, Mathias, "Fibre Boost for High Speed Copper," *Telecommunications*, Vol. 29, No. 6, 1995, pp. 81–82.

[20] Sanford, Curtis, "Casting the Access Net," *Telecommunications*, Vol. 30, No.1, 1996, pp. 41–43.

[21] Anderegg, J., et al., "Transparent Video Transmission in Contribution Networks," *Electrical Communication*, 3rd quarter 1993, pp. 227–234.

[22] Hoffmann, T., et al., "Video Compression Techniques for Multimedia Communications," *Electrical Communication*, 4th quarter 1993, pp. 407–410.

[23] Dampz, J., et al., "Multimedia Terminals: Advantages, Technology, Networking," *Electrical Communication*, 4th quarter 1993, pp. 387–393.

[24] Barezzani, M., et al., "Compression Codecs for Contribution Applications," *Electrical Communication*, 3rd quarter 1993, pp. 220–226.

[25] Hahn, Norbert, "Applying Videoconferencing," *Telecommunications*, Vol. 26, No. 8, 1992, pp. 27–30.

[26] MPEG-1 decoder chip, *Inside Multimedia IM*, Issue 113, Mar. 1996.

Copper Lines 3

Copper wire line transmission concerns transmission through

- Open wire lines (O/W lines);
- Twisted pair cable;
- Coaxial cable.

Occasionally, steel and aluminum conductors have also been used. Copper, however, is dominant; which is why this chapter focuses on copper conductors, with the exception of steel wire conductors for O/W lines. Applicationwise copper wire line transmission can be distinguished in

- Overhead lines;
- Underground cable;
- Submarine cable;
- Installation cable;
- RF feeder;
- Radiating cable.

The main characteristics of copper wire transmission lines are determined by the following properties of the lines:

- Conductivity of the conductor material;
- Diameter and spacing of conductors;
- Dielectric properties of the insulation between the conductors.

These properties determine the primary constants of transmission lines, which are usually defined per kilometer or mile as follows:

- Resistance (R), measured in ohms (Ω) and determined by the diameter and material of conductors;

- Conductance (G), measured in Siemens (S), depends on the quality of the insulation between the two conductors; the conductance is the reverse of resistance in ohm, in English-speaking countries, so "mho" is also used instead of S;
- Capacitance (C), measured in (micro)farad (F) and determined by the conductor distance and dielectric properties of isolation between conductors;
- Inductance (L), measured in (milli)henry (H) and determined by the conductor diameter, conductor material, and space relation between the conductors.

The locations of the primary constants in a homogeneous transmission line are indicated in Figure 3.1.

The transmission characteristics of the transmission line, or secondary constants, are basically calculated from the primary constants as the characteristic impedance

$$Z = \sqrt{(R + j\omega L)/(G + j\omega C)} \quad \text{(in } \Omega\text{)}$$

(independent of transmission line length) or propagation constant

$$\gamma = \sqrt{(R + j\omega L)/(G + j\omega C)}$$

in unit length. The propagation constant is also given by the formula $\gamma = \alpha + j\beta$, where α denotes the attenuation and is defined as

$$\alpha = \sqrt{0.5\{\sqrt{(R^2 + \omega^2 L^2)(G^2 + \omega^2 C^2)} + (RG - \omega^2 LC)\}} \quad \text{(in dB/unit length)}$$

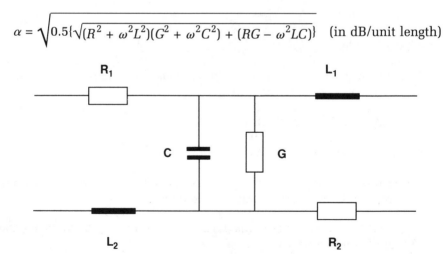

Figure 3.1 Primary constants in a homogeneous two-wire transmission line.

and β denotes the phase constant and is defined as

$$\beta = \sqrt{0.5\{\sqrt{(R^2 + \omega^2 L^2)(G^2 + \omega^2 C^2)} - (RG - \omega^2 LC)\}} \quad \text{(in radians/unit length)}$$

In the impedance formula for copper wire lines the resistance R dominates up to 20 to 100 kHz depending on the cable gauge. The inductance L dominates above 50 to 200 kHz. In between both are significant. In the range where R is significant, the cable impedance varies rapidly with frequency, although specific terminating impedances are often quoted. Above 100 to 200 kHz R is negligible compared with $j\omega L$ and the conductance is small in relation to $j\omega C$, so the characteristic impedance of transmission lines can be calculated using the simplified formula: $Z = \sqrt{L/C}$. Basically any increase in the distance between the two conductors of a transmission line results in an increase of inductance (as the effective inductance is proportional to the magnetic flux between the two conductors) and a decrease of capacitance (as the two conductors function as the two plates of an capacitor, in which the capacity decreases with distance between the plates), and finally an increase of characteristic impedance. Open wire lines thus have the highest characteristic impedance (600Ω to 2,000Ω), telephone cable 400Ω to 1,800Ω, transmission cable 120Ω to 200Ω, against typically 75Ω for coaxial cable.

The application of copper wire lines is, besides by the effect of crosstalk between the wires, limited by the attenuation of the line. This attenuation increases with distance and frequency. In the attenuation formula, the frequency is in $\omega = 2\pi f$. The frequency depends on the transmission capacity (bandwidth and number of telephone channels) of the multiplex system operated on the line. To compensate for the attenuation, repeaters are required at regular distances. Repeaters, however, have their limitations too. First, they amplify not only the desired signal but the input noise too; and second, their amplification is usually not constant over the whole bandwidth and, therefore, create a nonlinear distortion. Therefore, the maximum possible amplification is a compromise between cost and desired SNR. Figure 3.2 indicates the relation between the practical average repeater spacing and the transmission capacity of the three copper wire transmission media [1].

3.1 OPEN WIRE

Open wire lines were the first, and for economical reasons in many developing countries, still are the main transmission medium in the access network and are sometimes also used in the distribution network. The relatively modest investment for an open wire line defers costs due to

Repeater spacing

Km

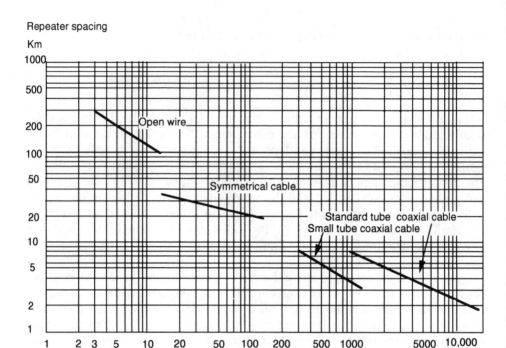

Figure 3.2 Relation between repeater spacing and transmission capacity of the three copper wire transmission media.

- The high vulnerability to damage by vandalism, theft, and natural calamities;
- Transmission quality, which is very much dependent upon the weather conditions—compared with dry weather, the line attenuation increases 60% to 70% by rain and 200% to 400% by ice and hoar-frost;
- The limited transmission capacity—future expansion comes at great expense;
- Maintenance costs, which are very high;
- Significant crosstalk and noise pick-up due to lack of shielding.

To obtain the lowest electrical resistance the open wire line conductors should be made from pure electrolytic copper. To withstand the strong mechanical stress, however, hard-drawn copper wires, bronze wires, steel wires, or copper-covered steel wires are used. The conductors usually have the following diameters:

- Copper: 3 mm, 4 mm, and 5 mm (occasionally: 1.5, 2, 2.5, 3.5, and 4.5 too);
- Bronze: 1.5 mm, 2 mm, and 3 mm (occasionally: 2.5, 3.5, 4, 4.5, and 5 too);
- Galvanized steel: 2 mm, 3 mm, 4 mm, and 5 mm;
- Copper-covered steel: 2.6 mm.

CCITT has defined, in G.311 and G.312, two line systems for operation on open wires: 3-channel and 12-channel systems. Both systems are defined for two-wire operation with go and return direction in separate nonoverlapping frequency bands. The major characteristics of the two systems are summarized in Table 3.1.

Instead of three VF-channels, one broadcast audio channel with a 50- to 10,000-Hz bandwidth can be carried. Instead of full-bandwidth the bandwidth of one or more of the telephone channels can be reduced to 300 to 2,700 Hz so that two VF-telegraph channels can be transmitted above those telephone channels (for example, at 2,940 and 3,100 Hz).

One wire pair can be shared by a 12-channel SOJ system, one 3-channel STO system, one service channel, and in some countries (such as in South Africa), still another 12-channel (SOX) system operating in an upperband above 143 kHz.

In order to reduce crosstalk and convert intelligible crosstalk, noninverted and inverted sidebands, as indicated in the table, are applied on systems operated in parallel on the same poles.

Table 3.1
CCITT 3- and 12-Channel O/W Systems

Characteristic System Designation	3-Channel System		STO-	12-Channel System		SOJ-
	kHz	*kHz*		*kHz*	*kHz*	
Line transmission range	4–16	18–30	A	N 36–84	I 92–140	A
	4–16	19–31	B	I 36–84	N 95–143	B
				N 36–84	N 93–141	C
				I 36–84	I 94–142	D
Pilot frequencies	16.11 or 17.80	31.11		40 and 80	92 and 143	
				47 and 97	107 and 157	
VF-band	300–3,400 Hz			300–3,400 Hz		
Number of VF-telephone channels	3			12		
VF-telegraph channels	4					
Typical repeater spacing on 3-mm copper wire	240 km			120 km		

Note: N = Noninverted sidebands; I = Inverted sidebands.

A basic block diagram of a 12-channel O/W terminal is given in Figure 3.3. The subscriber lines are connected to a two-wire/four-wire line termination and then to a 12-channel multiplexer that translates the 12 channels into a 60- to 108-kHz basic primary group. The basic primary group is translated to the 92- to 140-kHz line frequency in the transmit direction and derived from the 36- to 84-kHz line frequency in the receive direction. Two line pilots, 92 and 143 kHz, are added to the line transmission band in the transmit direction, whereas in the receive direction the two line pilots at 40 and 80 kHz are used for the automatic gain control of the receive amplifier to regulate the weather-dependent attenuation changes. An optional second highpass/lowpass filter is included to enable the simultaneous operation of a three-channel O/W system and a service channel on the same O/W pair.

To connect an O/W terminal or repeater to the overhead wires a number of precautions and conditions must be observed. First, special lightning protection is vital to safeguard subscribers and equipment. Second, for signaling (at 25 or 50 Hz) and fault location the whole line circuit should be dc-conductive; thus, line termination coils should be of the auto-transformer type, and lowpass

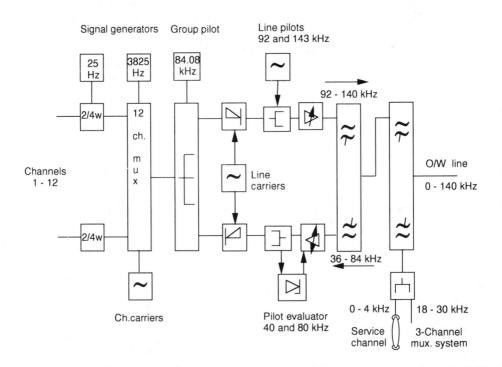

Figure 3.3 12-channel O/W terminal.

filters shall not block dc. To reduce crosstalk, special coils are inserted in the line with two windings in opposite directions to block the earth-unsymmetrical crosstalk whereas the mainly earth-symmetrical desired signal passes almost without loss. Figure 3.4 shows a typical arrangement for the connection of 12-channel and 3-channel systems to the O/W line.

The O/W conductors are supported by porcelain or glass insulators fixed to poles that are usually located at 50- or 60-m intervals. To reduce the mechanical stress the conductors are not fixed in a straight line, but they can adapt to the temperature with a certain dip. This dip typically varies in the middle of two poles between 20 cm at −25°C and 60 cm at +25°C with a 50-m pole interval and between 30 cm at −25°C and 75 cm at +25°C with a 60-m pole interval. Impregnated wooden poles (for a lifetime of typically 30 years) are used in lengths of 6m to 11m; beyond that length the poles are mainly made of concrete or steel. To keep the crosstalk between parallel routed conductor pairs within acceptable limits, a maximum of 16 to 20 12-channel systems can be operated on the same pole. The capacitance-related crosstalk between parallel systems is neglectable because of the neglectable capacitance coupling between the wires of parallel systems. The inductive crosstalk, however, is a serious problem and can be reduced, apart from applying frequency staggering and sideband inversion, by a regular crossing of the wires, for example, as shown in Figure 3.5 [2,3].

3.1.1 Carrier-Telephony on HT-Lines

A special application of O/W transmission is using a pair of HT-lines to transmit single-channel-carried telephone channels. Typically, up to six channels, each consisting of one speech and one data signal frequency modulated and in two-wire transmission, can then be accommodated with a 15-kHz spacing in a 35- to 150-kHz band for communication between various stations of a HT-grid. Larger systems can accommodate up to 24 channels within the 30- to 375-kHz transmission band. The carried signals are coupled to the HT-line by means of highly protective HT-capacitors. A carrier-frequency blocking filter inserted in the HT-line at the same point as the HT-capacitor, however, in the opposite direction of the desired transmission direction, ensures that the signals are transmitted between the desired stations only.

3.1.2 Aerial Cable

A special form of aerial installation of copper wires is the self-supporting aerial cable. This open air cable in the shape of an 8 consists of a main body with up to some 300 copper conductor pairs and high-strength-grade galvanized

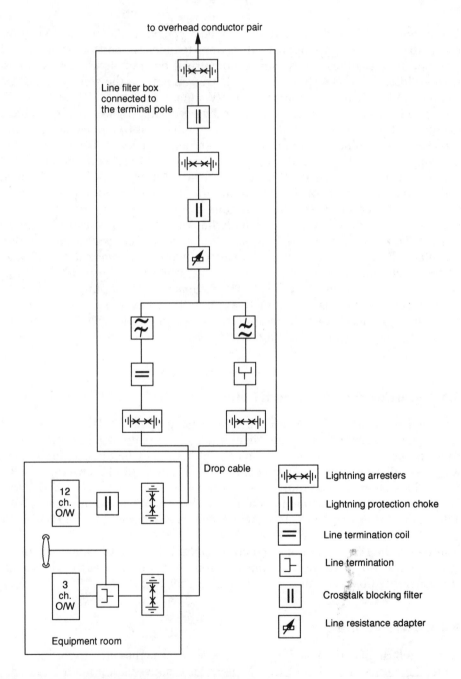

Figure 3.4 Typical O/W line-interface arrangement.

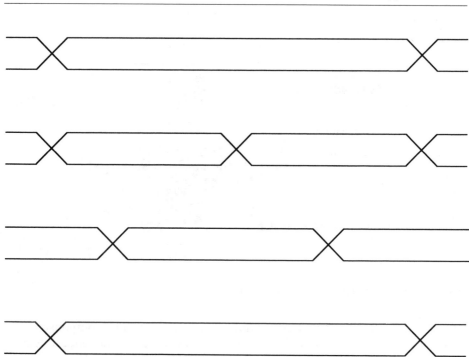

Figure 3.5 Wire crossing to reduce crosstalk.

steel wires as support messenger both covered by a common polyethylene sheath. Such cable is also used for installation on HT-power transmission grids or on overhead electricity distribution networks, thus taking advantage of existing support infrastructure but adequately protecting the telecommunication circuits from the electricity-carrying wires. The mechanical protection and electrical shielding is either provided by a corrugated (thus flexible) aluminum shield or a steel wire armoring. Figure 3.6 shows the cross section of an aerial telephone cable with 100 pairs of 0.5-mm copper conductors.

Apart from the 8-shaped aerial cable, occasionally special aerial cable with a flat-wire sheeting for pull-relief, or even normal earth cable attached to a steel-wire rope by means of clamps or a tie-wire are used. Aerial copper wire cable is mainly used for telephony but also is sometimes used as twisted pair symmetrical cable (see below) for the transmission of 12-, 24-, 60-, and 120-channel systems.

3.2 TWISTED PAIR CABLE

Contrary to O/W lines where the mechanical strength of the conductors is a major consideration, for underground telecommunication cable the mechanical

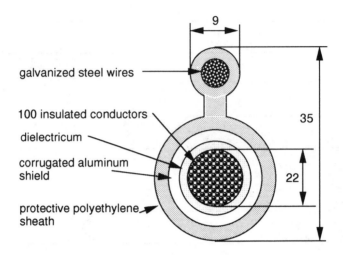

Figure 3.6 Suspended aerial cable cross section.

strength of the conductors is not as important since it can be compensated for by the protective cable shielding. The major concern in telecommunication cable construction is to minimize the attenuation, thus keeping R and C as low as possible. Therefore, the conductors are usually made from the purest electrolytic copper. To keep the capacitance low, the distance between the two conductors of a pair should be small. Hence, relatively thin (0.4- to 1.4-mm-diameter) conductors are used that are insulated with loosely wound nonhygroscopic tape (textile, impregnated paper or wood pulp) and thus with much air in the insulation or with a high dielectric strength plastic material like polyethylene, polypropylene, foam, or *polyvinyl chloride* (PVC).

Twisted pair cable can be divided into two groups: (1) *unshielded twisted pair* (UTP) telephone cable and (2) shielded twisted pair transmission cable.

The telephone cable is briefly described in Subsection 3.2.1 and the transmission cable in Subsection 3.2.3. Subsection 3.2.2 describes the new technologies to enhance the performance of telephone cable access networks.

3.2.1 Telephone Cable

The attenuation and the impedance of a twisted pair if used for a speech circuit can be calculated by approximating the simplified formulae

$$\alpha = \sqrt{(\omega RC)}/2 \qquad Z = \sqrt{R/j\omega C}$$

In addition to attenuation, crosstalk and thus inductive coupling is an additional matter to be mastered by specific means of twisting conductors and pairs.

Basically two conductors are twisted to form a pair, and neighboring pairs are twisted with different twist length. Two different approaches of twisting pairs together to a quad are applied as shown in Figure 3.7: the *Dieselhorst-Martin* (DM) quad (mainly applied in Germany) and the star quad. In North America quads were not used, just pair cables either built up in layers or in bundles called unit twins. Cables everywhere are now usually pair types.

With the DM quad two conductors (1 and 2) are twisted to form a pair, and a second pair (with conductors 3 and 4) is twisted with a different pitch around the first pair. In a star quad four conductors are twisted together and two diametrically opposed conductors (1 and 2, respectively 3 and 4) are used as a pair. The star quad twisting has the big advantage of requiring roughly 25% less volume than DM twisting, albeit at the cost of a higher attenuation.

Multiple pair telephone cable, to obtain an optimum use of the cylindrical shape, is constructed with a fixed hierarchy of pair allocation whereby a number of quads, usually five, are stranded into a basic bundle. Five basic bundles are then stranded to a main bundle, which thus has 4 × 5 × 5 = 100 wires, respectively, 50 pairs. A number of main bundles can be combined in a cable to yield a cable core with some 2,000 to 4,000 pairs.

The cable core is covered with mechanically and electrically protective shielding material (for example, lead or polyethylene, and copper foil), and the interstices between quads and shielding may be filled with a flooding compound (such as petroleum jelly) depending on the application. The cable sheath consists, depending upon the particular application, of simple PVC polyethylene, impregnated paper, nonhygroscopic tape, jute, steel tape, lead, and aluminum

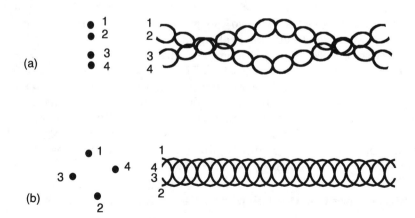

Figure 3.7 Twisting of cable pairs:(a) Dieselhorst-Martin twisting and (b) star twisting.

shield in various combinations. For further protection and maintenance accessibility, underground cable can be placed in special (polyethylene, PVC, or other material) cable ducts accessible from manholes at regular distances and buried at 0.8- to 1.2-m depth.

In soft soil the direct-buried cable can be plowed with cable plows into the appropriate depth of 0.5m to 1m.

Telephone cables have a relatively high capacitance and a very low inductance, which limits the unamplified transmission of telephone channels to only some 30 km. Loading coils—invented by Michael Odvorsky Pupin (1858–1935)—connected in series with the pairs at 1- to 3-km intervals can increase the unamplified distance to about 100 km, albeit at the price of limiting the upper frequency of the pair's transmission band. In the United Kingdom and North America standard loading coils were used with an inductance of 88 mH.

In order to further extend the maximum distance between an exchange and individual subscribers or to enable the use of thinner cable pairs, two electronic devices were introduced and usually installed in the exchange building as follows:

- NLT-amplifier (negative line in transistorized technology–amplifier), which compensates for the frequency-dependent attenuation of unloaded two-wire lines, thus providing a negative line impedance;
- Hybrid amplifier, which compensates for the frequency-dependent attenuation of pupinized lines.

The loading has to be removed in the case of digital transmission of an E1 or T1 signal on telephone cable, for example, for a cluster of subscribers connected to a line concentrator or for business subscribers with a PABX requiring a number of lines to the public exchange. Repeatered E1/T1 technology was developed in the 1970s to solve the problem of connecting increasing numbers of lines between telephone exchanges. The initial application of PCM with 24, respectively, 30 digitized telephone channels on two twisted pairs reduced or at least delayed the necessity of installing additional cables. The cable in the distribution network usually has pairs with 0.6- to 1.0-mm diameter requiring digital regenerating repeaters roughly every 2 km. In the access network there is usually a mixture of various types of cables with diameters below 0.6 mm, thus requiring E1/T1 repeaters at even shorter distances than in the distribution network. E1/T1 technology in the access network, therefore, is expensive and will be outdated by the xDSL technology.

3.2.2 Enhancement of the Telephone Cable Access Network

Copper wire-based telephone cables connect the majority of the worldwide 600 million telephone subscribers with their local telephone exchanges. Deregulation and privatization of network operation, which opened the door for competition, demands access network improvements in order to enable profitable new services. Billions of dollars have been invested in this enormous "copper mine"; and now, should it be replaced by an even more expensive glass-fiber infrastructure to connect those subscribers to the worldwide data highway? Eventually, yes, but evolutionarily and not revolutionarily and certainly not yet in this century.

Instead of replacing the copper wire cable by fiber, the data-capability of the local loop can be improved by adding appropriate electronic equipment, thus enhancing the largest asset of the telecommunication operators, the copper-based local loop, into a "gold-mine" with increased revenues over still many years to come.

In the past, the capacity of single copper wire pairs was increased by means of so-called *party line* and *pair-gain* systems that connected groups of subscribers to a single line. More recently, special digital subscriber line systems are being introduced that allow high-speed data, VOD, and multimedia on standard twisted copper pairs over the limited length of the local loop, which is typically between 2 and 5 km in Europe and 4 to 8 km in the United States. Those systems that enable telephone subscribers to communicate on the worldwide data highways are appearing under acronyms as:

- ADSL: asymmetrical digital subscriber line;
- SDSL: symmetrical digital subscriber line;
- VDSL: very high speed digital subscriber line;
- HDSL: high bit rate digital subscriber line;
- VHDSL: very high bit rate digital subscriber line.

In the following sections a short description is given of these subscriber line-enhancing systems.

3.2.2.1 Party Line Systems

Party line systems have been in use for many years mostly in rural areas. A party line system does not increase the capacity of a copper wire pair but allows automatic sharing by up to typically 10 to 15 separately located infrequent-calling subscribers. Subscriber secrecy, individual metering and ringing, recalling for busy line, and free calling between the party subscribers are typical features of this system. Party line systems are available for analog and for digital operation.

3.2.2.2 Pair-Gain Systems

Analog pair-gain systems have been widely used in both developing and developed countries. In the United States, in particular, an amplitude-modulated single-channel carrier *added-main-line* (AML) system has been in operation since the 1960s. A frequency-modulated version improved the SNR and crosstalk reduction performance. The most basic pair-gain system is the 1 + 1 system where one additional voice circuit is created on two carriers (one for the "go" and the other for the "return" direction) above the speech band. Figure 3.8 shows a principle diagram of a typical 1 + 1 system. Whereas the speech signal and signaling of the first subscriber continues to operate in the 300- to 4,000-Hz band, the speech signal and signaling of the second subscriber is modulated with a 28-kHz carrier in one direction and a 79-kHz carrier in the other transmission direction. An optional second telephone can be used by the second subscriber. In addition to the local power supply shown in the diagram,

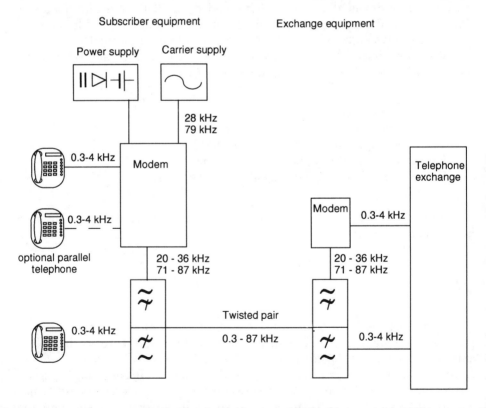

Figure 3.8 Pair-gain system 1 + 1 version (*After:* Alcatel STK Norway).

power feeding, and then battery charging from the line is possible for the second subscriber, too.

Analog pair-gain systems have been designed for two, four, and eight subscribers. These systems can serve a cluster of subscribers with a multisubscriber terminal, or they can serve separate subscribers along a line with each subscriber with the appropriate unit for his carrier frequencies. An analog pair-gain system typically can operate over 8 km via a 0.6-mm-diameter twisted pair; with additional repeaters this distance can be extended up to about 25 km.

Digital pair-gain systems have been on the market since 1987 using ISDN technology. The two ISDN 64-Kbps B-channels are used in digital pair-gain systems for two separate subscribers, whereas four subscribers can be served using subrate 32-Kbps ADPCM voice coding and splitting each B-channel into two subscriber channels. Digital pair-gain systems thus derived from ISDN provide the subscribers with one 64-Kbps voice and data communication channel and usually can be easily upgraded to full N-ISDN.

The transmission of speech and data and the DTMF signaling of each subscriber on one go and one return channel in full duplex operation on one single twisted pair is possible mainly thanks to a special echo cancellation technique that eliminates the signal reflections caused by the 2/4 wire interfaces and different diameters of the cable in the local loop. As an additional measure to overcome impairment from the UTP, a special baseband line code, 2B1Q, also derived from ISDN, is applied. The 2 binary 1 quarternary code is a baseband transmission technique equivalent to a four-level pulse amplitude modulation with a very good error-limiting performance. Digital pair-gain equipment usually includes power feeding of the subscriber equipment from the central office, thus eliminating power equipment and batteries at the subscribers premises.

Modern digital pair-gain systems are available in the following typical configurations:

- Two telephone lines each 64 Kbps plus two data lines each 19.2 Kbps for fax or modem;
- Four lines each 32 Kbps for either telephony or data.

The typical performance for a two-subscriber system is about 7 km on a 0.6-mm-diameter twisted pair [4,5].

3.2.2.3 ADSL

Asymmetrical digital subscriber line systems allow the use of an asymmetrical high-speed digital signal in addition to the normal telephone service on unconditioned twisted pair local loop cable. The unsymmetrical signal supports the bidirectional transmission of speech and data at speeds from 128 Kbps typically up to 1.544 Mbps, respectively, 2.048 Mbps (TI resp. E1) with a capability up to 6 to 9 Mbps in the direction to the subscriber (downstream) and upstream typically 16 or 64 Kbps and a capability up to 640 Kbps. With the data rate of 6 Mbps four, respectively, three video signals each MPEG-encoded at 1.5 Mbps, respectively, 2 Mbps can serve a subscriber's multimedia terminal over a 4- to 6-km local loop.

Two different modulation and line coding methods are still competing for worldwide standardization: *carrierless amplitude/phase* (CAP)—a passband transmission method similar to *quadrature amplitude modulation* (QAM; see Chapter 5) developed by AT&T Paradyne, and *discrete multi tone* (DMT)— developed by Amati Communications of Mountain View, California and already standardized by the *American National Standards Institute* (ANSI) under T1E1.4. The first ANSI-standard-based chips are expected mid-1996 in the following versions, each for loop length up to 3.7 km:

- 8 Mbps downstream and 220 Kbps upstream;
- 4 Mbps downstream and 640 Kbps upstream.

To promote the ADSL concept worldwide and facilitate the development of end-to-end network systems, an ADSL Forum was founded in 1994, registered in Palo Alto, California, and currently has over 60 member-companies. The Forum is currently working on standard definitions for providing systems that handle the transmission of signals from the subscribers set top box to the edge of the service provider's network. This includes work on standards for wiring the phone into the set top box, defining the methodology of controlling remote configuration, and describing requirements for network management. Standards are being developed for SDH and ATM interfaces. The major applications of ADSL are seen in VOD, interactive video (videoconferencing and multimedia), home shopping and home banking, education, and data network access (Internet). ADSL is conceived as a product for mass deployment; this deployment will be closely tied to the future of multimedia. If multimedia meets its expectations, ADSL will become available roughly at the price of this book [6,7].

3.2.2.4 SDSL

Symmetrical digital subscriber line systems are similar to ADSL but support the full-duplex bidirectional transmission of speech and data at speeds from 128 Kbps to 2.048 Mbps.

3.2.2.5 VDSL

Very high speed digital subscriber line systems will support both symmetrical and asymmetrical data transmission in the local loop for bit rates up to about 50 Mbps, covering the last few hundred meters on copper in an otherwise optical fiber access network.

3.2.2.6 HDSL

High bit rate digital subscriber line systems are mainly conceived for SOHO subscribers. HDSL, developed by Bell Comm. Research Inc., in fact, is a transparent replacement for repeatered E1/T1 circuits in the access network. Repeatered E1/T1 technology is expensive, and HDSL will outdate that technology. Figure 3.9 illustrates the basic differences between E1/T1 repeatered transmission and HDSL technology. The drawing shows an example based upon a 2-Mbps E1 signal on two twisted pairs with 0.5-mm (24 AWG) diameter. In the traditional repeatered version one pair is used for the upward direction and the other pair for the downward direction. With HDSL, to reduce the required bandwidth, half the 2-Mbps, respectively, 1.5-Mbps signal is sent in a bidirectional mode over one pair and the other half over the second pair. After splitting, for example, the 2.048-Mbps signal into two 1.024-Mbps signals, 144 bits for synchronization and monitoring are mapped into each half signal resulting in a 1.168-Mbps signal, respectively, 24 bits are added for T1 resulting in a 784-Kbps signal. The 1.168-Mbps, respectively, 784-Kbps signals are transmitted in the same frequency band, with the transmit and receive signals being separated by a hybrid. A complex echo cancellation technique eliminates the residual signal of the transmitter in the receive path and echoes coming from the far end of the line caused by heterogeneity of the line. A digital signal processor calculates the reflections of its own transmit signal. The calculation results are used by an adaptive filter to emulate the echo as precisely as possible. The emulated echo signal is then subtracted from the receive signal; the reflected components of the signal are, in effect, canceled out. Thanks to this procedure, a high line quality is not a prerequisite of HDSL; on the contrary, HSDL can be installed in the local loop without the conventional complex and time-consuming measurements, bridge taps (leftover cable from previous subscribers) removing, and selection of the better pairs within a cable.

Similar to ADSL, two line codes are used: 2B1Q and CAP. Unlike 2B1Q (derived from ISDN as mentioned within the discussion of digital pair-gain), CAP, as mentioned in conjunction with ADSL, is a baseband transmission method similar to QAM that for HSDL is combined with trellis coding to further improve the bit error performance into the range that is usual for optical fiber

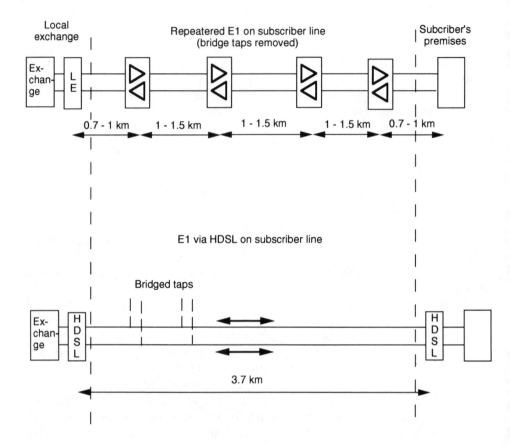

Figure 3.9 Comparison of HDSL and E1/T1 technology (*After:* Alcatel).

operation. For HDSL with trellis-coded CAP, a BER $> 1 \times 10^{-11}$ per 24-hr interval is state of the art for operation on two cable pairs as follows:

- 0.4-mm cable diameter: > 4 km;
- 0.8-mm cable diameter: > 10 km;
- 0.9-mm cable diameter: > 11 km.

HDSL interfaces with SDH and ATM and is also available at fractional E1/T1 bit rates (for example, 384 Kbps for videoconferencing) for operation on one single twisted copper pair. ETSI has specified the use of HDSL via three twisted pairs separating the 2.048-Mbps E1 signal into three bidirectional 784-Kbps signals in order to cope with the high variety—concerning length and other physical parameters—of copper wire-based local loops in Europe.

3.2.2.7 VHDSL

Very-high bit rate digital subscriber line systems have their origin in UTP-LAN networks, where data rates of 10 and 16 Mbps (10BaseT Ethernet and Token Ring) and now also rates of 100 Mbps as *copper-distributed data interface* (CDDI) and 155 Mbps as SDH STM1 are required over short distances of some 100 meters between data processing equipment and wiring hubs. Typical configurations for a distance of some 200m are

- 51.84 Mbps (SONET STS-1) downstream and 2 Mbps upstream;
- 622 Mbps (SDH STM-4) downstream and 155 Mbps upstream.

Beyond using VHDSL in LAN cabling, another application is in the local loop in combination with optical fiber in a so-called *fiber-to-the-kerb* (FTTK; see Chapter 4) configuration. The major length of the local loop is served by optical fiber cable with a passive optical network termination at a kerb near the subscribers premises; so the existing UTP, although not for the "last mile," can still be used for the last hundred meters, thus serving a subscriber with B-ISDN without a need for digging up the subscribers premises [8,9].

3.2.2.8 Data Transmission on Newly Installed UTP

To minimize the problems inherently connected with data transmission on UTP in such a way that UTP can still be used economically on various applications instead of optical fiber, special attention needed to be paid to the UTP cable plant within buildings, corporate campus, and LANs. The *Electronic Industry Association* (EIA) and the *Telecommunication Industry Association* (TIA) of the United States, therefore, defined minimum electrical and physical characteristics for UTP cable and accessories. Major limiting factors for data transmission on UTP are bandwidth and electromagnetic radiation. The electromagnetic radiation mainly concerns

- Electromagnetic fields from the power supply grid (for example, steep high-voltage pulses generated during power switching) by electrical (such as welding machines) and electronic equipment (for example, laser printer), fluorescent lamps, and electromagnetic switches causing an interference voltage on the data signals;
- Lightning causing low-frequency high-energy pulses on the line;
- Electrostatic discharge from surrounding persons or apparatus.

Upper bandwidth limits together with correspondingly low degrees of susceptibility to electromagnetic influence [*electromagnetic interference* (EMI) or *electromagnetic compatibility* (EMC)] and generally performance capabilities have been defined in standard EIA/TIA-568A for five categories of UTP cable as summarized in Table 3.2.

Although not explicitly stated in EIA/TIA-568A, categories 3, 4, and 5 roughly correspond to data rates up to 10 Mbps (e.g., Ethernet), 16 Mbps (e.g., Token ring), and 100 Mbps (such as VOD and multimedia), respectively. In the meantime "super" category 5 UTP cable is on the market, enabling the transmission of ATM 155-Mbps signals.

Improper cable installation, such as too close proximity to an EMI source, can easily result in a breakdown of computer systems working at high data speed. The basic issues of installation practice are therefore addressed in EIA/TIA-569. This standard focuses on such practical matters as maintaining the noise level within acceptable levels by installing the cable at appropriate distances from EMI sources such as electrical cabling and maintaining twist rates to within 13 mm (0.5 in) of the connections.

Application of the EIA/TIA-568A/569 standards—which are being accepted worldwide in a number of incarnations such as ISO, GOSIP (*Government OSI Profile* of the United Kingdom), and Australian Standard (3080)—may result in an initial investment increase in cable plant but, in the long run, would be offset by savings on the subsequent transmission equipment investment, especially for future applications.

Meeting the stringent radiation requirements of category 5 might not always be possible with UTP cable. This has resulted in the application of a thin aluminum foil around the twisted pairs in FTP or *shielded twisted pair* (STP) cable, together with the development of shielded components and patch fields for FTP cable. The cable shielding can be applied as individual shielding around each pair or as global screening around the cable core. The following cable designations are used:

- STP: pairs are individually shielded;
- S/STP: pairs are individually shielded and the cable core is shielded;
- S/UTP: the cable core is shielded, but not the individual pairs.

Table 3.2
UTP Cable Categories According to EIA/TIA-568A

Category	Bandwidth
1	analog voice and vf-telegraphy
2	digital voice, low speed data, and N-ISDN
3	up to 16 MHz
4	up to 20 MHz
5	up to 100 MHz

Conventional backbone cabling design uses multipair UTP for voice and optical fiber for data. Whereas optical fiber itself is becoming less and less expensive, it still costs slightly more than UTP and is more susceptible to buckling, bending, and tensile stress than twisted copper wire cable. Optical fiber components and fiber-based data equipment are significantly more expensive than their copper counterparts. UTP and FTP cable, therefore, and since they are now even capable of supporting ATM, will find a niche application for data transmission over short distances in an industrial environment. Hence, for LANs, a cable with optical fiber pairs and UTP might be cost effective to interconnect copper-based data equipment [10–12].

3.2.3 Twisted Pair Transmission Cable

The attenuation and impedance of twisted pairs are very much dependent of the frequency. L decreases to about 70% of its initial value as the frequency increases from 50 kHz to 1 MHz and is stable beyond. G is very small for plastic-insulated conductors but increases roughly at a proportional rate for nonhygroscopic tape (textile, impregnated paper, or wood pulp) insulation. R, which is approximately constant over the voice band, increases at higher frequencies proportional to the square root of frequency due to skin and proximity (eddy current) effects.

Therefore, for the transmission of multiplexed signals, attenuation and crosstalk on ordinary twisted pair cable are too high. When new telecommunication cables need to be installed in the distribution network, optical fiber cables are used exclusively. With certain cable modifications, however, transmission of 24- to 120-channel multiplexed analog signals, or broadcast audio and video signals on twisted pairs, is possible and has been widely applied for many "pre-optical-fiber" years. Those cable modifications include

1. Application of better dielectricum;
2. Careful twisting with exact symmetrical arrangement of the pairs;
3. Adjacent quads with different unrelated twisting pitches;
4. Application of shielding;
5. Two separate cables for go and return direction.

In applying cable modification 1 to 3, symmetrically balanced twisted pair cable, called symmetrical or balanced pair cable, have been developed for the transmission of higher frequencies in the following combinations:

- Tape insulation for frequencies up to 252 kHz;
- Styroflex insulation for frequencies up to 552 kHz.

The typical characteristics of the analog transmission systems that were used on this cable are summarized in Table 3.3.

Shielding was applied for frequencies above 552 kHz and generally for symmetrical pair cable if additional electromagnetic protection was required, for example, if buried near railway tracks or electricity grids. The shielding surrounded the conductors with a uniform spacing along the entire length and was grounded so that the capacitance of the conductors was uniform throughout the length of the cable and balanced to ground.

Digital transmission over symmetrical pair cable has been applied for E1/T1 links to interconnect exchanges. Unlike with analog transmission, for E1/T1 transmission four-wire operation on two pairs of the same cable is possible. Intermediate regenerative repeater are then required at 1- to 3.5-km intervals, depending on the cable characteristics and the number of parallel systems within the same cable. The total length that can be covered depends on the diameter of the symmetrical cable conductors and is typically as follows:

- 30 km on 0.4 mm;
- 55 km on 0.8 mm;
- 80 km on 1.2 mm.

Time-consuming measurements concerning crosstalk distribution along the line and between the various quads of the cable are necessary as a base for careful planning of the regenerator locations and for selecting most suitable pairs. Significant electromagnetic interference came from the electromagnetic telephone exchanges, so the distance between line terminal equipment and the first and last regenerators had to be reduced to about 50% to 60% of the normal distance between two regenerators.

The regenerators are usually power-fed and include performance-monitoring and dc- and pulse-fault detection facilities. As power feeding a constant dc current of typically 60 mA flows via the phantom circuit of the star quads.

Table 3.3
Analog Transmission Systems Operated on Twisted Pair Transmission Cable

Number of Channels	Number of Cables	Transmission Band (kHz)	Conductor Diameter (mm)	Repeater Spacing (km)
12	1	go 6–54 return 60–108	1.4	32
24	2	6–54 + 60–108	1.4	35
60	2	12–252	1.2	30
120	2	12–252 + 312–552	1.3	20

Power-feeding units with an output of 100V to 200V, depending on the number of regenerators within a power-feeding loop, are located in the terminal equipment or in aboveground regenerator stations. Figure 3.10 shows the typical arrangement of an E1 line operated on symmetrical pair cable.

The transmission of 8-Mbps signals over symmetrical cable is also possible, but in that case two separate cables are required for go and return direction. Distances up to 9 km can be covered when using pairs with 1.3-mm conductor diameter [1].

3.3 COAXIAL CABLE

3.3.1 General

At frequencies around 1 MHz the attenuation even in a symmetrical pair cable reaches prohibitive values mainly due to the skin effect in the conductors, so the coaxial cable is used in such cases. A coaxial cable consists of a center conductor surrounded by a concentric outer conductor, both made out of copper. The supporting dielectric material between the inner and outer conductors is either a solid low-dielectric-loss foam or air, in which case a polyethylene spiral or adequately spaced disc keep the inner conductor exactly in the center.

The primary constants of coaxial cables are under better control and are less frequency-dependent than those of the twisted pair cables because of the inherently more consistent mechanical structure and the shielding from outside influences provided by the outer conductor.

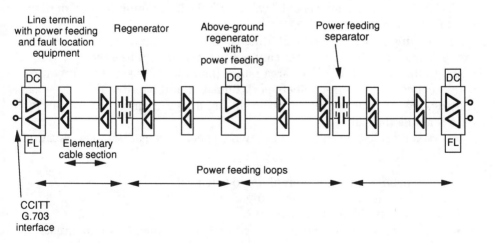

Figure 3.10 E1 line on symmetrical pair cable.

The capacitance, C, is independent of frequency and is a function of the conductor diameter ratio and the permittivity (dielectric constant) of the dielectric.

The inductance, L, is also practically independent of frequency over the normal frequency range used for coaxial cables. In fact, the inductance increases at frequencies below 60 kHz, but the shielding effect at those frequencies is insufficient; so coaxial cable is hardly used below 60 kHz. The shielding, on the other hand, increases with frequency, and coaxial cable offers an excellent crosstalk attenuation especially at higher frequencies. As a consequence, contrary to symmetrical cable, both transmission directions of a signal can be accommodated in the same coaxial cable.

The conductance, G, is a function of the dielectric used between the inner and outer conductors; for the usual air insulation, this conductance is negligible over the whole frequency range applied on coaxial cable.

The resistance, R, is the major performance-determining constant. Because of the skin effect, R increases at the square root of frequency. The attenuation formula for coaxial cable can be simplified as

$$\alpha = (R/2)\sqrt{C/L} \approx R/2Zo \quad \text{(dB/km)}$$

Thus, the loss is directly proportional to R and, hence, directly proportional to the square root of frequency. Figure 3.11 shows the typical attenuation versus frequency over the frequency band of 60 kHz to 5 MHz for the CCITT standardized (G.352) small-diameter coaxial cable with the inner/outer conductor diameters of 1.2/4.4 mm.

Doubling the cross-sectional dimensions of the coaxial conductors halves the resistance (as the skin effect causes the resistance to be inversely proportional to the conductor surface area rather than the cross-sectional area). As a consequence, the loss in decibels per kilometer, and bandwidth and repeater spacing, are inversely proportional to the conductor diameters provided their ratio remains the same.

Coaxial cable is used for the following specific transmission applications:

- Long-distance transmission of analog signals;
- Long-distance transmission of digital signals;
- Cable-TV distribution networks;
- RF-feeders between transmitting/receiving equipment and antennas;
- RF-radiating cable for extension of broadcast and mobile radio into electro-magnetically shielded areas (tunnels, mines, buildings).

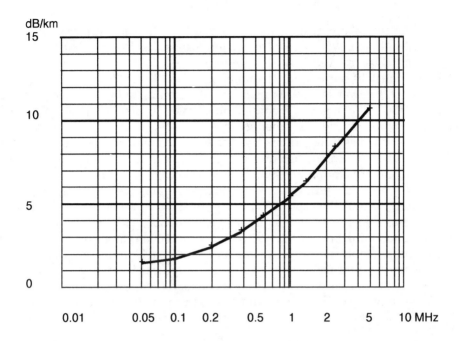

Figure 3.11 Attenuation versus frequency for 1.2/4.4-mm coaxial cable.

3.3.2 Long-Distance Transmission of Analog Signals on Coaxial Cable

For long-distance transmission CCITT has defined three types of coaxial cable as summarized in Table 3.4.

Coaxial cable initially was conceived for high-performance long-distance transmission of high-capacity analog transmission systems. A number of prerequisites must be fulfilled in order to achieve the high performance. To com-

Table 3.4
CCITT Defined Coaxial Cable

Name	Designation	Conductor Dimensions Inner (mm)	Outer (mm)	Typical Attenuation at 50 MHz and 10°C (dB/km)
Standard	2.6/9.5	2.6	9.5	8
Small-diameter	1.2/4.4	1.2	4.4	18
Mini-tube	0.7/2.9	0.7	2.9	32

pensate for the relatively high attenuation of coaxial cable, numerous complex underground repeaters are required at short distances along the line, similarly as shown in Figure 3.10. Those repeaters are power-fed from their line terminals at a very constant current, for example, 60 mA ±1% for 300- and 960-channel systems. The power feeding is made via the inner conductors of the coaxial cable. At the end of a power-feeding loop a power-feeding termination galvanically interconnects the two inner conductors of the same cable (go and return direction), whereas transformers provide the interconnection between the two cables for the high-frequency analog signals. Those transformers also provide a high-voltage separation between the two feeding loops to prevent lightning and power grid calamities from affecting adjacent power feeding loops and generally to reduce line noise. Typically up to 20 repeaters of a 960-channel coaxial line system can be fed from one terminal. With two power-feeding sections a total link of 2×4 km $\times 20$ repeaters $= 160$ km can be bridged between two terminal stations. For longer distances an above-ground repeater station with power-feeding equipment can add two power-feeding loops and thus double the maximum possible distance between two terminal stations.

The repeaters for analog transmission basically have the following functions:

- The pilot controlled linear amplification of the weak received signal;
- To equalize the frequency-dependent attenuation;
- To equalize the temperature-dependent attenuation (for example, day and night difference, which typically is 0.5°C; and seasonal difference, which typically is ±10% in Western Europe; attenuation variation is 2‰/°C);
- Line supervision and fault reporting.

Depending on regional climatic conditions, the actual depth of the buried coaxial cable and the total length of the cable, not all of the above functions are combined in all of the underground repeaters of a link, for example, temperature equalization in every third or fourth repeater.

Coaxial cable usually has a few normal plastic insulated copper wires, *interstice pairs,* that fill up the cylindrical shape of the cable. These interstice pairs are then used for supervision and service channel operation.

The basic data of the various analog transmission systems (that used to be) operated on coaxial cable are summarized in Table 3.5 [1].

3.3.3 Analog Submarine Coaxial Cable Systems

Coaxial cable for submarine application has only one coaxial pair that is used for two-wire transmission with go and return direction in separate nonoverlapping frequency bands. The maximum diameter of a coaxial cable, and thus its attenua-

Table 3.5
Analog Transmission Systems Operated on Coaxial Cable

System-Capacity Telephone Channel	Frequency Band (kHz)	Coaxial Cable Type	Repeater Spacing (km)
300	60–1,363	1.2/4.4	8
600	60–2,792	1.2/4.4	6
960	60–4,287	1.2/4.4	4
1,260	60–5,680	2.6/9.5	9.5
2,700	300–12,435	2.6/9.5	4.65
		1.2/4.4	2
10,800	4,322–59,684	2.6/9.5	1.55

tion and transmission capacity, is determined by the bending radius of the cable. This radius must be large enough so that the outer conductor neither buckles on the compression side nor ruptures on the tension side. Larger diameters, on the other hand, increase cost and space requirements on the cable-laying ship. Thus, there is a relationship between cable diameter, transmission capacity, and number of repeaters. Consider the major characteristics of TAT-7—the last transatlantic coaxial cable, which was inaugurated in 1983—as summarized in Table 3.6.

For submarine operation not only special coaxial cable but also special analog transmission systems have been developed. The world's longest coaxial cable submarine structure is the SEA-ME-WE I (South-East Asia–Middle East–Western Europe submarine cable one).

SEA-ME-WE I was inaugurated in 1986 and, like most other submarine systems, was designed for a operational lifetime of 25 years, so the cable might

Table 3.6
Characteristic Data of TAT-7

Characteristic	Value
Route length	5,500 km
Outer diameter	52 mm
Bending radius	1.025m
Number of circuits	4,200
Repeater spacing	10 km
Power feeding:	
Voltage	5,150V
Current	657 mA

serve another 15 years! The submarine cable (see Figure 3.12), with a total length of 13,585 km, connects three continents: Asia, Africa, and Europe. The Suez-Alexandria section is constructed with conventional terrestrial coaxial cable installed along the Suez Canal.

The shallow-water cable, with a diameter of 1 to 1.5 in is armored with steel wires as protection against fishing boats and trawlers. The deep-water cable (about 85% of the route) is unarmored and has a diameter of 1.5 to 1.7 in.

Two different multiplex systems operate on the cable: a 2,580-channel system with a bandwidth of 25 MHz and a 1,260-channel system with a bandwidth of 12 MHz. The 2,580- and 1,260-channel capacities refer to the normal channel spacing of 4 kHz. For applications at 3-kHz channel spacing the capacity can be increased to 3,440/1,680; whereas with analog circuit multiplication equipment (for both techniques please see Chapter 2) and 4-kHz spacing a transmission capacity of 5,000, respectively, 2,500 channels is available. Figure 3.13 shows the special channel arrangement of the two submarine multiplex systems.

The line equipment used for this cable includes terminal equipment, submarine repeaters, and submarine equalizers. The terminal equipment incorporates power feeding for the repeaters and equalizers, applying a constant current of 545 mA and a maximum voltage of 6,500V. The repeaters located in

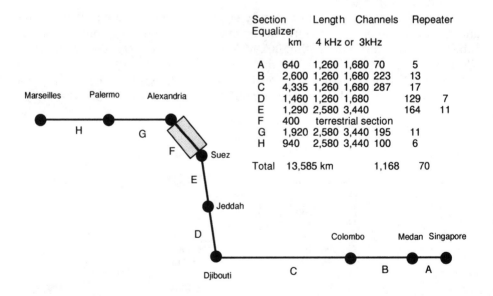

Section Equalizer	Length km	Channels 4 kHz or 3kHz			Repeater
A	640	1,260	1,680	70	5
B	2,600	1,260	1,680	223	13
C	4,335	1,260	1,680	287	17
D	1,460	1,260	1,680	129	7
E	1,290	2,580	3,440	164	11
F	400	terrestrial section			
G	1,920	2,580	3,440	195	11
H	940	2,580	3,440	100	6
Total	13,585 km			1,168	70

Figure 3.12 Route of SEA-ME-WE I.

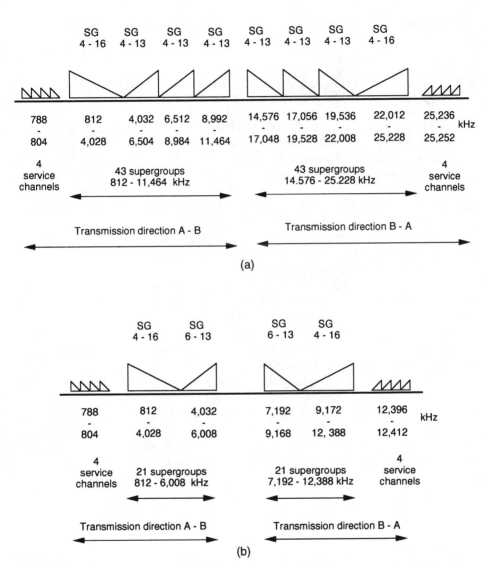

Figure 3.13 Analog submarine multiplex systems: (a) 25-MHz system and (b) 12-MHz system.

shallow waters are equipped with temperature-regulated amplifiers to compensate for the temperature changes, which, of course, are not required at the stable temperatures in the over 4,000-m-deep sea water. In addition to the about 1,170 repeaters, some 70 equalizers are used that are remotely controlled to enable adaptation of the cable equalization to the actually used bandwidth. The route is engineered for an overall average noise level of 1 pW/km and a maximum of 2 pW/km only [13,14].

3.3.4 Long-Distance Transmission of Digital Signals on Coaxial Cable

Digital transmission systems have replaced most of the analog transmission systems operated on coaxial cable. The regenerator spacing for the digital systems has been adapted to the repeater spacing of the analog systems to enable easy upgrading from analog to digital without digging and cutting the existing cable. The digital signals have a dc-free line signal (for example, 4B3T or MS43) that enables them to pass through the transformers of the power-feeding terminations.

The regenerators are equipped with amplifiers and equalizers to compensate for the frequency-dependent and temperature-dependent attenuation before regenerating the digital signals.

Digital transmission systems operating on coaxial cable have been developed for transmission capacities of 480, 1,920, and 7,680 telephone channels. Systems for higher transmission capacities have not been developed because emerging optical fiber systems offered both technically and economically better solutions. The basic data of the digital transmission systems operating on coaxial cables are summarized in Table 3.7.

Thanks to the signal regeneration without line noise amplification, significantly longer distances can be covered between two terminal stations than with analog systems. In fact, the maximum length is not limited by the signal quality but by practical considerations in connection with the dimensioning of the system-incorporated fault location system. For 34-Mbps systems, applied mainly between adjacent exchanges, the maximum distance between two line

Table 3.7
Digital Transmission Systems Operated on Coaxial Cable

System Capacity Telephone Channels	Transmission Speed (Mbps)	Coaxial Cable Type	Repeater Spacing (km)
480	34	0.7/2.9	2
		1.2/4.4	4
		2.6/9.5	9.5
1,920	140	1.2/4.4	2
		2.6/9.5	4.5
7,680	565	2.6/9.5	1.55

terminals is typically a few hundred meters; whereas for 140-Mbps systems this distance is typically 1,000m on 1.2/4.4 cable and 2,000m on 2.6/9.5 cable.

3.3.5 Cable-TV Distribution Networks

Although the deployment of analog long-distance transmission systems on coaxial cable has been stopped in favor of digital systems and no additional long-distance coaxial cable will be laid for the transmission of digital signals, coaxial cable laying for cable-TV (CATV, originally abbreviated from "Common Antenna TeleVision") networks continues, although with a decreasing trend due to emerging optical fiber systems.

A coaxial cable-based CATV network basically consists of a head end and a tree-and-branch configured coaxial cable distribution network. Figure 3.14 shows the typical layout of such a network.

In the head-end the various broadcast and TV signals are received by satellite and professional antennas and via radio-relay systems and terrestrial cable from satellite, TV transmitters, TV-distribution nodes, studios, and from outside-broadcast vans. The received signals are then amplified and without modulation, thus in their original VHF/UHF frequency band, distributed via the coaxial cable network. In Figure 3.15 the frequency allocation of a CATV system as used in Germany is indicated.

The transmission band below 47 MHz is used for two upstream TV channels and for data monitoring and service channels. TV band I contains the two national (ARD and ZDF) programs plus one regional program on their genuine frequency. The 87.5- to 108-MHz band contains up to 30 UHF stereo-broadcast programs, and the 111- to 125-MHz band up to 16 digital satellite broadcast channels. The 174- to 230-MHz band accommodates Band III national TV programs at their genuine frequency; whereas the lower, upper, and extra-sideband carry other regional and foreign TV programs, up to a total of 45 TV channels for the whole system. The system can cover an area of approximately 30 km^2.

3.3.5.1 CATV for Multimedia

After 50 years of CATV in the United States with coaxial cable lines running past 97% of U.S. homes, this coaxial cable network may, rather than making place for optical fiber, experience a resurgence thanks to emerging multimedia services. CableLabs, the American cable TV industry's research and development consortium, recently announced the standardization of high-speed modem connections with CATV systems. These standards will specify:

- Consumer RF interface connecting the CATV network to the home;
- Connection to the subscriber's PC;

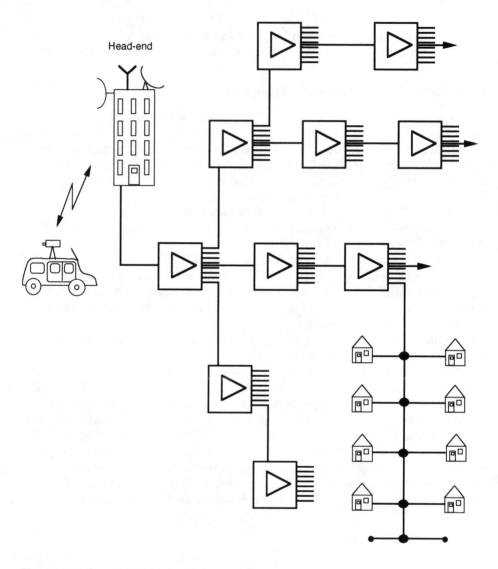

Figure 3.14 Coaxial CATV network.

- Interfaces within the head-end or distribution hub, including router and operation support system interfaces;
- Head-end to WAN interface, to include Internet and other global data networks.

A set-top box based on those standards for two-way operation, if mass produced, would offer an economical solution and immediately bring the inter-

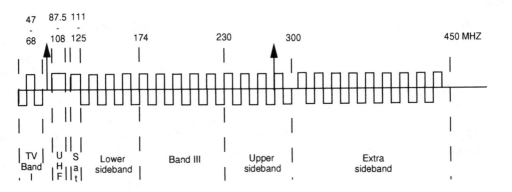

Figure 3.15 Frequency allocation of a German CATV system.

national data highway to millions of homes, provided sufficient protection against prevailing interferences on the CATV infrastructure from CB radio, police scanners, and others can be achieved. Efficient multimedia and videoconferencing via the existing CATV network could well find a place in competition with ADSL and ISDN [15].

3.3.6 RF Feeders

Coaxial cable is used as RF feeders between antennas and the transmitter/receiver of radio-relay systems, mobile radio systems, and earth stations and for radio and TV broadcast transmitters and CATV networks operating between 30 MHz and 3 GHz. Coaxial cable for RF feeders, unlike the long-distance coaxial transmission cables, are not standardized but are available in a large variety of sizes and types that can be divided into two groups: (1) air-dielectric and (2) foam-dielectric coaxial cable.

3.3.6.1 Air-Dielectric Coaxial Cable

The air-dielectric coaxial cables have copper wire or copper tubes as inner conductors. In large coaxial cables corrugated tubes are used as inner conductors. The outer conductor is a thin-walled longitudinally welded and corrugated solid copper tube that makes the cable semiflexible, gas-tight, and mechanically stable. To separate the inner and outer conductors a high-density polyethylene is used with particularly low-dielectric loss, so the amount of dielectric material is limited to the minimum required for safe centering of the inner conductor. The outer conductor is usually covered with an abrasion-resistant polyethylene outer jacket. Since the outer conductor is gas-tight, the cable can be gas-pressurized to increase power rating, facilitate supervision, and prevent ingress of moisture.

3.3.6.2 Foam-Dielectric Coaxial Cable

The foam-dielectric coaxial cables use a nonhygroscopic low-dielectric loss polyethylene foam with very small and closed cells that reliably centers the inner conductor and hermetically closes the space between inner and outer conductors. The inner conductors are again copper wire, copper tubes, or corrugated copper tubes. For highly flexible cable copper-clad aluminum inner conductors are also used. The outer conductor is the same as for the air-dielectric cable.

The air-dielectric coaxial cable is used for such applications where stringent transmission conditions prevail; whereas the cheaper foam-dielectric coaxial cable is suitable for more moderate applications, especially for smaller stations where pressurization is not available or required.

3.3.6.3 RF-Performance

The performance of coaxial cable as RF feeders is determined by additional characteristics, including

- Relative propagation velocity and delay;
- Electrical length;
- Reflections factor, respectively, VSWR;
- Power rating;
- Maximum operating and cut-off frequency.

3.3.6.4 Relative Propagation Velocity and Delay

The relative propagation velocity is defined as

$$v_r = (v_\psi/c_o) \cdot 100 = (l/l_o) \cdot 100 \quad (\text{in } \%)$$

in which v_ψ denotes the propagation velocity in the cable, c_o is the propagation velocity in free space $(300 \cdot 10^3 \text{ km/s})$, and l and l_o denote the geometrical and electrical length, respectively, in meters. Delay is defined as

$$t_\psi = 333.6/v_r = 10^8/(v_r \cdot c_o)$$

The propagation velocity is frequency-dependent due to the skin effect. As the frequency decreases, the velocity decreases and thus the delay increases. The propagation velocity is measured at frequencies around 200 MHz as standard.

3.3.6.5 Electrical Length

The electrical length is defined as

$$l_e = (100/1)/v_r \quad \text{(meters)}$$

The electrical length of RF cables is dependent upon temperature and, in case of air dielectric cables, on the pressure and humidity of contained air. The influences are quite small but must be taken into account for such cases where the cables are very long compared to the operating wavelength.

3.3.6.6 Reflections Factor, Respectively, VSWR

The reflections factor sums up the effects of all the impedance variations within the cable and at its ends at a certain frequency. It is the ratio between the (vector) addition of all the reflections and the incident signal, measured at the near end of the cable. It is also customary to use the term *voltage standing wave ratio* (VWSR), based upon the standing wave, that the cable under test would produce in a homogeneous transmission line connected to its near end and having its nominal characteristic impedance.

The VSWR is $s = (1 + r/100)/(1 - r/100)$. The reflections factor $r = (s - 1)/(s + 1) \cdot 100$, expressed as a percentage.

Instead of the reflections factor, the term *return loss* is also used and is defined as

$$A_z = 20 \log(100/r) \quad \text{(dB)}$$

3.3.6.7 Power Rating

The power rating is the lower of the peak power rating and mean power rating. The peak power rating is the input power for which the peak RF voltage rating is reached when the cable is operating in its matched condition. It is defined as

$$P_p = 500(U_p^2/Z_c) \quad \text{(kW)}$$

in which U_p is the RF voltage rating peak value in kilovolts, and Z_c is the characteristic impedance.

The peak power rating is independent of frequency, can be increased considerably by operating the cable under an overpressure, and decreases with altitude in the case the inner cable is allowed to assume the pressure of the environment.

The mean power rating is defined as

$$P_{max} = (0.8686 \cdot P_v)/(2 \cdot \alpha_t) \quad \text{(kW)}$$

in which P_v denotes the maximum admissible power dissipation in Watts per meter, and α_t is the attenuation under operation condition in decibels per hundred meters.

The mean power rating is the input power at which the inner conductor reaches an agreed temperature for the applied dielectric material (typically 85°C for foam and 150°C for Teflon in air dielectric) and is usually defined for installation in still air of 40°C without direct solar radiation and under normal atmospheric pressure.

3.3.6.8 Maximum Operating and Cutoff Frequency

Energy transmission in a coaxial RF cable takes place in the normal coaxial wave mode. Above cutoff frequency, which is a function of cable dimensions, other wave modes can also exist and the transmission properties are no longer defined. The approximate value of the cutoff frequency can be computed as

$$f_c = (1{,}91 \cdot v_r) \cdot (D_i + d_i)$$

in which v_r is the relative propagation velocity, D_i is the inner diameter of the outer conductor in millimeters, and d_i is the outer diameter of the inner conductor in millimeters.

In addition to cutoff frequencies, maximum operating frequencies are usually stated that include a certain safety margin. For some cables the maximum operating frequency is determined by other constructional criteria and may then significantly deviate from the cutoff frequency [16].

3.3.7 Radiating Cable

Radiating coaxial cables assure reliable communication within tunnels, subways, buildings, and mines. They function like antennas over their entire length and can be used to transmit or receive RF energy to or from their immediate surroundings. This effect is produced by openings in the outer conductor. The RF signal can be routed wherever local signal coverage is required. In addition to conventional application within confined areas, surface applications are also emerging, such as beside motorways for carrying radiotelephone traffic with a minimum of environmental interference and disruptions, beside railway lines to present a highly dependable carrier for automatic steering of trains, and in streets to improve performance of CT and PCN networks.

Radiating cables operate in the frequency range of 30 MHz to 2 GHz. The cable is connected to a base station that usually will obtain its input signal from an outside cellular radio, or broadcast repeater, or terrestrial cable. Power splitters distribute the RF signal on the coaxial cable; whereas, for larger installations, additional RF-repeaters compensate for the coaxial cable attenuation. Figure 3.16 shows a typical installation for a private two-way communication system for a large office building with a garage within a campus.

The system operates at 150 MHz, and the base station has an output power of 1W.

Radiating cable, if state of the art, is fireproof, does not produce toxic and corrosive compounds if exposed to flames, contains no halogens or fluorine, and gives off very little smoke to meet stringent environment protection requirements inside buildings.

Radiating coaxial cables are usually with foam dielectric and can be divided into the two groups, slotted and leaky cables.

The *slotted cables* have a continuous end-to-end longitudinal slot and are most cost effective for all applications where salt rather than dirt coupled with moisture is expected to deposit on the cable surface. Radio paging systems in buildings and radio systems in tunnels are typical examples for the application of slotted cable.

Leaky cables have individual apertures at regular spacings along the length of the cable. The leaky cable, which is more elaborate to manufacture, has more constant electrical characteristic even under adverse conditions. A special version of leaky cables is a double-slotted coaxial cable with two rows of apertures separated by 180 degrees on the circumference of the cable. Cables of this design are less sensitive to the effects of the mounting surface. Another version of leaky cable applies angled slots, rather than small holes, into the sheath for better performance.

The world's largest radiating cable network is in operation in the tunnel under the English Channel. Some 250 km of radiating cable are installed in three separate networks.

Network I is an operation and maintenance network including the overall service communications in the three tunnels, gangways, and service stations.

Network II is installed in the two main rail tunnels for operational communication between the TGV high-speed trains and control posts.

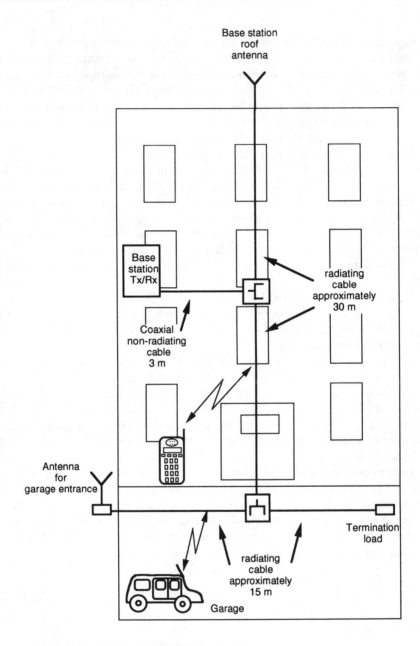

Figure 3.16 Radiating cable network (*After:* Andrew).

Network III is installed in the two main tunnels, also, however, to serve the passengers. All traffic in the Channel Tunnel is carried by rail; the vehicles are loaded onto special wagons and the passengers will remain inside during the 30-minute passage. Network III broadcasts entertainment and information to the passengers through their own car radios. Each wagon is equipped with leaky cable to retransmit in the FM-VHF band the signals that are picked up, and amplified, from the leaky cable track in the tunnel [16,17].

3.4 SPECIFIC APPLICATIONS

Copper line initially was the only medium for transmission. Technical progress, however, successively created the radio, satellite, and optical fiber media with better transmission merits, leaving virtually void any domain where investments on new copper lines are still economically justifiable. The only applications where new copper lines still offer advantages are

- Minor extensions of existing copper line networks;
- Coaxial cable in access networks for the last few hundred meters between an optical fiber network and the subscriber home terminals, to prevent the subscribers from needing relatively expensive transitions from optical fiber to their copper-based telecommunication equipment;
- RF feeder connecting antennas with radio equipment;
- Radiating cable in confined areas where radio transmission is not satisfactory and for surface applications such as beside motorways for carrying radiotelephone traffic with a minimum of environmental interference and disruptions, beside railway lines to present a highly dependable carrier for automatic steering of trains, and in streets to improve performance of CT and PCN networks;
- UTP and STP cable pairs in optical fiber cable to ease interconnection of copper-based data equipment directly without expensive optical fiber converters.

Upgrading existing copper lines with capacity-extending equipment, however, may very well be justifiable with such equipment as

- ADSL, SDSL, and VDSL to enable data transmission, VOD, and multimedia via ordinary subscriber telephone cable;
- HDSL and VHDSL to enable high bit rate and very high bit rate, respectively, communication via ordinary subscriber telephone cable.

References

[1] Members of the Technical Staff, *Transmission Systems for Communications*, Winston-Salem: Bell Telephone Laboratories, Inc., Revised 4th ed., Dec. 1971.

[2] Sarkowski, H., Editor, *SEL-Taschenbuch*, Stuttgart: Standard Elektrik Lorenz AG, 1962.

[3] Ludolph, Ir. G. L., *Technisch Vademecum Elektro- en Radio- Techniek*, Haarlem-Antwerpen-Djakarta: N.V. De Technische Uitgeverij H. Stam, 1953.

[4] Subscriber Carrier-Frequency System BF 1+1, Data sheet of Alcatel STK Norway.

[5] Hills, Tim, and David Cleevely, *Rural Telecoms Handbook Volume 2*, Cambridge, UK: Analysis Publications Ltd., 1992.

[6] Joshi, Bhagvat, and Tony Cooper, "Planning for Broadband Services," *Telecommunications*, Vol. 29, No. 9, 1995, pp. S161–S180.

[7] Gage, Beth, "Where In The World Is ADSL?" *Communications International*, Vol. 23, No. 1, 1996, pp. 50–54

[8] Moons, Marc, "HDSL: A New Lease of Life for Copper?" *Telecommunications*, Vol. 27, No. 11, 1993, pp. 67–72.

[9] Heidelberger, Christof, and Daniel Rieger, "The Copper Revolution," *Telecommunications*, Vol. 28, No. 9, 1994, p. 57–60.

[10] Mak, Max, "Twisted Pair Cable for Future Data Services," *Telecom Asia*, Oct. 1995, pp. 38–40.

[11] Andres, Paul, "A Universal High Performance Cable," *Telecommunications*, Vol. 27, No. 7, 1993, pp. 49–51.

[12] Biederstedt, Lutz, "The Advantages of Shielded Cabling Systems," *Telecommunications*, Vol. 28, No. 6, 1994, pp. 49–50.

[13] Paul, D. K., "Communications via Undersea Cables: Present and Future," *Fiber Optics, Short-Haul and Long-Haul Measurements & Applications*, SPIE, Vol. 559, 1985.

[14] SEA-ME-WE Submarine Cable Project, Marketing leaflet of Submarcom, Clichy, France.

[15] Lopez, Summer, "Modem Challenge," *Communications International*, Vol. 23, No. 2, 1996, p. 11.

[16] Transmission Line and Antenna Systems, Catalogue of Radio Frequency Systems Hannover, Division of kabelmetal electro GmbH, Edition 8, No. 1192.150.03.

[17] Radiax Slotted Coaxial Cable, Bulletin No. 1058G (12/91) of Andrew Corp., Orland Park, IL 60462 USA.

Optical Fiber 4

4.1 INTRODUCTION

The principle of guiding light through a transparent conductor was physically explained and demonstrated as early as 1870 by British physicist John Tyndall (1820–1893). Tyndall demonstrated that light could be guided along a curved path in water. Ten years later, in 1880, A. G. Bell—four years after he discovered the telephone—invented and patented a "Photophone," which he used to transmit speech signals at distances of a few hundred meters. Bell used sunlight sound-intensity-modulated by and reflected from one mirror to another and there detected with a selenium device. Weather dependence and still insensitive selenium photocells hindered practical application.

The invention of the laser (which stands for "light amplification by stimulated emission of radiation"), in which light could be converted into electricity and vice versa, by T. Maiman (1927–) in 1959, inspired research efforts to uncover a low-loss, well-controlled guided optical medium.

The breakthrough occurred in 1966 when K. C. Kao and G. A. Hockham—working at *Standard Telephone Laboratories* (STL) in the United Kingdom—predicted that fibers drawn from extremely pure glass would be an ideal support for the transmission of modulated light waves [1]. Optical-electrical transmission as proposed by Kao and Hockham consists basically of (see Figure 4.1) a light source, a cabled optical fiber, and a light detector. The light source is either an inexpensive *light-emitting diode* (LED) or a complex but nowadays also inexpensive laser. The light source emits light as a function of an electrical input signal (modulated with the information to be transmitted) at a wavelength suitable for transmission through the optical fiber. At the receiving end the detector converts the light back to the original electrical (modulated) signal. The detector usually is a fast pin-photo diode or an avalanche photo diode.

At that time the lowest loss in glass fibers still was in the order of 1,000 dB/km; and although Kao and Hockham predicted that a loss of

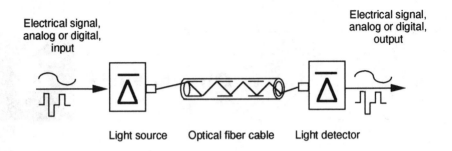

Electrical signal,
analog or digital,
input

Electrical signal,
analog or digital,
output

Light source   Optical fiber cable   Light detector

Figure 4.1 The principle of optical transmission.

20 dB/km should be attainable, the STL management refused to submit their patent application. So it happened that the German scientist M. Boerner— arriving at similar ideas as Kao and Hockham while researching for Telefunken—still in 1966 filed a patent in Germany. From the German Patent Office, he obtained patent No. DBP 1254.513 for a "transmission system using a semiconductor laser, a glass fiber, and a photodetector for transmission of PCM-signals." Boerner obtained similar patents in the United Kingdom and the United States and died soon thereafter.

F. D. Kapron, D. B. Keck, and R. D. Maurer of the Corning Glass Works of the United States achieved the 20-dB goal with industrially produced doped-silica clad fiber in 1970. A vapor phase process resulted in the required purification by a factor 1,000 or more with respect to unwanted metal ions. Optical fiber with attenuation below 1 dB/km appeared on the market, successively improved, currently to even less then 0.2 dB/km.

Optical fiber was first introduced as a replacement of copper cable on interexchange and other long-distance links. The first commercial optical fiber system was put into operation in 1973 with a 2-Mbps system operating on a 24-km link between Frankfort and Oberursel in Germany. In 1987 the world's first commercial long-distance submarine deep-water cable with a capacity of 280 Mbps went into service between the French mainland and Corsica. The world's longest terrestrial optical fiber link operating a 2.5-Gbps SDH system went into operation in Australia between Brisbane via Melbourne to Perth in 1996.

Another major application of optical fiber cable appeared in the access network. The first major test using optical fiber in the access network began in 1978 in Japan with two-way video service. A second large-scale four-year test started in 1986 in Berlin with an optical fiber access network operating at 140 Mbps and including ATM exchanges. It was with such and many other projects that user acceptance of broadband services was tested and provided a sound base for the present implementation of *fiber in the loop* (FITL) systems [2,3].

4.2 OPTICAL FIBER TECHNOLOGY

4.2.1 Refractive Index

The underlying principle of transmission in optical fibers is that light—in an encoded sequence of pulses—is reflected or refracted on boundaries of two different optical media constituting a waveguide for light transmission. The density of an optical media is given by the index of refraction n. This refractive index indicates the factor by which light travels slower in that medium than in a vacuum (in which $n = 1$). An optical fiber for transmission (see Figure 4.2),

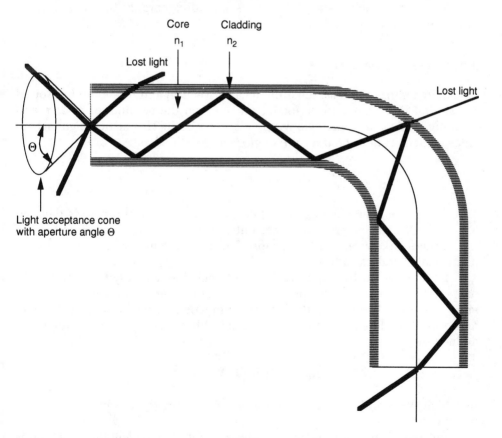

Figure 4.2 Propagation of light in an optical fiber.

therefore, consists of a light-conducting cylindrical core of high-grade silica glass with a refractive index n_1 surrounded by a concentric sheath (cladding) of silica glass with a refractive index n_2 slightly lower than the index of the core. The different refractive index is obtained by doping the pure silica glass—for example, with boron, which reduces the refractive index, or germanium, which increases the refractive index. A different refractive index would also be available with a simple glass-air interface; in practice, however, this interface would be unstable due to the following:

- Surface damages, such as scratches or inclusions;
- Surface contamination, such as grease or acid;
- Air instability, such as changing humidity and contamination.

To prevent such instability, the wave guiding core is surrounded by an exactly defined cladding [4,5].

4.2.2 Aperture Angle

To obtain a minimum loss, the difference between n_1 and n_2 should be around 1% so that light rays will be totally reflected on the boundaries of core and cladding if injected at an angle within a critical limit. This critical light acceptance angle is called the *aperture angle* Θ and defined as

$$\Theta = \text{arc } \sin \sqrt{(n_1^2 - n_2^2)}$$

The sine of the aperture angle is called the *numeric aperture* and is denoted $NA = \sin \Theta$. Light entering the fiber outside this acceptance angle is lost for transmission; likewise, light can get lost if the fiber is bent beyond the maximum permissible bending radius.

Depending on the aperture angle of each incident light ray a specific path—called *mode*—will be followed through the fiber. The wider the core the more modes (each with slightly different path length) can simultaneously exist: hence multimode and single or monomode fibers. The number of modes is defined by

$$N_m = 0.5(d_c \cdot NA \cdot \pi/\lambda)^2$$

in which N_m stands for the number of modes and d_c for core diameter in micrometers.

Related to the core diameter is also the *cut-off wavelength*, λ_c, which is defined as the wavelength below which propagation modes, other than the basic mode, first appear. The spectrum of the transmission source must therefore lie at a value above λ_c to enable single-mode transmission. The cutoff wavelength can be calculated by the formula [5]

$$\lambda_c = \pi d_c (1000/2.42)\sqrt{(n_1^2 - n_2^2)}$$

4.2.3 Transmission Windows

Silica glass offers three windows of minimum attenuation located between the boundaries of ultraviolet and infrared absorption in the 800- to 1,600-nm range. The theoretical minimum attenuation between the two borders is determined by the Raleigh scattering, which is a physical material property. Local glass density changes cause a steady fluctuation of the refractive index that cannot be eliminated and limits the lowest spectral attenuation of the fiber to low value of about 0.13 dB/km at 1,550 nm. Further attenuation, however, is caused by material impurities such as *hydroxyl* (OH) ions that enter the fiber during the production process and evoke peaks of attenuation at 950, 1,240, and 1,380 nm. Single-mode fiber with an attenuation of 0.17 dB/km is in the experimental production stage, whereas single-mode fiber from regular production with values slightly below 0.2 dB/km are standard. Figure 4.3 shows the typical attenuation of the current single-mode fiber versus the wavelength of the light used for transmission through the fiber. The range of low (almost Raleigh) attenuation between peaks of absorption attenuation are called "windows." As shown in this figure, optical fiber has three windows of minimum attenuation.

The first window is at 850 nm. The initial fiber optic systems operated in this window using LEDs and *gallium-arsenide* (GaAs) *laser diodes* (LDs). FITL systems (explained in Subsection 4.5.3) also operate in this window because unrepeatered operation up to 5 km is economical with the relatively low cost 850-nm components.

Constant improvement of both LEDs and LDs enabled the use of the second window at 1,300 nm (1.3μm) from 1976. Finally, for the third window at 1,550 nm (1.5 μm), at which Raleigh scattering loss and infrared absorption is minimized, appropriate LDs were developed in the early 1980s and is now generally used for long-distance optical fiber transmission systems.

Figure 4.3 illustrates the enormous transmission capacity of optical fiber and indicates the approximate bandwidth per window. This bandwidth can be calculated as

$$f = \text{speed of light/light wavelength}$$

For the first window, with the approximate range of 800 to 950 nm,

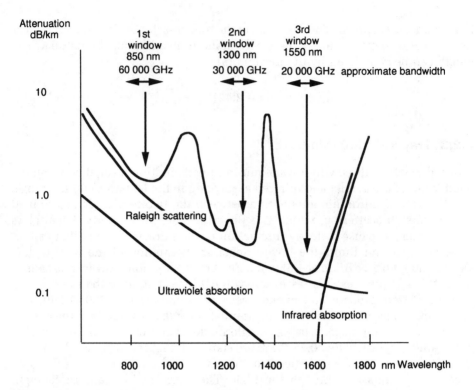

Figure 4.3 Windows of minimum attenuation for optical fiber transmission.

$$f(\text{GHz}) = (3 \cdot 10^8/800 \cdot 10^{-9}) - (3 \cdot 10^8/950 \cdot 10^{-9}) = 60{,}000 \text{ GHz}$$

For the second window, with the approximate range of 1,200 to 1,350 nm,

$$f(\text{GHz}) = (3 \cdot 10^8/1200 \cdot 10^{-9}) - (3 \cdot 10^8/1350 \cdot 10^{-9}) = 28{,}000 \text{ GHz}$$

For the third window, with the approximate range of 1,450 to 1,600 nm,

$$f(\text{GHz}) = (3 \cdot 10^8/1450 \cdot 10^{-9}) - (3 \cdot 10^8/1600 \cdot 10^{-9}) = 19{,}500 \text{ GHz}$$

The total capacity of the three windows thus amounts to almost 110 THz, which corresponds to some 15 million color TV channels [4,6–8].

4.2.4 Bandwidth

For practical applications the bandwidth of optical fibers is defined as the product of the bandwidth in GHz, in the conventional understanding, and the possible unrepeatered distance (in km) for a maximum optical power loss of 3 dB. The bandwidth thus represents the frequency analogy to pulse scatter, usually called *dispersion,* in the time range; the lower the pulse dispersion on the transmission route, the greater the bandwidth.

This bandwidth expressed in GHz · km is given by the formula

$$B = 0.44/\delta T$$

in which δT is the dispersion in nanoseconds, which can be calculated as [5]

$$\delta T = \sqrt{(t_2^2 - t_1^2)}$$

where t_1 and t_2 are the 3-dB values of the transmitted and received pulses, respectively.

4.2.5 Dispersion

The dispersion δT (in nanoseconds) is determined at the half amplitude width (50% value of the pulse amplitude or 3-dB point). Figure 4.4 shows the dispersion with an input pulse and the amplified output pulse after passing through 1 km of fiber and the corresponding bandwidth (1 GHz · km) of the fiber. The dispersion is essentially composed of modal dispersion and chromatic

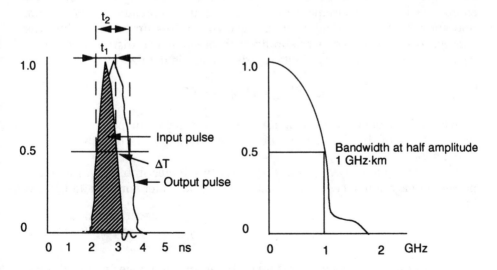

Figure 4.4 Signal dispersion and bandwidth of optical fiber.

dispersion, whereas at very high transmission bit rates polarization mode dispersion and power-dependent dispersion create transmission limitations [5].

4.2.5.1 Modal Dispersion

As mentioned in Subsection 4.2.2, depending on the aperture angle of each incident light ray, a specific path will be followed through the fiber. The wider the core the more modes each, with slightly different path lengths possible. A typical multimode fiber with a 50-μm core may have several hundred modes. The broadening of the optical pulses caused by the different path length of the different modes is called the modal dispersion. This modal dispersion can be reduced by reducing the core diameter, thus reducing the number of possible simultaneous modes. Hence, the monomode fiber with a 8- to 10-μm core diameter has zero modal dispersion [4].

4.2.5.2 Chromatic Dispersion

The chromatic dispersion is due to the color spectrum of the light and is composed of material dispersion and wave guide dispersion. The light pulses usually have a small wavelength spectrum and thus consist of different colors that travel at slightly different speeds through the fiber, thereby creating a chromatic material (while depending on the fiber material) dispersion, also called *group velocity dispersion* (GVD). The narrower the spectrum of the light source, the lower the chromatic dispersion. Figure 4.5 illustrates this phenomenon showing the broadening of the received pulses up to the point where consecutive pulses start overlapping each other, hence distorting the signal transmission. The chromatic material dispersion is defined by

$$\delta T_{\mathrm{mat}} = D_{\mathrm{mat}} \cdot \delta\lambda \cdot L$$

in which D_{mat} denotes the material dispersion factor in ps/nm $\cdot$ km, $\delta\lambda$ is the spectrum of the light source in nanometers, and L is the fiber length in kilometers.

The chromatic wave guide dispersion originates from the difference in power propagation speed in the center and at the outer surfaces and is given by

$$\delta T_{\mathrm{wg}} = D_{\mathrm{wg}} \cdot \delta\lambda \cdot L$$

The total chromatic dispersion is given by adding the material and the wave guide dispersions

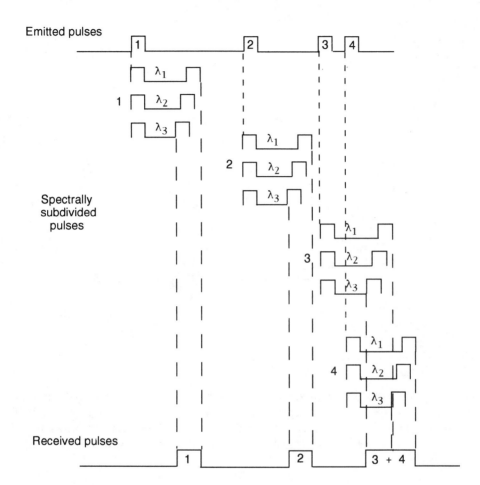

Figure 4.5 Chromatic dispersion.

$$\delta T_{\text{crom}} = (D_{\text{mat}} + D_{\text{wg}}) \cdot \delta\lambda \cdot L \tag{4.J}$$

Figure 4.6 shows the approximate relation between chromatic dispersion and wavelength for both the material and the wave guide dispersion. Beyond approximately 1,300 nm the material and the wave guide dispersion have different signs, thus the total chromatic dispersion can become zero. This *zero-dispersion point* (ZDP) in pure silica glass occurs at 1,270 nm thus in the second window. A signal with a 10-nm spectrum (typically for a laser), therefore, will hardly experience any chromatic dispersion if transmitted in the second window but will be broadened by roughly 160 ps after 1 km if transmitted in the third window. In order to take advantage of the lower attenuation in the third window,

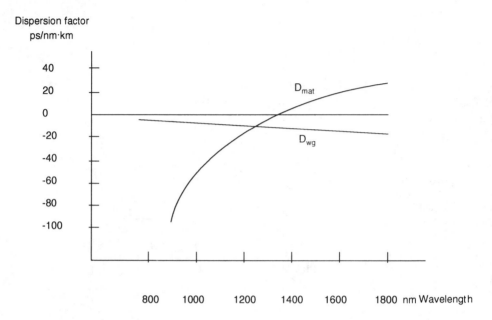

Figure 4.6 Chromatic dispersion versus wavelength.

special *dispersion-shifted fiber* (DSF) with ZDP at 1,550 nm has been developed by adding dopant material in the fiber. The characteristics of single-mode DSF cables are specified in CCITT Recommendation G.653. Dispersion-flattened fibers are developed that are optimized for operation in the second and third windows [4,8].

4.2.5.3 Polarization Mode Dispersion

Polarization is a property of light relating to the direction of its vibrations. Light rays traveling in a fiber can be depicted as vibrating entirely in the x-axis, entirely in the y-axis, or even vibrating in both axes. Each axis represents a polarization mode. The *polarization mode dispersion* (PMD) arises from the fact that the single-mode fibers are essentially bimodal from the polarization point of view. Due to fiber imperfections such as random core ellipticity and anisotropic static stress, delay time occurs between the two polarization modes, resulting in a pulse broadening and thus in a transmission performance degradation. The PMD is measured in $ps\sqrt{km}$, and a maximum level of $0.5\ ps\sqrt{km}$ is considered standard for single-mode fiber. The impairment of optical transmission due to PMD increases rapidly with a bit rate beyond 2.5 Gbps. Special low-PMD fiber is manufactured, mainly for submarine application, in a manufacturing process that spins the fiber as it is drawn and by changing the fiber's orientation in the cable with respect to external pertubations [9,10].

4.2.5.4 Self-Phase Modulation

A further transmission limitation is given by the power-dependent behavior of the fiber. This phenomenon, which is called the *Kerr effect,* creates an intensity-dependent phase shift leading to a spectral broadening that results in self-phase modulation and thus degradation of the pulse shape. This Kerr effect increases at higher bit rates due to the reduced bit interval and the increased spectral density at higher bit rates.

The Kerr effect stands for the phenomenon that when light is present in a medium an intensity-dependent change of the refractive index can occur. At normal pulse launch power levels this change is small, but over the typically long fiber lengths the effect is accumulative and significantly alters the phase of the light. A pulse of light can be considered a distribution of a range of intensities; and as the phase change is proportional to the intensity, each part of the pulse experiences a different phase shift. In the frequency domain this leads to a frequency shift or *chirp.* When such a chirp is present, the frequencies in the leading half of the pulse are lowered while those in the trailing half are raised, resulting in a broadening of the spectrum. Chirp builds up in direct proportion to the length of fiber traversed, with subsequent reductions of the transmission span as a function of the transmission speed [9,10].

4.2.6 Amplification of Optical Signals

For the amplification and handling of optical signals in terminal and repeater equipment there are two basic solutions available:

- *Opto-electronic* (O/E) components;
- Optical amplifiers subdivided into *semiconductor optical amplifiers* (SOA) and *erbium-doped fiber amplifiers* (EDFA).

4.2.6.1 Opto-Electronic Components

Opto-electronic components basically provide a conversion between optical and electrical energy virtually without energy loss. They consist of light sources and light detectors. The light sources are mainly LEDs and GaAs laser diodes—the less expensive LEDs used for short-distance (for example, FITL systems) and low-capacity systems, and higher powered LDs for long-distance high-capacity systems. LDs of the Fabry–Perot *buried heterostructure* (BH) and *distributed feedback* (DFB) types are widely used. Fabry–Perot BH LDs produce oscillation by means of mirrors at the ends of the amplifying medium, whereas in DFB LDs oscillation is caused by a diffraction grating situated in the amplifying medium. DFB LDs produce virtually monochromatic light, thus limiting chromatic dispersion. A laser diode is considered monochromatic if the ratio of the principal mode to the lateral modes is greater than 25 dB across the operating temperature range. High-speed LDs for transmission speeds up to 25 Gbps, or direct modulated with analog signals up to 860 MHz for TV distribution, and a spectrum of below 6 nm with a linewidth between 1.2 and 0.25 nm, tunable over the third window range, combined with a drive current of 10 mA, and 50-mW output power mark the state of the art (for example, *double channel-double channel planar buried heterostructure,* DC-DCPBH; or *semi-insulated buried heterostructure,* SIBH; and *Multi Quantum Well GRaded INdex Separate Confinement Heterostructure,* MQW GRINSCH, LDs).

The light detectors, which convert photons to electrons, are pin-diodes or *avalanche photo diodes* (APD). A pin-diode basically consists of an intrinsic-pure semiconductor—very lightly doped layer sandwiched between p and n layers. In an APD a photon of light entering a depletion region between p and n layers generates an avalanche of electrons that results in an internal amplification of the photo current and hence an increased sensitivity. Both the pin-diodes and the APDs are available for transmission speeds up to 20 Gbps.

Opto-electronic components are emerging toward integrated *optical and electrical circuits* (OEICs) composed of active and passive components, combining transistors (FET, DHBT, and HEMT transistors), photo detectors, LEDs, and LDs. An O/E amplifier, for example, basically consists of a signal processor and a signal output part. The signal processor processes the incoming signal, including filtering, bit synchronization, signal management, distribution, and, of course, the signal amplification. The signal output part takes care of the signal adaptation and or conversion toward the interfacing equipment and metallic or fiber line.

The integration efforts in opto-electronics have not yet reached the level of maturity of their electronic counterpart, in fact leading to a technological electronic bottleneck since fibers can transmit more information in terms of intrinsic bandwidth than present O/E circuitry can handle. The limitations of the O/E component circuitry cause the transmission speed of single wavelength fiber-optic systems to become increasingly difficult near 20 Gbps, hence WDM systems (as described in Section 2.4) [4,11].

4.2.6.2 Optical Amplifiers

An optical amplifier amplifies a light beam without converting the signal or message it conveys into electrical form. The nature, type of modulation, bandwidth and bit rate of the message, and even the number and positions of the electrical or optical carriers are therefore immaterial, at least superficially, to the behavior of the amplifier and its ability to amplify the information-carrying light beam. Of the various types of optical amplifiers introduced over the last three decades, there are only two that are increasingly finding application for transmission: SOAs and EDFAs. Both of them use the same physical principle; that is, an incident light beam is amplified by stimulated emission (an expression created by Einstein) in a medium that causes amplification by the injection of energy (also called energy pumping). Whereas SOAs use special pumped lasers and apply electrical pumping, in EDFAs the amplifying medium is a short section (typically a few tens of meters) of optical fiber doped with rare earth ions inserted into the optical line circuit with optical pumping applied.

For amplification at 1,550 nm the fiber is doped with erbium ions (Er3+, Nr. 68 in the periodic system of elements, a so-called rare earth element). For the energy injection a powerful light beam generated by a laser diode emitting at a wavelength below the wavelength of the optical signal, for example, at 980 or 1,480 nm, is pumped into the doped fiber section and excites the energy carriers, the erbium ions, to the higher energy level. When population inversion has been achieved (that is, when the energy level is more populated than the quiescent level), the carriers can return to the quiescent level, emitting a photon as they do, by one of two methods: stimulated emission due to the action of an incident photon or spontaneous emission. The stimulated emission is the basis for the optical amplification of an incident photon. Spontaneous emission can in turn be amplified by stimulated emission; this then is undesired amplified noise. Figure 4.7 depicts the principles of optical transition and optical amplification.

EDFAs exhibit almost-ideal characteristics with small-signal gain in excess of 40 dB, maximum saturated output power around +20 dBm, and noise figures close to the quantum limit (3 dB). Furthermore, they are highly compatible with the fiber environment, relatively temperature-independent, absolute linear, and weakly dependent on the state of the optical polarization of the incoming signal. The EDFAs slow-gain dynamics prevent saturation-induced crosstalk and intermodulation distortion effects. However, the optical amplification is inherently obtained at the expense of noise (due to the spontaneous emission) added to the signal and amplified by succeeding EDFAs.

To operate in the second window the optical amplifiers use fluoride fibers doped with the rare earth praseodymium (Pr, Nr. 59 in the periodic system of elements) and have similar characteristics as the EDFAs (logically they should be called PDFAs, but more conveniently the nomination *fiber amplifier* might become the common name for EDFA and PDFA).

SOAs generally have a lower gain and lower saturated output power than fiber amplifiers. On the positive side, however, are their lower noise, low

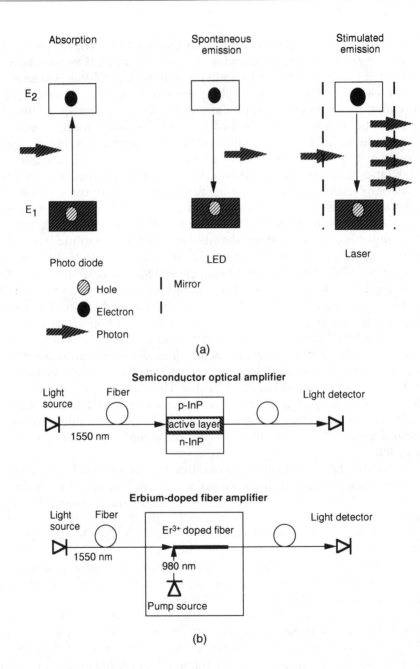

Figure 4.7 The principle of (a) optical transition and (b) optical amplification.

electrical consumption, and their availability for the second and third windows [9,12].

4.2.7 Soliton Transmission

Soliton is the name for a light pulse that either does not change its shape or changes its shape along the length of the fiber, thereby periodically returning to its original shape—shapewise, a soliton thus acts as a *perpetuum mobile.* This ideal pulse behavior ironically is the result of two detrimental nonlinear effects of optical transmission: the chromatic material dispersion (GVD) and the normally likewise unwanted *self-phase modulation* (SPM) of pulses, which are both explained in Subsection 4.2.5.

As mentioned in Subsection 4.2.5, with SPM the frequencies in the leading half of the pulse are lowered while those in the trailing half are raised; whereas due to GVD above the zero dispersion point, thus in the range where the GVD is negative, the actual group velocity (not the GVD itself) increases with frequency. Hence, the GVD acts contrary to the SPM and, in effect, can correct the SPM; in other words, the self-phase modulated pulse pulls itself back to the original shape after a certain distance called the *soliton distance.* This soliton distance can be calculated with a complex mathematical formula and can range from less than 1 km to over 100 km on standard single-mode fiber. Figure 4.8 in a highly simplified form depicts the principle of soliton transmission with pulse shape distortion through SPM and subsequent pulse shape correction through GVD.

To produce solitons, a power threshold (for creating SPM) needs to be surpassed. Keeping the pulse shape, however, does not eliminate the fiber attenuation. Thus, whereas with soliton transmission the pulsewidth remains constant, the pulse-amplitude decreases as a function of the fiber attenuation. Signal regeneration thus is not required, but regular optical amplification is of utmost importance to sustain soliton production—hence, the emerging combination of soliton transmission with carefully spaced EDFAs.

The aforementioned 1- to 100-km soliton distance can be further increased (at least tenfold) by lengthening the input pulse (thus lowering the transmission speed), hence WDM, and by reducing the fiber core diameter. The significance of increasing the soliton distance is that there is a corresponding reduction in the intensities needed for the soliton production. Thus, soliton transmission requires a compromise between transmission speed, fiber core diameter, and laser power.

The pumping of laser power by the fiber amplifiers causes the fiber medium to interact with the incident light, resulting in so-called *Raman scattering.* This means that from the pumping frequency (for example, pumping at 980 nm) a small amount of energy is scattered to another frequency, that of the

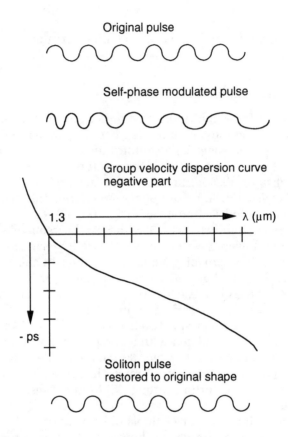

Figure 4.8 Soliton transmission.

solitons (at 1,550 nm), resulting in an amplifying effect and thus compensation of the fiber attenuation so that the soliton can continue on its way without loss over the soliton distance, after which again laser power needs to be injected. The maximum practical soliton distance, thus EDFA spacing, is around 50 km, which is a suitable distance for submarine amplifier spacing. Experimental soliton transmission has been achieved up to a length of 12,000 km using EDFAs every 25 to 40 km. On the other hand, longer unrepeatered optical fiber spans (over 300 km) have been obtained with soliton transmission using high-power lasers at the terminal stations.

Soliton transmission thus is genuine optical transmission with a transmission performance beyond the present technological limits of ultrafast electronics. Figure 4.9 clearly demonstrates the superior performance of soliton transmission [9,10,13].

Maximum distance without signal regeneration

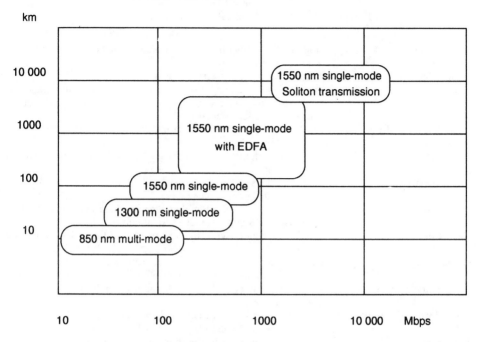

Figure 4.9 Evolution of optical fiber performance (*After:* Alcatel).

4.2.8 Principal Advantages of Optical Transmission

The transmission characteristics concerning attenuation, bandwidth, and cross-talk of *pure silica glass* (SiO_2) are far superior to those of copper lines, so fiber soon will be the main transmission medium. The raw material, sand, used for the fiber production is abundantly available, although extensive purification and processing is needed. A summary of the principal advantages of optical transmission is given in Table 4.1 [4,7].

Disadvantages of transmission on optical fibers include the difficulty of exact fiber jointing, the high price of connectors, and the impossibility of power transmission.

4.2.9 Actual Limits of Optical Fiber Transmission

In spite of the significant advantages outlined in Table 4.1, and although the intrinsic transmission capacity of optical fibers are considered to be virtually

Table 4.1
Principal Advantages of Optical Fiber for Transmission

High bandwidth	Huge available fiber bandwidth around 50,000 GHz in the two low attenuation windows
High reliability	Transmission quality with a BER in the order of 10^{-10}
Low loss	Fiber attenuation of typically 0.2 dB/km enables terminals and repeaters to be widely spaced; for example, 300-km unrepeatered link at 622-Mbps operation
Electromagnetically immune	No electromagnetic interference, hardly any crosstalk, and very difficult to tap
Fire resistant	The melting point of silica fiber is about 1,900°C (versus 1,100°C for copper)
Fire protective	Photonics does not generate sparks and, thus, is safe for operation even in explosive atmospheres
Small size	Typical two-pair cable outer diameters are: fiber 2 to 5 mm, twisted-copper 3 to 6 mm, coaxial 12 to 16 mm
Light weight	Fiber cable weight typically 10% to 30% of the weight of copper cable
Oxidation-free	Glass is chemically stable and thus can endure adverse environments (such as the ocean bottom or mines) better than metal cables
Electrically isolated	Fiber is completely safe for high-voltage, monitoring, and control applications
High physical flexibility	Optical fiber cable can easily be bent, allowing quick installation in already used conduits
Resources	Contrary to copper, silica is abundantly available

unlimited now that electronic circuitry and O/E components enable transmission speeds around 10 Gbps, both the components and the fiber itself are presenting practical limits. The noise amplification in optical amplifiers, well known from the analog technique, can become a problem especially on very long (submarine) links. The attenuation compensation is limited by the available output power of the optical transmitters and the sensitivity of the receivers. Although a 0.2-dB/km fiber attenuation is the state of the art, still another 0.1 dB/km has to be added in practical systems for splicing, connector losses, and system margins. The approximate limits for state-of-the-art high-speed transmission systems on single-mode fiber are indicated in Table 4.2 [9].

System assumptions are as follows:

- Dispersion: SMF 1300 and DSF 1550 $D = 3.5$ ps/nm $\cdot$ km, SMF 1500 $D = 17$ ps/nm $\cdot$ km;
- Transmitter power: −3 dB;
- Receiver sensitivity: −20 dB at 10 Gbps, −17 dB at 20 Gbps, 14 dB at 40 Gbps;

Table 4.2
Approximate Optical Fiber Link Line Limitations

Bit Rate (Gbps)	Dispersion Limits (km)			Attenuation Limits (km)	
	1,550 nm		1,300 nm		
	SMF	DSF	SMF	1,550 nm	1,300 nm
10	60	280	400	50	30
20	15	70	100	40	25
40	4	20	25	30	20

SMF: single-mode fiber; DFS: dispersion-shifted SMF.

- Fiber attenuation installed: 0.45 dB/km at 1,300 nm, 0.3 dB/km at 1,550 nm;
- System margin: 3 dB.

The approximate theoretical limits of 1,550-nm SMF are indicated in Figure 4.10.

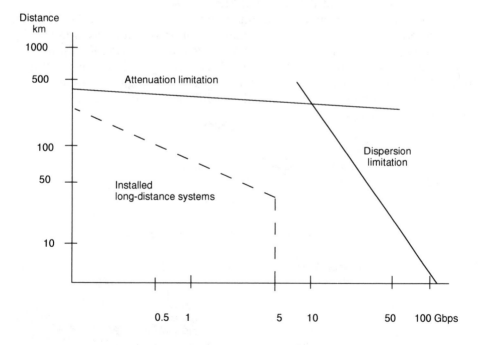

Figure 4.10 Theoretical limits of SM fiber at 1,550 nm.

4.3 TYPES OF OPTICAL FIBER

Optical fiber in this book concerns silica glass fiber only, as this is the major fiber for telecommunications. Other fibers are plastic and fluoride glass fibers.

Plastic fibers, although introduced in the early 1980s, have only found application in short data links, illumination, automobiles, and production facilities. The advantages of plastic fiber compared with glass fiber are higher mechanical strength and flexibility, easier connection (due to wide cores of 100 to 1000 μm), and lower cost. The major disadvantage is the high attenuation combined with a low operating wavelength, as shown in Table 4.3 [4].

Fluoride glass fiber is commercially available too. The attenuation is still very high in the order of 25 dB/km, but 1 dB/km has already been achieved with experimental fluoride fibers. Different from silica glass fibers, fluoride glass fibers do not adsorb the infrared wavelength; so operation is possible in the so-called mid-infrared band between 2,000 nm and 5,000 nm. Fluoride glass fibers are made of *zirconium fluoride* (ZrF_4) and *barium fluoride* (BaF_2). Thus far, the fluoride glass fibers are mechanically and environmentally more vulnerable than silica glass fibers. They appear to be very promising, however, in respect to an ultralow attenuation around 0.05 dB/km and subsequent unrepeatered links in the range of thousands of kilometers.

The available telecommunications silica glass fibers can be subdivided into two groups depending on the signal transmission mode as, specifically, *multimode fibers* (MMF) and *single-mode fibers* (SMF), whereby a mode (as explained in Subsection 4.2.2) is defined as a group of light rays bouncing through the fiber at a common incidence-reflectance angle. In respect to the physical construction of the fibers, three different types of fiber have been successfully developed:

- Step-index fiber;
- Graded-index fiber;
- Single-mode fiber.

Table 4.3
Major Characteristics of Plastic Fiber

Wavelength (nm)	Attenuation (dB/km)
400	400–600
500	250–523
600	200–300
700	550–600
800	1,500–1,600

The step-index and the grade-index fibers were first developed and allowed multimode transmission only. The single-mode fiber, in fact, is also a step-index fiber, however, with an extremely small core diameter that eliminates multimode transmission. Figure 4.11 shows the major characteristics of the three types, and short descriptions of each type are given in the following three subsections [7,8].

4.3.1 Step-Index Fiber

The first available fibers were step-index fibers, which have a uniform refractive index n in the core of the fiber, which is higher than that of the fiber cladding.

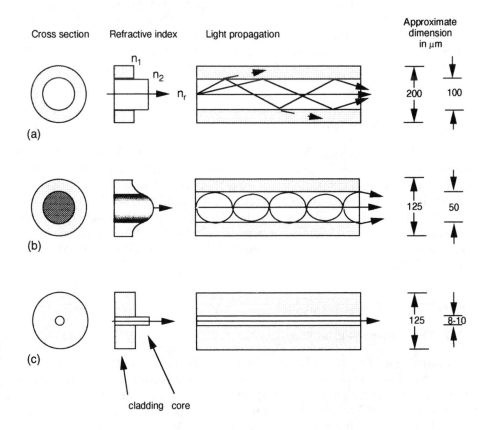

Figure 4.11 The three types of optical fiber: (a) step-index multimode fiber, (b) graded-index multimode fiber, and (c) single-mode fiber.

The refractive index transition at the boundary between the core and the cladding is therefore shaped like a step. Since the wavelength of the light is short relative to the core diameter, modal dispersion limits the transmission bandwidth of an optical fiber cable with step-index fibers to about 100 MHz · km.

4.3.2 Graded-Index Fiber

In graded-index fibers the profile of the core refractive index is parabolic, with the maximum value at the core center, and reducing to the same value as that of the cladding at the cladding/core boundary. The relative refractive index difference between the core center and the cladding is about 1%. As a result of this parabolic index profile, the light rays now follow a sinusoidal or spatial helical path along the fiber axis, thereby reducing the spread of the transmit times for different modes. With an optimum refractive index profile, modal dispersion can be minimized and a transmission bandwidth of 1 GHz · km achieved. The graded index fiber has been standardized in 1980 in CCITT Recommendation G.651, with a 50-μm core diameter and 125-μm cladding diameter.

4.3.3 Single-Mode Fiber

Single-mode fiber is state of the art. The core diameter of the fiber is sufficiently reduced so that the light is guided in an almost straight line and thus, only a single mode can be propagated. In this case there is no longer any path-dependent transit time difference (thus no modal dispersion) and the fiber has a very wide transmission bandwidth around 10 GHz · km. Since the fiber attenuation is lowest, single-mode fibers are now generally used for long-distance optical fiber transmission. CCITT has standardized the characteristics of single-mode fibers in G.652 in 1984 [5,7].

4.4 OPTICAL FIBER CABLES

4.4.1 The Optical Fiber Production

The production of optical fibers, in order to ensure the highest purity of the fiber, takes place in special clean rooms with a controlled atmosphere and in a virtually uninterrupted production process to ensure a homogeneous noncontaminated product. Whereas an uncontrolled atmosphere easily contains some 2.5 million dust particles (at the size of 0.5 to 5 μm) per cubic meter, the clean room should have less than 35,000—corresponding to the cleanness of an surgery room, and special confined areas should have even less than 500 dust particles per cubic meter. The fiber production starts from a highly pure silica glass tube in which special chemicals are deposited in several different methods to obtain the different refractive indices of the core and the cladding. The tube is modified to a preform, from which the fiber will be drawn. The different methods are

- CVD: Chemical vapor deposition;
- PCVD: Plasma chemical vapor deposition;
- MCVD: Modified chemical vapor deposition;
- VAD: Vapor axial deposition;
- OVD: Outside vapor phase deposition;
- IMCVD: Intrinsic microwave heated chemical vapor deposition.

The MVCD method is most frequently used in the production of monomode fibers and, therefore, will be briefly described.

In the MCVD fiber production method for monomode fibers (see Figure 4.12) the silica glass tube is rotated over a slowly moving burner while the chemical gasses flow into the tube. For a controlled oxidation of the chemical gasses helium and very pure oxygen are added to the chemical gas mixture. Both burner temperature and movement and gas distribution are PC-controlled. At an internal temperature of about 2,150°C the chemical gasses oxide and deposit on the inside of the tube. After deposition of the chemicals the burner temperature is increased to about 2,400°C and then the burner movement significantly slowed so that the glass tube collapses into a solid glass rod whereby the deposited material constitutes the material for the fiber core. During the collapsing period pure oxygen and pure chloride are pumped through the tube to minimize the number of OH ions and thus to minimize the optical attenuation. Preforms used to have a fiber-length capacity of 80 to 100 km of fiber and are emerging toward capacities of 200 to 300 km of fiber.

The preform is then placed in a vertical fiber drawing tower, the height of which determines the fiber pulling speed—typically 5 m/s for a 10-m tower and 15 m/s for a 20-m tower. At a temperature of 2,200°C the preform melts and is carefully drawn into a fiber with a diameter of, say, 125 μm at an accuracy of ±0.5 μm. The fiber is first cooled down in a helium cooler and then conducted through two resin baths each followed by an oven for baking the protective coating. With the protective coating the fiber has a diameter of 245 to 250 μm. After further testing for diameter control the fiber passes through a "dancing trio," which has a buffering function to prevent mechanical vibrations and is stored on a pulling spool at a regular length of, for example, 6.4, 12.6, or 25 km. The speed of the pulling spool is controlled by the fiber diameter control tester to obtain the required ±0.5-μm accuracy [4,14,15].

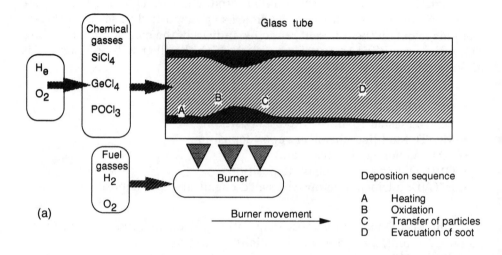

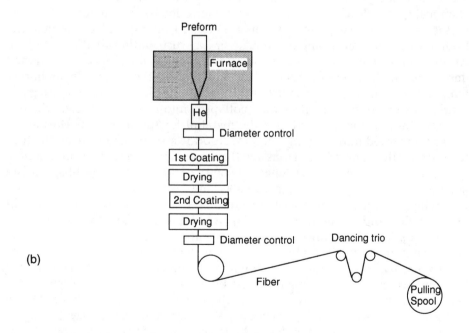

Figure 4.12 MCVD (a) preform and (b) fiber production.

4.4.2 The Optical Fiber Cable Characteristics

Optical fiber cable basically differs from copper cable in that there is no need for electromagnetic protection and a higher need for mechanical protection. Fully metalfree optical fiber cable can be used in explosive areas, on HT grids, for highly sensitive medical and computer environments, and does not attract lightning-strokes.

Contrary to copper wire cable, twisting of optical-fiber pairs (to reduce crosstalk) is not required, but the fiber pairs require sufficient motional freedom in order to prevent tensile stress. Therefore, optical fibers are packed into a cable with a cable core in one of the following versions (see Figure 4.13):

- Fibers more or less loose buffered inside protective tubes;
- Fibers in a slotted core;
- Fibers encapsulated in flat ribbons.

4.4.2.1 Fibers Loose Inside Protective Tubes

With this construction 1, 2, 4, 10, or even up to 144 fibers are loosely located inside a plastic tube that is usually jelly-filled to keep out any moisture. A number of such tubes are combined into a cable.

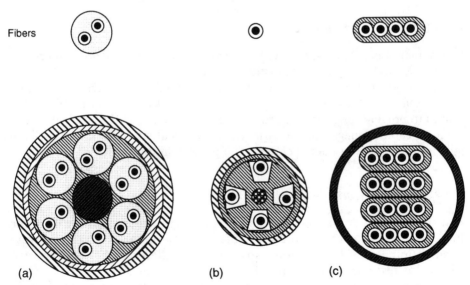

Figure 4.13 Optical fiber cable versions: (a) tube cable, (b) slotted cable, and (c) ribbon cable.

Special versions of steel "minitube" cable have been constructed for submarine application. Up to 48 fibers are placed inside a steel minitube with a typical external diameter of 2.3 mm and a wall thickness of 0.2 mm. The steel tube, which provides a high level of protection against compression and crushing, is for additional fiber protection against moisture and longitudinal abrasion and is filled with a thixotropic jelly. To protect the fibers from longitudinal constraints, mainly during installation, they have a controlled excess length of 2 to 3 ‰.

The external structure of the minitube cable depends on the application. Cables for amplified/regenerated routes requiring power feeding usually obtain one or two layers of steel wires as vault around the minitube. A copper ribbon is then placed and lengthwise welded tightly around the steel wires so that copper ribbon and steel wires together form a compact electrical conductor for the power feeding and resist 1,000 isostatic pressure bars and 60 kN of traction at breaking point. To insulate the whole of this metallic structure from the sea water a typically 5-mm layer of polyethylene is extruded around the copper ribbon. This insulation has to stand a dc voltage level of 15 kV during its 25-year service life.

For repeaterless links a similar vault is applied, however without copper ribbon, and a 3-mm-thick jacket protects the cable.

4.4.2.2 Fibers in a Slotted Core

One or more fibers are inserted with a degree of slack in helical grooves arranged around the axis of a thermoplastic core. The fibers can move within these grooves to accommodate cable stretching and to obtain zero-stress on the fibers. For submarine cables the helical grooves are filled with a water-repellent compound to prevent longitudinal propagation of sea water within the cable in the event of cable damage. To increase the mechanical strength, for example, for submarine cable, the plastic core can be pressed around a steel wire rope.

4.4.2.3 Fibers Encapsulated in Flat Ribbons

Flat ribbon cable is the preferred solution for high-density cable with several hundreds or even several thousands of fibers, for example, for FITL networks. A small number of fibers are encapsulated in a plastic ribbon. Several such ribbons placed one above another form the core of the cable.

For tension relief, threads made of tension-resistant plastic and/or silica are stranded around the cable core, which is then covered by a cable jacket made of polyethylene, polyvinyl chloride, or a composite jacket. If rodent protection is more important than lightning protection, a strip metal wrap can be placed over the jacket. This metal wrap is then covered by another plastic sheath [14,16].

4.5 OPTICAL FIBER TRANSMISSION SYSTEMS

Optical fiber transmission systems, also called optical fiber line systems, are used to transmit analog and digital signals through specific optical fiber cables. Such a system consists of line-terminating equipment connecting the electrical or optical signals with the fibers; repeaters compensating the line losses and regenerating the digital signals; optical signal splitters distributing optical signals in various directions; power feeding for underground or submarine repeaters; and fault-locating, supervision, and network management equipment. Applicationwise, optical fiber systems are divided into the following four groups:

- Terrestrial transport systems;
- Submarine systems;
- FITL systems;
- CATV systems.

4.5.1 Terrestrial Optical Fiber Transport Systems

Terrestrial optical fiber transport systems are available for the various digital PDH and SDH multiplex hierarchy bit rates. The low attenuation of optical fiber allows repeater section's exceeding 100 km in length depending on transmission capacity and type of cable, compared with coaxial systems that require repeaters every 1.5 to 10 km. Moreover, optical fiber repeaters do not need to equalize a temperature-dependent attenuation, as is the case with coaxial cable repeaters. Figure 4.14 shows the relation between repeater spacing and transmission speed on optical fiber systems operating on standard single-mode optical fibers.

With the long (and flexible) repeater spacing for terrestrial optical fiber systems, the repeaters normally are located in telecommunication buildings instead of underground at fixed short intervals along the route. It is no longer necessary, therefore, to power feed the repeaters through the cable; so optical fiber cables can be completely metalfree (thus no lightning danger). Figure 4.15 shows a typical arrangement for an optical fiber transport system with a capacity of 565 Mbps operating on single-mode fiber at a wavelength of 1,550 nm. Regenerative repeaters can be flexibly located at 50- to 100-km intervals, enabling a total digital line path up to a length of 10,000 km. The multiplex equipment for multiplexing four 140-Mbps bit streams into a 565-Mbps line signal is integrated in the line terminal equipment. Similar arrangements exist

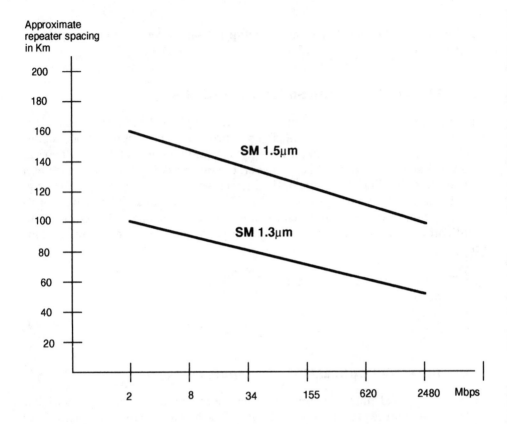

Figure 4.14 Repeater spacing versus transmission speed for optical fiber systems.

for 4- by 155-Mbps SDH systems with a 622-Mbps line signal or even 16- by 155-Mbps SDH systems for a 2,488-Mbps (2.5 Gbps) line signal. With the line equipment for 622-Mbps and 2.5-Gbps SDH, the line signal usually can be assembled from various tributaries. The 2.5 Gbps, for example, can be assembled from 16 STM-1 bit streams (155 Mbps each), or 16 PDH 140-Mbps bit streams, or any coherent mix of STM-1 and PDH tributaries.

The aforementioned systems with up to 100 cascaded regenerators, of course, are for long-haul operation. Less sophisticated systems are existing for short haul operation. Table 4.4 summarizes typical systems for long- and short-haul operation for the PDH and SDH CEPT multiplex hierarchies [17,18].

4.5.2 Submarine Optical Fiber Systems

The development of the submarine optical fiber transmission systems is well ahead of the development of terrestrial optical fiber systems to meet the require-

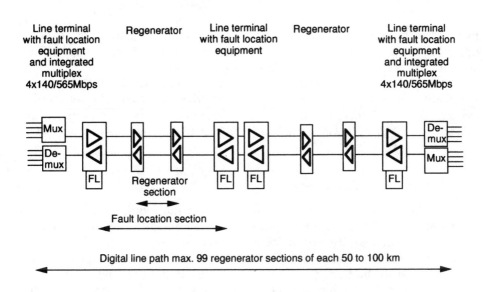

Figure 4.15 Optical fiber transport system operating at 565 Mbps (*After:* Siemens).

Table 4.4
Typical Terrestrial Optical Fiber Transport Systems (for PDH and SDH CEPT-Hierarchy)

Description	Bit Rate (Mbps)	Approximate Regenerator Section Length (km) SM 1,300 nm	MM 1,300 nm	SM 1,550 nm
Low capacity				
Short haul	2	60–70		
Long haul	2	80–100	65–75	115–130
Short haul	8	50–65		
Long haul	8	80–90	60–70	105–120
Medium capacity				
Short haul	34	20–30	10–15	
Long haul	34	70–80	50–65	90–105
High capacity				
Short haul	140/155	20–30	10–25	
Long haul	140/155	40–70		50–130
Short haul	565	10–20		
Long haul	565	35–60		50–115
Short haul	622	10–30		10–30
Long haul	622			20–45
Short haul	2,488	10–15		
Long haul	2,488			10–50

ments for transmission systems with the highest transmission capacity operating almost uninterrupted over 25 years of service, with repeaters/regenerators on the bottom of the sea at depths up to 9,500m.

Submarine optical fiber transmission systems are available for the classical application of bridging oceans and as repeaterless systems. The repeaterless systems are used between neighboring continents, between continents and offshore islands (or platforms), between islands, and along coastal lines instead of landline systems.

4.5.2.1 Submarine Equipment

Optical fiber submarine systems consist of terminal equipment, submerged regenerators, submerged branching units, and the optical fiber undersea cable.

The terminal equipment provides the transmission interface between the land transmission network and the optical bit stream on the undersea cable and is similar to the terrestrial optical fiber terminal equipment apart from a higher degree of protection and redundancy to meet the 25-year service life.

Associated with the terminal equipment is the remote power-feeding equipment that supplies a dc current to the submerged cascaded regenerators. At both terminal stations the power-feeding equipment is connected to a copper conductor of the cable—usually a longitudinally welded copper tube shrunken on to the steel wire vault of the cable. This conductor tube is isolated from the sea water by a polyethylene jacket. The two regenerators in the middle of the route are connected with the sea, which thus acts as the return path for the dc current. In the case of a breakdown in the power feeding at one terminal, the power feeding at the other terminal can usually power the whole link.

The submerged regenerator units include a two-way regenerator per pair of fibers. Figure 4.16 shows a basic diagram of such a two-way regenerator. The digital light signal arriving from terminal A, respectively B, is detected by a photo diode, amplified and reshaped, resynchronized, and retransmitted by a laser to direction B, respectively A. Power supply and supervision and looping circuits communicating with both terminals are common for both regenerators. A high degree of redundancy is usually incorporated in the regenerators to meet the 25-year service life.

Submerged branching units separate the fibers and route transmitted information to several destinations. Various options are possible for their internal structure, ranging from simple splicing of fibers to more complex systems including regenerators, remote-controlled optical interfiber switching, or electrical switching between power-feeding conductors. Figure 4.17 demonstrates a theoretical application with two branching units located in coastal areas on a transatlantic cable with landing points in the United Kingdom and France on the European side and in Canada and the United States on the North-

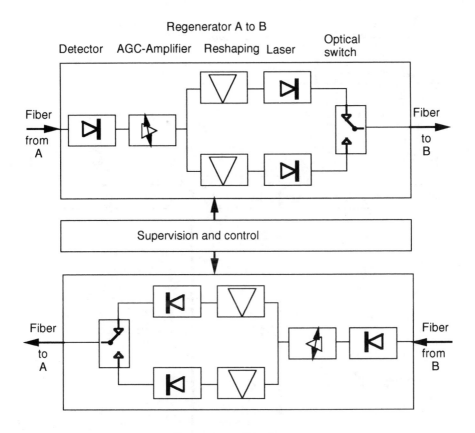

Figure 4.16 Two-way submarine regenerator.

American side. The two branching units form a complete telecommunications network in which traffic can be switched between the four terminals depending on traffic load distribution and the effect of breakdowns on individual fiber pairs can be minimized. A similar solution, however, with only three landing points (in France, the United Kingdom, and the United States), is applied on the first optical fiber transatlantic cable (TAT-8) [16,19–21].

4.5.2.2 Submarine Systems

The first-generation submarine optical fiber line systems carry 280-Mbps and 420-Mbps signals on SM fiber operating at 1,300-nm wavelength using regenera-

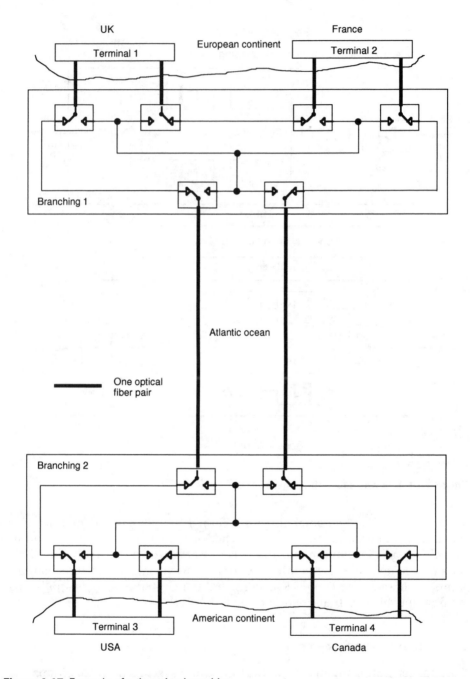

Figure 4.17 Example of submarine branching.

tors at average intervals of 50 km. Second-generation systems carry 560 Mbps and 622 Mbps operating at 1,550 nm with regenerators at typical 100-km intervals. The latest submarine system with a transmission speed of 5 Gbps uses EDFAs at 45-km intervals. The system applies soliton transmission operating on dispersion-shifted SM fiber at 1,550 nm, thus eliminating the need for signal regeneration.

The world's first commercial long-distance submarine deep-water cable went into service in 1987 between the French mainland and Corsica. The 390-km link carries two systems each with a capacity of 280 Mbps using 65-km repeater spacing and operating at 1,300 mm on single-mode fibers.

One year later the first transatlantic optical fiber cable—TAT-8—operated on 2 × 280 Mbps with 1,300 nm on single-mode fibers, too. TAT-8 applied for the first time an underwater branching unit positioned on the continental shelf providing dual shore end connections into both the United Kingdom and France. The transmission capacity of TAT-8 with 7,680 telephone channels was fully exhausted within one and a half years. Therefore, TAT-9 was installed in 1991 with double the capacity (2 × 560 Mbps) and booked out in a few months!

Currently, TAT-12/13 with two pairs each operating at 5 Gbps are being installed in a ring configuration consisting of TAT-12 connecting the United Kingdom with the United States and TAT-13 connecting France with the United States whereby the two landing points of TAT-12 and TAT-13 in the United States will be interconnected by a third submarine cable. A fourth submarine cable interconnects the two landing points in France and the United Kingdom. Each cable has two pairs of dispersion-shifted monomode fibers. EDFAs are located at 45-km intervals on the transocean cables and at 74-km intervals on the cable between the United Kingdom and France. This ring system with automatic rerouting in case of interruptions thus can carry over 300,000 telephone circuits between the two continents without any submerged signal regeneration.

Such a 5-Gbps system, too, is being installed (in 1996) on a 2,800-km route called "JASURAUS" between Jakarta (Indonesia) and Port Hedland (Australia).

The world's longest optical fiber submarine cable system ever laid, SEA-ME-WE 2, was inaugurated on October 18, 1994 (see Figure 4.18). Stretching 18,190 km from Singapore to Marseilles, SEA-ME-WE 2 links 13 countries on three continents: Singapore, Indonesia, Sri Lanka, India, and Saudi Arabia on the Asian continent; Djibouti, Egypt, Tunisia, and Algeria on the African continent; and Turkey, Cyprus, Italy, and France on the European continent. The cable has two single-mode fiber pairs. Each pair operates a system with a wavelength of 1.5 μm and a capacity of 2 × 560 Mbps; thus, the equivalent of a total of 15,600 telephone channels on both pairs. The system includes 160 submerged repeaters and branching units (SEA-ME-WE 1 operates 820 analog

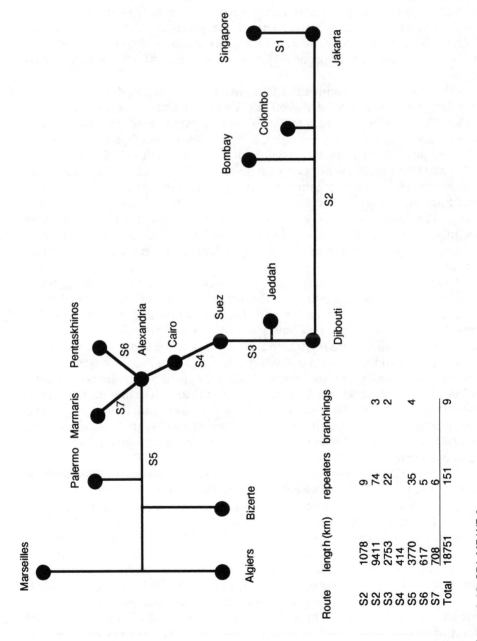

Route	length (km)	repeaters	branchings
S2	1078	9	3
S2	9411	74	2
S3	2753	22	
S4	414		4
S5	3770	35	
S6	617	5	
S7	708	6	
Total	18751	151	9

Figure 4.18 SEA-ME-WE 2.

repeaters!) and serves 52 telecommunication operators. Shortly after the inauguration of the SEA-ME-WE 2, a MoU was signed in Singapore by 16 international
telecommunications administrations regarding plans for SEA-ME-WE 3, which
will carry commercial traffic on two fiber pairs each operating at 4 × 2.5 Gbps
by December 1998.

By the time this book is published SEA-ME-WE 2 most likely will not be
the longest optical fiber submarine cable anymore: FLAG will be the champion.
FLAG, which stands for Fiber-optic Link Around the Globe, will then cover
27,300 km and operating at 5 Gbps on two fiber pairs carry the equivalent of
120,000 telephone circuits. FLAG will interconnect Europe with Japan through
the Mediterranean and Indian Ocean with 14 in-between landing points. Different from current practice, FLAG is not planned and financed by telecommunication operators but by a private enterprise consortium.

Moreover, FLAG will for only a short time be number one if "Africa ONE"
is realized. Africa ONE, as proposed by AT&T, would be a 32,000-km submarine
ring around the African continent. The ring would have some 30 landing points
around the coast of Africa and connect to the noncoastal African countries by
satellite, radio relay, and optical fiber cable. Connections with the PANAFTEL
terrestrial network, the RASCOM satellite network, SEA-ME-WE 2, and FLAG
are also planned.

The ocean's deepest submarine cable was laid in 1994 between Australia
and Guam. This so-called "PacRim West" optical fiber cable with a length of
7,080 km, a capacity of 560 Mbps, and 53 optical repeaters lay at depths of
8,000m to 9,000m. This cable was the final portion of a 16,500-km system
installed in the South Pacific and connecting the region to Asia, North America,
and Europe [7,22].

4.5.2.3 Repeaterless Systems

Repeaterless systems can cover distances in the range of a few hundred kilometers between two terminals. Those systems are available for transmission speeds
of 1.5- to 565-Mbps PDH signals and 155-Mbps, 622-Mbps, and 2.5-Gbps SDH
signals. The average distance between two terminals is typically 150 to
200 km. To eliminate the need for submerged regenerators on even longer
distances, optional optical amplifiers are added on or near to the output side
of the terminal and eventually on or near to the input side of the terminal
stations too, resulting in repeaterless spans of 200 to 300 km.

The world's longest unrepeatered route, put into operation in late-1995,
for example, covers 309 km between the Spanish mainland and the Island of
Majorca, operating at 622 Mbps on one pair of SM fiber (supplied by Alcatel
and operated by the Spanish telecommunications operator Teléfonica). An
error-correcting code is used in this system to lower the signal detection threshold at the receiving end and to achieve errorfree operation on standard SM
fiber [20,21].

4.5.3 FITL

"Fiber in the loop" abbreviated FITL is the terminology commonly used for the application of optical fiber transmission for interactive and distributive services in the access network.

Telecommunication access networks already present an investment of some 40% of the total telecommunication network, and this figure is expected to increase beyond 50% within the next 10 years. The access network thus presents a heavy investment for which a service lifetime of at least 30 to 50 years is to be expected. Therefore, an access network installed today needs to meet the requirements of tomorrow. The needs of today concern the transmission of N-ISDN and CATV. The needs of tomorrow concern the transmission of B-ISDN integrated with High-Definition CATV. Thus to meet the requirements of tomorrow the present copper access network will need to be replaced— in the course of the next 20 to 50 years—by a broadband access network based mainly upon optical fiber and partially upon radio (where fiber cable deployment is too expensive or comes late).

This transition from copper to fiber within the access networks will be evolutionary rather than revolutionary and will go in heavy competition with the copper cable enhancement equipment (xDSL as described in Chapter 2), but undoubtedly optical fiber will dominate in the end. The deployment of fiber in the access network is going on in a multitude of different approaches known under a variety of acronyms in the order of increasing local loop coverage:

- FTTZ: fiber to the zone;
- FTTR: fiber to the remote unit;
- FTTC: fiber to the curb;
- FTTK: fiber to the kerb;
- FTTP: fiber to the pedestal;
- FTTB: fiber to the building, basement, or business;
- FTTF: fiber to the floor;
- FTTO: fiber to the office;
- FTTD: fiber to the desk;
- FTTA: fiber to the apartment;
- FTTH: fiber to the home;
- FTTT: fiber to the terminal.

Among those acronyms FTTT might become the ultimate common term replacing the distinction between FTTH for residential subscribers and FTTO, FTTB, respectively, FTTD for business subscribers. FTTB is confusingly used for FITL service to small offices or to residential buildings as well as generally to a business area. FTTZ is an expanded concept of FTTC whereby a zone may serve a few hundred or even a few thousand subscribers who used to be served by remote switching units. FTTK and FTTP are the same as FTTC and refer to fiber up to the cable distribution box nearest to the subscriber.

Generally the acronym indicates the location of the *optical network termination* (ONT) that terminates the fiber coming from the local exchange and or CATV head-end. The continuation from the ONT to the final destination at the subscriber's premises, usually called "the-last-mile" although in some case covering a few tens of meters only, for cost reasons, is either by existing twisted pair or coaxial cable or, if such cable is not existing or of inferior quality, the quickest and cheapest solution might be by radio.

Some of the major FITL solutions are shown in Figure 4.19. An FITL system usually starts with so-called *optical subscriber access* (OSA) equipment located at or nearest to a local PSTN and/or ISDN exchange and/or a CATV head-end. The key functions of the OSA are

- Electrical interface with the public communication and CATV distribution networks;
- Optical interface with the optical fiber cable network;
- Software-controlled configuration of the FITL network—including cross-connecting and grooming, for example, of leased lines and nonlocally switched services;
- Operation, maintenance, and management of the FITL network.

A *passive optical network* (PON), as the name implies, contains only passive optical components, such as optical splitters, that connect an incoming fiber to a number (typically 16, 32, or 128) of outgoing fibers.

The ONT terminates the optical fiber, provides the optical/electrical conversion of the signal, and interfaces with the subscriber unit(s) in the FTTH, FTTT, and FTTO applications or with the last mile cable(s) or radio in all other applications.

The *radio in the loop* (RITL) solution shown in Figure 4.19 as one of the access solutions with the *radio distribution point* (RDP) and *radio from the curb* (RFTC) is described in Subsection 5.3.5. The FITL cable network can be in a multistar configuration as shown in Figure 4.19 or in a ring configuration as shown in Figure 4.20. The ring configuration might be more expensive than the star configuration but provides higher flexibility in adding future subscribers and enables alternative routing for increased reliability. In Figure 4.20 an STM-1 signal operates on the ring for N-ISDN operation; for future B-ISDN operation the capacity can be increased to STM-4.

For the transmission of the optical signal either two fibers are used, one for each direction, or only one fiber for both directions thereby using *time*

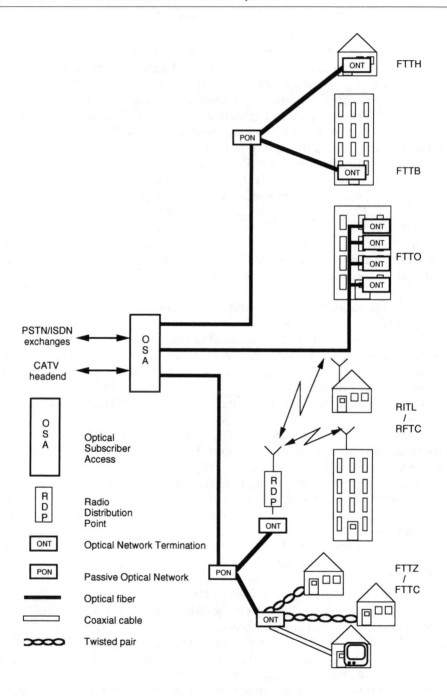

Figure 4.19 Various FITL solutions.

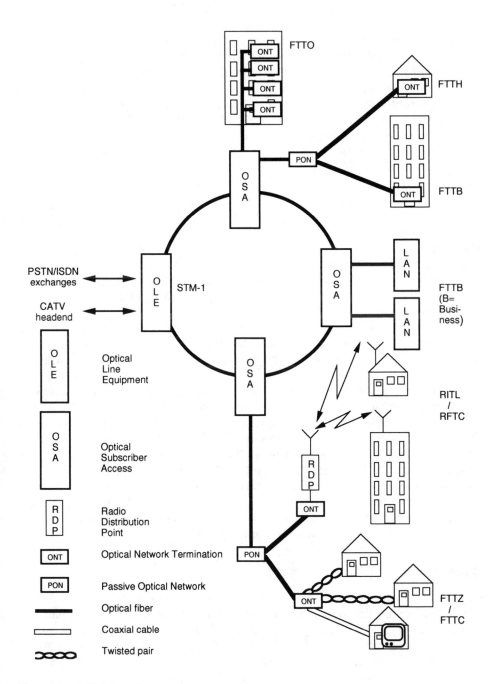

Figure 4.20 FITL in a ring topology.

compression multiplex (TCM), also known as "Ping-Pong"—in which case the downstream signals (toward the subscribers) are transmitted using a conventional TDM and in the upstream direction a TDMA scheme, similar to that used in satellite transmission, is applied. In the case of narrowband operation with public communication and CATV, WDM is applied with the public communication operating on 1,300 nm and CATV on 1,550 nm. For the migration to broadband service WDM can be applied too.

The first major FITL deployment was made in 1990 in Germany where, as a result of the German reunification, a unique opportunity was given to replace the over 60-year-old telecommunication network in the previous German Democratic Republic of some 2,000 villages without any telephone at all by a modern network based upon optical fiber. Over 5 million potential subscribers—many of them who had been waiting more then 10 years for a telephone connection—were connected to a modern broadband-prepared network, with FTTC/FTTB for 1.2 million subscribers within less than four years. Over 50,000 subscribers for whom cable laying was too expensive or no rights of way could be obtained were connected with their local exchange by RITL equipment.

In Japan, also in 1990, *Nippon Telegraph and Telephone Corporation* (NTT) announced an ambitious plan according to which all Japanese households would have a broadband access in a national wholly optical network by the year 2015.

FITL deployment (as demonstrated in Figure 4.21) typically starts with the most profitable service to the business area with FTTO, followed within a few years by servicing dense residential areas with FTTC. A few years later FTTH will be deployed beginning mainly in the newly developed residential areas. Gradually the deployment of FTTC will decrease and in many areas a migration from FTTC to both FTTH and FTTO will take place. In metropolitan areas with loop lengths around 1 to 3 km a 50% penetration might be achievable within a deployment period of 10 years. For urban areas with loop length of 3 to 5 km, and even more in rural areas with loop length of 5 to 20 km, naturally longer deployment periods will be required [23].

4.5.4 CATV

Optical fiber cable applied instead of coaxial cable for CATV provides the big advantage that no line repeaters are required and much more TV channels can be distributed. Moreover, the application of EDFAs as power boosters enable the passive distribution of CATV in a network with a radius of typically 30 km around the head-end to millions of subscribers, instead of hundreds so far. Laboratory experiments showed the feasibility of distributing digital HDTV channels to over 8 million subscribers in a 10-Gbps passive optical network using cascaded EDFAs. Optical fiber CATV will speed up the ongoing evolution

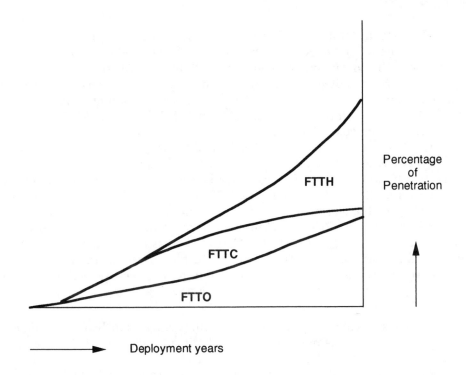

Figure 4.21 FITL deployment.

from basic broadcast CATV to interactive services ranging from access-controlled channels (premium, pay-per-view), *near-video-on-demand* (NVOD), true VOD, and eventually to *video dial tone* (VDT).

The migration from coaxial-based to fiber-based CATV networks currently happens in hybrid fiber-coax configurations, whereby the fiber is mainly used for the signal transport and the coax for the ultimate distribution to the subscribers premises. The fiber will move closer to the subscriber as a function of a price decrease of the electro-optic components and the attractiveness of new services.

The migration of CATV networks from coax to fiber follows—similarly as described for the evolution of FITL—different approaches depending on existing infrastructure, offered services and degree of integration with voice and data telecommunication. Beyond the solutions known under the listed acronyms generally for FITL, the following apply specifically for fiber deployment in CATV systems:

- FTTF: fiber to the feeder;
- FTTSA: fiber to the service area;
- FTTLA: fiber to the last amplifier.

FTTF stands for the basic hybrid fiber-coax architecture in which the fiber is used as the trunk between the head-end and the network hubs. The hubs then "feed" the coaxial branch-and-tree distribution networks.

In FTTSA architecture, the fiber is deployed up to ONTs that feed a servicing area, typically of 500 to 2,000 passings (a passing being a CATV subscriber access point). A small cascade of three to four coaxial cable amplifiers is required in the coax plant behind the ONT.

In FTTLA architecture, the fiber is pushed deeper into the network and ONTs serve as the last amplifier, which is then followed by a purely passive coax distribution network serving typically 100 to 300 subscribers. With the fiber nearer to the subscriber than in a FTTF and a FTTSA, a FTTLA considerably facilitates the subsequent addition of voice and data services. A CATV network applying the FTTLA architecture is shown in Figure 4.22. A fiber pair is used from the head-end up to the ONTs, whereas the ultimate distribution to the subscribers is made through coaxial cable. EDFAs in the hub provide the power to extend the length of the trunk and to feed several subscriber branchings. Typical characteristics of such a network are

- Transmission of up to 50 analog [AM-VSB (vestigial side band) modulated] and digital TV channels, up to 30 FM and 16 digital broadcast channels within a frequency band of 45 to 862 MHz;
- Reverse channel for upstream data and video in a 5- to 30-MHz band;
- Optical splitting in hub 8X and in PON 16X;
- Distance between hubs 15 km;
- Length of coaxial chain 250m;
- Optical wavelength: downstream 1,550 nm, upstream 1,300 nm.

Voice, data, and video signals can be transported on the same fiber pair; a full integration of CATV into B-ISDN, therefore, would be the logical solution. Such systems are being introduced, too, with typical characteristics as follows:

- Optical wavelength, both directions: 1,300 nm;
- N-ISDN: 2 Mbps;
- TV upstream: 5 to 55 MHz;
- TV downstream: 80 to 862 MHz.

Traditionally, however, in most countries CATV is not operated by the telecommunication operators but by independent operators usually possessing their own network infrastructure. It can be expected, therefore, that despite the emerging B-ISDN, independent (stand-alone) CATV networks, increasingly interactive, either will continue to operate over many years to come or will

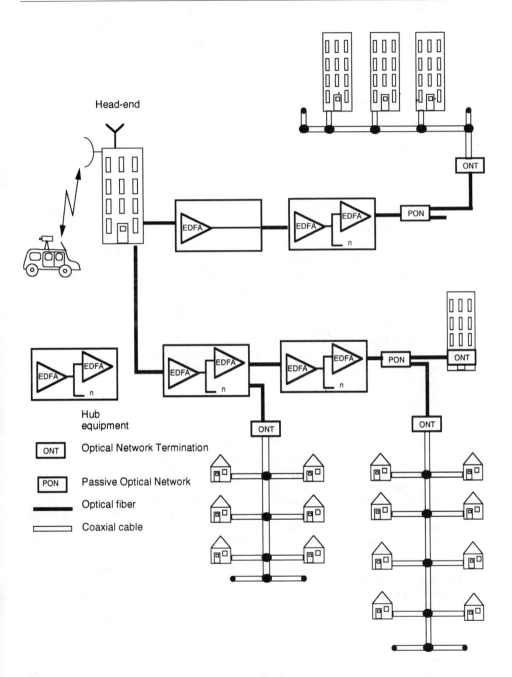

Figure 4.22 CATV network applying the FTTLA architecture.

integrate B-ISDN themselves and operate in competition with the established telecom operators [24–26].

4.6 SPECIFIC APPLICATIONS

Optical fiber transmission, because of its striking advantages and despite being the newest of the transmission media, has become the standard solution for the following applications:

- Undersea systems;
- High-capacity terrestrial systems, except where adverse terrain such as jungle, marshes, and rocky areas has to be crossed;
- Low- and medium-capacity systems replacing open-wire systems;
- Access systems in newly developed areas (in various FITL solutions);
- CATV networks with large numbers of subscribers and large distances;
- Standard conditions with regard to project implementation schedule, security, distances, and rights of way;
- Easy extension to meet high-traffic growth, and frequent drop and insert;
- Permanent location of the stations;
- Easy access to isolated stations and no existing infrastructure such as towers and power plant that otherwise would favor radio-relay or satellite transmission;
- Areas with high electromagnetic radiation.

References

[1] Kao, K. C., and G. A. Hockham, "Dielectric-Fiber Surface Wave-Guides for Optical Frequencies," *Proc. IEEE*, Vol. 113, July 1966, pp. 1151–1158.

[2] Schwartz, Morton I., "Optical Fiber Transmission—From Conception to Prominence in 20 Years," *IEEE Communications Magazine*, Vol. 22, No. 5, 1984, pp. 38–47.

[3] Boerner, Manfred, Mehrstufiges Übertragungssystem für in Pulsecodemodulation dargestellte Nachrichten, DBP-Nr. 1254.513 vom 21.12.1966.

[4] Minoli, Daniel, *Telecommunications Technology Handbook*, Norwood, MA: Artech House, 1991.

[5] "Optical fibre cables," *Fibre Optic Communications*, Cologne: Philips Kommunikations Industrie A.G., Publication No. 34.2e27.1.2.

[6] Tosco, Federico, "Optical Transmission," *Telecommunication Journal*, Vol. 58, XII/1991, pp. 890–895.

[7] Huurdeman, Anton A., *Transmission: A Choice of Options*, Paris: Alcatel Trade International, June 1991.

[8] Rittich, Dieter, et al., "Physikalische Grundlagen der optischen Nachrichtentechnik," *Nachrichtentechnische Berichte*, ANT Nachrichtentechnik, Vol. 3, 1986.

[9] Chesnoy, J., et al., "Ultrahigh Bit Rate Transmission for the Years 2000," *Electrical Communication*, 3rd quarter 1994, pp. 241–250.

[10] Lees, Clare, "The Soliton Solution," *Telecommunications*, Vol. 21, No. 12, 1987.

[11] Refi, James, " A Recipe for High capacity Fibre Transmission," *Telecommunications*, Vol. 29, No. 12, 1995.

[12] Gabla, P. M., and O. Scaramucci, "Long-Haul, High Bit Rate Optical Fibre Links with Optical Amplifiers," *Communication & Transmission*, Vol. 3, 1992, pp. 67–74.

[13] Carballes, J.-C., "Einfluß der optischen Kommunikation," *Elektrisches Nachrichtenwesen*, 4th quarter 1992, pp. 4-10

[14] Kiesewetter, Willi, and Eduard Schlauch, "Herstellung von Glasfasern für die optische Nachrichtenübertragung in drei Processschritten," *Maschinenmarkt*, Vol. 95, No. 27, 1989, pp. 68–72.

[15] Jocteur, R et al., "Evolution of Optical Fibre and Cable Technologies," *Electrical Communication*, 3rd quarter 1994, pp. 275–280.

[16] Reinaudo, Ch., "Undersea Cables: A State-of-the-Art Technology," *Electrical Communication*, 1st quarter 1994, pp. 5–10.

[17] Communication Transmission Systems, Product Catalog 1991/1992, No. A42020-S166-A1-7680, Siemens Munich.

[18] Alcatel Network Systems Product Booklet, 1994, No. 3CL 00078 0001 TQZZA Ed. 01, Alcatel, Paris.

[19] The Submarine Link, Product information from Alcatel Submarcom, No. 9337/3, Alcatel Submarcom, Clichy (France).

[20] Newsletter of the International Telecommunication Union, No. 9, 1994, p. 23.

[21] Blanc E. et al., "Unterwasser-Glasfaser-Übertragungssysteme," *Elektrisches Nachrichtenwesen*, 4th quarter 1992, pp. 45–50.

[22] Newsletter of the International Telecommunication Union, No. 3, 1995, p. 30.

[23] Nassar, Farès, and Michel Triboulet, "FITL: The New Communications Highway?" *Telecommunications*, Vol. 28, No. 9, 1994.

[24] RVS 1550: Ericsson Raynet Video System, Technical Description 02/1995 of Ericsson Raynet.

[25] Bellermann, Jan, et al., "Cable TV—the Multimedia Messenger?" *Telcom Report International*, Siemens Telecommunications, Vol. 19, No. 4, 1996, pp. 30–33.

[26] Heidemann, Rolf, "The IVOD Berlin Projekt: Access Technology for Service Provisioning," *Alcatel Telecommunications Review*, 3rd quarter 1996, pp. 196–200.

Radio Relay 5

5.1 INTRODUCTION

The beginning of radio-relay transmission can be clearly traced back to a series of famous experiments carried out by Heinrich Hertz in 1888. Heinrich Rudolf Hertz (1857–1894), professor of physics at the Technical University of Karlsruhe (Germany) from 1886 to 1889, conducted a series of experiments during that period to prove the validity of a theory formulated in 1873 by the Scottish physicist James Clerk Maxwell (1831–1879). Maxwell's theory stated that electromagnetic waves are of the same nature as light. Heinrich Hertz made his decisive experiment, proving the validity of Maxwell's theory, with two parabolic antennas, one connected to a spark bridge and the other, at a distance of about 15m, with a resonator circuit connected to a spark gap. Once the spark bridge was activated, electromagnetic energy received at the second antenna produced minuscule sparks in the spark gap.

Before practical radio-relay transmission could be realized, however, considerable technological progress was still required, such as the discovery and development (chronologically) of the coherer, Marconi's short-wave radio, the diode, the triode, and the HF-generator. Finally the VHF generator developed by Heinrich-Georg Barkhausen (1881–1956) in 1920 resulted in the development of velocity-modulated tubes such as the klystron in 1930, the magnetron in 1934, and the *traveling wave tube* (TWT) in the late 1940s.

With klystron available in 1930, radio-relay equipment could be developed, and the world's first commercial radio-relay system, which was *amplitude modulated* (AM) using a klystron-producing 1-W RF output power and operating at 1.7 GHz, was put into commercial operation across the English Channel covering a distance of 56 km. The date of initial operation, January 26, 1934, marks the beginning of the radio-relay transmission era!

The information contained in this chapter is almost entirely taken from Anton A. Huurdeman, *Radio-Relay Systems,* Norwood, MA: Artech House, 1995.

This first commercial radio-relay link carried one telephone and one telegraph channel simultaneously. The link remained in operation until 1940 when service was interrupted by World War II.

Radio-relay development continued in the United States and resulted in a 100-telephone channel 4-GHz *frequency-modulated* (FM) system put into operation on a link between New York and Boston in 1947.

The emergence of television around 1950 gave a big impetus to radio-relay transmission. A 4-GHz radio-relay network in the region of New York, Boston, Philadelphia, Baltimore, and Washington supported the live broadcasting of President Harry S. Truman's inauguration in 1949. In January 1953, some 75 million people watched on TV the inauguration of President Dwight D. Eisenhower thanks to a 12,000-km radio-relay network throughout the United States from New York to San Francisco.

In Europe radio-relay equipment manufactured in the United Kingdom, France, Germany, and Italy was installed just in time to enable the simultaneous *"Eurovision"* TV broadcasting in Western Europe of the coronation of Queen Elizabeth II of Britain on June 2, 1953. The air-interface across national borders became possible thanks to the standardization work of the ITU and subsequent CCIR and CCITT Recommendations.

With the advent of microwave semiconductors around 1960 it was possible to replace klystrons and TWTs in the RF power output stages by FET devices, and so it became technically and economically possible to manufacture all-solid-state radio-relay equipment.

All-solid-state technology substantially reduced the power consumption (by about 80%) and equipment size (by about 50%) and subsequently reduced the cost and dimensions of primary power supply (diesel generators, rectifiers, and batteries) and equipment accommodation (shelters or prefabricated buildings for repeater stations on isolated sites). As a result, radio-relay transmission became an increasingly flexible, economical, and reliable means of transmission for low, medium, and high capacities.

The transmission capacity of radio-relay systems gradually increased to 960, 1,260, 1,800, 2,700, and even 3,600 telephone channels per RF channel. Analog radio-relay transmission proved itself in long routes for trunk networks, where many centers had to be linked across areas that are difficult to cable. Today, with digital transmission making it feasible to use the less crowded higher frequency bands, radio-relay transmission has become equally suitable for short hops where users are close together but separated by densely packed streets and buildings, as is the case in major towns and cities all over the world.

The world's longest radio-relay link was taken into operation in early 1996 in Russia. The 7,600-km link from Moscow to the city of Novosibirsk built by Siemens, and from that city to Chabarovsk built by NEC, was completed in only 18 months. The link operates at 155 Mbps and connects Western Europe through Russia with Korea and Japan.

The radio-relay equipment mentioned thus far is all for *point-to-point* (P-P) operation. Around 1980, the advent of cost-effective integrated circuits—such as *very large scale integrated circuits* (VLSI), *application-specific integrated circuits* (ASIC), and *monolithic microwave integrated circuits* (MMIC)—made it economically possible to use radio-relay transmission in a *multiple access* (MA) and *point-to-multipoint* (P-MP) mode of operation, in a domain so far left at the fringe of telecommunications, that is, the vast rural areas of the world.

The latest, and very promising, addition to radio-relay transmission are the emerging *radio in the loop* (RITL) also called *wireless local loop* (WLL) systems that serve many subscribers with N-ISDN, and soon B-ISDN and multimedia, quicker, or at lower cost than otherwise would be possible at their location with optical fiber.

5.2 RADIO-RELAY TECHNOLOGY

Radio-relay transmission is the technology of generating, modulating, amplifying, and directing very high frequencies through the atmosphere for subsequent selective receiving, amplifying, and demodulating. In other words, radio-relay technology essentially shapes modulated, very short electromagnetic waves into rays or beams for propagation at the velocity of light through the atmosphere in a specific direction where it will be received practically undisturbed by other waves that might be simultaneously in the air. The waves used in radio-relay transmission have a frequency ranging from 300 MHz to about 100 GHz. They behave, as Hertz confirmed, like light and can thus only be propagated in an almost straight line and lose their intensity as a function of the distance covered. Consequently, radio-relay transmission is limited in two ways: to the direct line of sight between stations and by the attenuation in the atmosphere. As a consequence of these limitations and depending on the actual topology between the terminal stations of a link, a number of repeater stations might be required. A repeater station amplifies the weak RF signals received from the transmitters of adjacent stations and, if necessary and the signal is digital, regenerates the signal and retransmits it to the adjacent stations. Double-terminal stations, in addition to performing the same functions as a repeater, give access to the information carried on the link so that information can be extracted and added.

Radio-relay technology, beyond standard electronics, mainly concerns

- Modulation;
- Radiation;
- Propagation.

In contrast to transmission via copper or optical fiber cable, which have homogeneous characteristics, radio-relay uses the atmosphere as transmission medium, which is by nature nonhomogeneous both in distance and time. To overcome the disadvantages of this nonhomogeneity, beyond careful planning of the path between the terminal stations, specific radio-relay techniques are required, such as diversity, equalization, crosspolarization, interference cancellation, and error correction.

5.2.1 Modulation

Modulation is the process by which certain characteristics of a wave are modified in accordance with the characteristics of another wave. In radio-relay technology, this means that the information to be transmitted from one place to another—called the baseband signal—by a radio-relay system is modulated onto the RF carrier at the transmit side and demodulated again at the receive side.

The RF carrier, like any electromagnetic wave, is a sine wave, which is defined by

$$U = U_{max} \cdot \cos(\phi + 2\pi ft)$$

As indicated in Figure 5.1, modification of the RF carrier by the baseband signal is possible in the following three modes:

- By changing the amplitude U, resulting in AM;
- By changing the phase ϕ, resulting in PM;
- By changing the frequency f, resulting in FM.

Modulation always adds some noise to the original signal (modulation noise). Therefore, to optimize the modulation performance with respect to noise, selectivity, and system planning flexibility (for example, no demodulation required on repeater stations), the baseband signal is normally in heterodyne mode and first modulated on an *intermediate frequency* (IF), which is then modulated on the RF carrier. In the case of digital modulation, the heterodyne mode is of less importance for the following reasons:

- Digital regeneration can only be performed on the digital baseband signal. The IF signal, even if modulated by a digital signal, remains an analog signal that can be amplified but not regenerated.

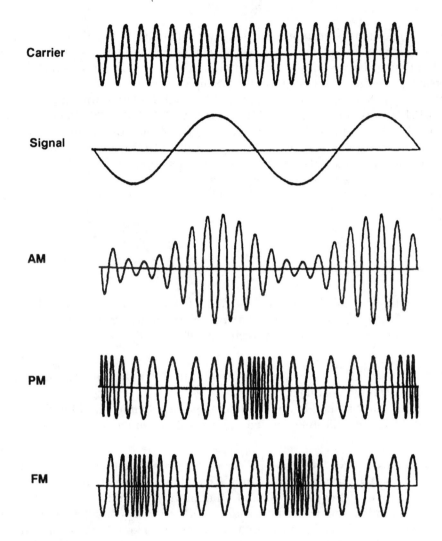

Figure 5.1 Modulation options.

- Digital regeneration does not add any noise to the signal, so the number of modulation stages is of no importance for the signal quality.

Direct modulation of the baseband on the RF carrier used to be applied for low-cost analog systems with low and medium capacities. However, it is now reappearing in modern highly integrated digital radio-relay equipment. So far RF frequency generation—usually in a *phase-locked loop* (PLL) circuit—

and modulation have been performed by two separate circuits. Recent improvements in circuit architectures, however, enable RF frequency generation and modulation to be performed by a single circuit, thereby eliminating the IF, increasing modulation quality, and reducing equipment size and cost. This is particularly advantageous for compact outdoor radio-relay equipment operating in the EHF bands (30 to 300 GHz). Consequently, CCIR has not defined a specific IF for digital systems but has recommended that heterodyne modulation, if used with radio-relay systems operating below 15 GHz, should preferably use the same IF as standardized for analog systems so that equipment can be converted from analog to digital operation simply by replacing the modulation equipment (provided the radio equipment and antenna system are still in good condition).

5.2.1.1 Analog Systems

Analog radio-relay systems are usually frequency modulated primarily because FM is less susceptible to distortion than AM. Heterodyne systems usually apply, in compliance with CCIR Recommendation 403, an IF of

- 35 MHz (below about 1 GHz) or 70 MHz (about 1 GHz) for 12, 24, 60 and 120 telephone channels;
- 70 MHz for basebands up to 1,800 telephone channels and for TV;
- 140 MHz for radio-relay systems above 1,800 telephone channels.

When using FM, the IF and RF carriers deviate from their nominal values as a function of the baseband frequency. The relation between frequency deviation and baseband frequency is called the modulation index. The higher the modulation index, the better the *signal-to-noise ratio* (SNR) but also the larger the bandwidth required in the RF frequency spectrum (and in the IF, as already shown). Therefore, CCIR defined (in Recommendation 404) an effective channel deviation for each baseband at the values shown in Table 5.1.

Noise in an RF circuit increases with frequency, so the SNR of an FM signal would decrease within the baseband (nonlinear frequency response). To compensate for this noise increase with frequency, it is common practice to apply amplitude pre-emphasis on the baseband signal and thus to increase the level of the top frequencies. Hence, they are transmitted with a higher frequency deviation than the bottom frequencies, thereby increasing their modulation index and the SNR. After demodulation, the baseband response is equalized again by corresponding de-emphasis.

Table 5.1
Relation of Channel Deviation and IF With System Capacity

Telephone Channels	Baseband (kHz)	Channel Deviation (at RMS in kHz)	IF (MHz)
12	60–108	35	35/70
24	(6)12–108	35	35/70
60	60–300	50/100/200	35/70
120	12–552	50/10/200	35/70
300	60–1,300	200	70
960	60–4,188	200	70
1260	60–5,680	140/200	70
1800	312/316–8,204	140	70
2700	312/316–12,388	140	140

5.2.1.2 Digital Systems

The objective of digital modulation is to bring the baseband signal onto the RF carrier using the minimum bandwidth. Furthermore, the transmission capacity for digital signals has to be accommodated in existing frequency plans that were originally defined for analog transmission. This bandwidth economy, which is known as "spectral efficiency," is defined in bits per Hertz (transmission capacity/RF carrier bandwidth). Spectral efficiency depends not only on the selectivity and frequency response of the radio-relay equipment, but also largely on the modulation mode.

Digital signals have basically two amplitude states ("0" and "1") corresponding to phases of 0 degrees and 180 degrees. In the simplest digital modulation mode, this two-state condition is keyed on the RF carrier by shifting the phase of the carrier. It is therefore known as *phase-shift-keying* (PSK) modulation (see Figure 5.2).

In two-state PSK, or 2 PSK, shifting the carrier phase by 180 degrees requires 1 Hz of the carrier frequency for each bit of the baseband, so the spectral efficiency is 1 bit/s/Hz. A 2-Mbps baseband modulated with 2 PSK thus requires an RF carrier with a bandwidth of 2 MHz. A first improvement is obtained by using 4 PSK or *quaternary* PSK (QPSK) modulation. In this case, the binary signal is converted into a quaternary signal and the four possible phases of the quaternary signal are keyed onto the RF carrier, shifting the carrier phase in 90-degree steps, resulting in a spectral efficiency of 0.5 bits/s/Hz. Thus a 1-MHz RF carrier bandwidth is needed to transport a 2-Mbps signal. The next logical improvement is 8 PSK, shifting the carrier in 45-degree steps and resulting in a spectral efficiency of 3 bits/s/Hz. In this case, a 0.67-MHz RF carrier bandwidth would be sufficient for a 2-Mbps signal.

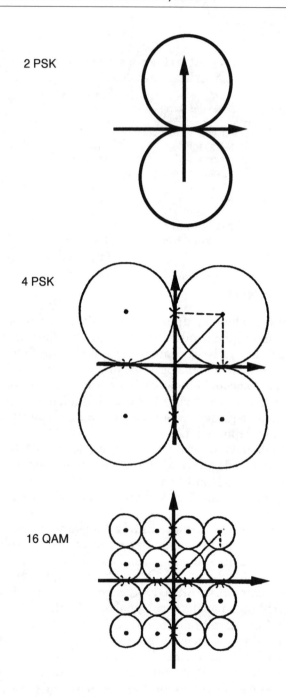

Figure 5.2 Digital modulation modes.

Each higher PSK modulation mode requires a better SNR performance, which is difficult to achieve. Consequently, instead of 16 (or higher) PSK modulation, *quadrature amplitude modulation* (QAM) is used. This is a combination of phase shifting and amplitude modulation of the carrier. Two carriers 90 degrees out of phase—hence quadrature—are amplitude modulated by a digital signal (baseband) with a finite number m of amplitude levels and are subsequently added to one another. It is thus known as m-QAM. With 16 QAM, 16 different signal states are detected and amplitude and phase shift modulated on the RF carrier, resulting in a spectral efficiency of 4. A 140-Mbps baseband thus only requires a bandwidth of 140/4 = 35 MHz on the RF carrier, which corresponds to the 40-MHz RF channel, for example, for the 4-, 6.7-, and 11-GHz bands. However, 16-QAM modulation cannot be used to transmit 140 Mbps in the 2-, 4-, 6.2-, and 8-GHz bands with an RF channel spacing of 29/30 MHz. The next logical step to 64-QAM modulation had to be made, resulting in an spectral efficiency of 6 and enabling the transmission of a 140-Mbps signal in a 23-MHz bandwidth, which fits with sufficient selectivity in the 29/30 -MHz RF channel spacing. Implementations of 128 QAM with a spectral efficiency of 7 bit/s/Hz and 256 QAM with a spectral efficiency of 10 have already been realized. At this time, 512 and even 1,024 QAM (which would provide a spectral efficiency of 24 bit/s/Hz) are being studied, but such highly complex modulation schemes are vulnerable to fading and interference.

The introduction of SDH posed another challenge, namely, to fit the STM-1 level (155 Mbps) and the STM-4 level (622 Mbps) into existing RF frequency plans. This, together with the availability of highly integrated digital processing components and the constant need for further performance improvements (in view of the excellent performance of optical transmission equipment), led to the choice of a modulation mode consisting of a combination of QAM and signal coding.

Various signal coding solutions have been investigated, such as *block coding modulation* (BCM), *modulation matched coding* (MMC), and *trellis coded modulation* (TCM). It appears that TCM is generally accepted as the preferred method and is therefore briefly explained here.

Trellis coding assumes a logical continuation of the baseband signal so that on the modulation side only changes in the logic are coded. On the demodulation side disharmonious signals, which do not correspond to a logical signal continuation and are not announced by coding, are eliminated and replaced by the logically expected signal. To this purpose a *maximum likelihood sequence estimation* (MLSE), with a Viterbi algorithm, is implemented. The convolutional positioning of the signal states are compared with a choice of four predetermined positions per signal state on a convolutionally sloped grid (hence the name Trellis coding), as shown in Figure 5.3.

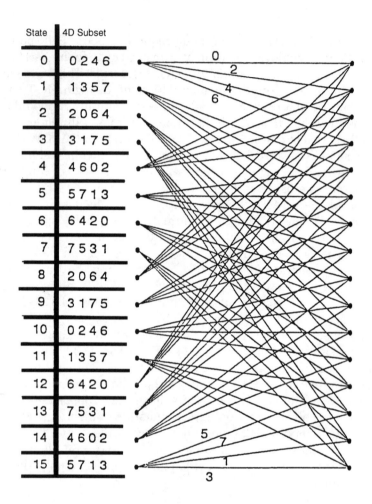

Figure 5.3 Trellis coding.

TCM is thus QAM-improved with error correction. Compared with 64 QAM, it has only a slightly improved spectral efficiency of 6.5 for 128 TCM but a substantially improved BER. The modulation mode 128 TCM is the present state of the art for 155-Mbps modulation for transmission within 29/30-MHz RF channel plans. For the transmission of STM-1 within 40-MHz RF channel plans (in the 4-, 6.7-, and 11-GHz bands), either 32 TCM (with a spectral efficiency of 4.5) or 64 TCM (with a spectral efficiency of 5.5) can be used, depending on the required quality.

Taking advantage of the error-correcting feature of TCM, this modulation scheme developed for high-capacity systems in the meantime is applied for (mostly rural) P-MP systems, too. Due to the error correction, namely, a lower receiver reception threshold can be accepted; thus, a lower transmit power is possible, resulting in lower power consumption and amplifier cost and thus less investment in primary power. Instead of a lower RF power, smaller antennas can be used with subsequent lower cost for antennas, towers, and tower foundations.

A summary of the digital modulation modes with their spectral efficiencies and signal configurations is given in Table 5.2.

5.2.2 Radiation

The electromagnetic waves generated and modulated/demodulated in the radio-relay equipment are radiated into and received from the atmosphere by antennas. The antenna is thus the interface between the radio equipment and the free space transmission medium. The connection between the radio equipment and the antenna is generally called the feeder; it is constructed using coaxial cable or hollow waveguide.

5.2.2.1 Antennas

In its simplest form, an antenna is a metallic conductor, the length of a full wavelength of the wave, to be radiated. One end of the conductor is connected to the radio equipment, the other end radiates freely. The RF current flowing

Table 5.2
Summary of Digital Modulation Modes

Type of Modulation	Spectral Efficiency (bits/s/Hz)	Approx. RF Bandwidth for 140-Mbps Baseband (MHz)
2 PSK	1	140
4 PSK	2	70
8 PSK	3	47
16 QAM	4	35
32 QAM	5	28
64 QAM	6	23
128 QAM	7	20
256 QAM	8	14
512 QAM	9	10
32 TCM	4.5	31
64 TCM	5.5	25
128 TCM	6.5	22

along the conductor establishes an electric field. The distribution of the current along the conductor corresponds to a standing wave, and therefore, any change in this current generates a magnetic field whereas an arriving magnetic field generates a current in the conductor. Electromagnetic wave propagation is based on this interaction between electric and magnetic fields at the metallic conductor. Such a basic antenna radiates an electromagnetic field spherically around the conductor and is therefore called an isotropic radiator, which is taken as a standard for the definition of certain antenna characteristics.

Antennas can be classified according to their radiation pattern as either omnidirectional or directional. The theoretical isotropic radiator is an omnidirectional antenna. In radio-relay transmission, omnidirectional antennas are only used for central and sometimes for the repeater stations of P-MP systems. Directional antennas are used for all other applications. Moreover, most antennas can be used with dual polarization, that is, with one group of RF channels using horizontal polarization and a second group simultaneously using vertical polarization.

The major electrical and mechanical antenna characteristics, which are important for radio-relay transmission, are as follows:

- Gain: defined as the relation between the RF energy radiated in the directed beam and the energy radiated by an isotropic antenna (taken as a reference) in the same direction and expressed in decibels;
- Half-power beamwidth: the beamwidth of the main lobe at which the radiation intensity drops to half the maximum value of the main lobe. Normally in addition to the main beam (or main lobe), there are several undesired smaller sidelobes, and even small lobes in the opposite direction to the main lobe;
- *Front-to-back* (F/B) ratio: defined as the ratio of its maximum gain in the forward or intended direction to its maximum gain in the backward direction;
- *Radiation pattern envelope* (RPE): showing the antenna radiation intensity in relation to the direction;
- Return loss: refers to the power that cannot effectively be used because of a mismatch between the antenna and feeder. This mismatch, or reflection coefficient, is measured in terms of *voltage standing wave ratio* (VSWR), which is the ratio of the amplitude of the reflected wave to that of the incident wave;
- *Crosspolarization discrimination* (XPD): the ratio of the power radiated in the desired polarization to the undesired power converted to the orthogonal polarization;
- Interport isolation: isolation between the two inputs of dual-polarized antennas;

- Intermodulation: occurs when two signals are coupled to the same antenna, as a result of the nonlinear behavior of the metal to metal contact (for example, feeder-antenna connector);
- Wind and ice resistance: radio-relay antennas are designed for outdoor installation, and radio-relay transmission has to continue even (or especially) under very bad weather conditions. Antennas must thus be able to withstand extreme wind and ice conditions. The windload of an antenna normally is specified for operational wind speed and survival wind speed.

The operational wind speed is defined as the wind speed at which the antenna axis deflection is less than one-third of the half-power beamwidth, which depends on the climatic conditions, which are usually specified between 110 and 180 km/h.

The survival wind speed is the maximum wind speed that the antenna should withstand without permanent damage. Again, depending on the station location, survival wind speeds are specified between 180 and 250 km/h, although they may be as high as 320 km/h in regions prone to hurricanes.

Antennas installed in cold weather regions have to withstand substantial ice loads. Ice loads are typically specified at a maximum of 3 cm of radial ice or 900 kg/m^3.

Several antenna constructions can be used for radio-relay transmission. For frequencies up to 2.7 GHz the following antennas may be used:

- Yagi antennas with directing and reflecting rods;
- Corner reflector antennas, used as a simplified Yagi type;
- Antennas with reflector plates;
- Helical antennas, with one or more coil-like radiator(s);
- Log-periodic or "end-fire" antennas;
- Parabolic reflector antennas.

Parabolic reflector antennas are used for frequencies above 2 GHz. Reflector diameters start at 60 cm and normally go up to 3.6m. Exceptionally, for transhorizon transmission they may go up to 40m. Reflector antennas normally have a solid metal or a surface-metalized fiberglass reflector. Grid antennas, however, instead of a solid reflector, use a number of metallic rods arranged in the shape of a parabola, thus substantially reducing weight and windload without significantly compromising the F/B ratio. Grid antennas can only be used for frequencies below 3 GHz.

Reflector antennas use one of the following types of radiators in front of the reflector:

- Dipole radiator: mainly for frequencies below about 3 GHz usually mounted on a "swan neck";

- Horn radiator;
- Cassegrain radiator (using a feed horn that radiates into a secondary reflector mounted at the focus of the main reflector);
- Offset radiator (using a feed and subreflector placed outside the main beam, thus reducing interference).

Both the Horn and Cassegrain radiators can be used for dual-polarized operation, whereas the dipole radiator only operates on one polarization. Thus two dipole radiators are required for dual polarization.

Parabolic reflector antennas are available in various constructions that are suitable for different applications. Table 5.3 summarizes the major reflector antenna versions indicating their performance with a 3-m reflector when operating in the lower 6-GHz band (5,925 to 6,425 MHz).

5.2.2.2 Feeders

The feeder is the interconnection between radio and antenna. RF energy is "fed" from the transmitter to the antenna or from the antenna to the receiver. Thus, a feeder is a transmission line for electromagnetic waves. The efficiency of a feeder is expressed as a VSWR, which typically is around 1.02. Coaxial cable can be used at frequencies below about 3 GHz. At higher frequencies the attenuation of coaxial cable is too high and hollow waveguides must be used.

In a waveguide, electromagnetic waves propagate as two different fields: a transverse electrical field termed the E mode or TE wave and a perpendicular magnetic field termed the H mode or TH wave. To ensure efficient reflectionfree propagation of these two fields, the waveguide must be dimensioned according to the frequency band being used; that is, the higher the frequency, the smaller the waveguide.

Table 5.3
Summary of Reflector Antennas

Antenna Type	Gain (dB)	F/B Ratio (dB)	XPD (dB)
Standard	43	51	30
Improved performance (e.g., "focal plane" or "deep")	43	65	30
High performance	43	71	33
Ultra high performance	43	76	33 (or 40)
Cassegrain	43	62	30
Offset	43	75	36
Horn reflector	43	95	40

Waveguides are available as rigid waveguides (for all frequency bands) or as flexible waveguides for the frequency bands from 2 to 23 GHz. Flexible waveguides are available as either flexible corrugated copper waveguides or semiflexible aluminum waveguides.

5.2.2.3 Passive Repeaters and Reflectors

Passive repeaters and reflectors are RF beam direction changers. They are used to overcome obstacles in an otherwise direct line of sight between two radio-relay stations.

A passive repeater consists of two parabolic antennas, each facing one of the two radio-relay stations and interconnected by a waveguide. The RF energy received by one antenna is thus reradiated by the other without electronic amplification of the RF signal. In many applications where the angle of transfer of the RF signal exceeds 40 degrees, a single flat reflector can be used rather than a combination of two antennas.

5.2.3 Propagation

The propagation of electromagnetic waves can be classified in three different types:

- Ionospheric propagation: which is possible for short waves at frequencies between 1 and 30 MHz for very long distance (around the world) transmission;
- Tropospheric propagation: which is possible at frequencies between 30 MHz and about 5 GHz for transmission over the horizon (also called over the horizon or scatter radio-relay transmission);
- Atmospheric, or ground wave, propagation: which is possible at frequencies from 300 MHz to about 300 GHz for line-of-sight radio-relay transmission.

Electromagnetic waves travel at the speed of light, which is about 300,000 km/s (the exact value is 299,792 km/s). The frequency is defined in hertz (Hz), where the unit 1 Hz stands for one cycle per second. Thus the relation between wavelength and frequency is given by

$$\text{wavelength (m)} = \text{velocity of light (m/s)} / \text{frequency (Hz)}$$

As an example, for a 2-GHz RF carrier the wavelength is

$$300,000,000 \text{ m/s} / 2,000,000,000 \text{ Hz} = 0.15 \text{ m}$$

The propagation of such centimetric waves in the atmosphere is, to a first approximation, governed by the laws of free space propagation, which define the free-space attenuation (or loss) F for a radio-relay path between two stations at a distance d km and an RF carrier wavelength of λ m as

$$F = 20 \, \log(4\pi d/\lambda)$$

The path attenuation in free space thus increases not only, as aforementioned, as a function of the distance but also with the inverse of the wavelength and, therefore, with frequency. This relation between free space and distance and frequency is given in Figure 5.4. The achievable hop length of a radio-relay link thus decreases as the frequency increases, all other parameters being constant. Doubling the distance at a given frequency, on the other hand, increases the free-space attenuation by only 6 dB. This relatively small increase of free-space attenuation compared with the increase of distance might give the impression that long hops easily can be obtained by simply increasing the transmitter output power, the receiver sensitivity, or the antenna gain. Unfortunately, however, this is not so easy because the total path attenuation, in addition to the free-space attenuation, is determined by a number of influences, which are connected with

- The propagation path;
- Atmospheric attenuation;
- Fading;
- The frequency allocation.

5.2.3.1 The Propagation Path

For wave propagation the surface of the earth, free-space, and atmospheric attenuation give only an approximate value for the actual loss, because loss can be markedly affected by reflections from the ground between the stations as well as by ray diffraction and refraction in the atmosphere. In so far as the line-of-sight between the two antennas is not obstructed, ground reflections are the major source of propagation disturbances. Assuming a low-loss reflecting ground, the transmitted RF energy reaches the receiver in two ways: directly along a straight line and indirectly by reflection (see Figure 5.5). The reflected ray arrives with a phase delay at the receiving antenna, composed of the 180-degree phase shift at the reflection point plus the phase shift caused by the path length difference. With a path difference of half a wavelength, corresponding to a 180-degree phase shift, the direct ray and reflected ray arrive in phase and are thus added. However, with a path difference of a full wavelength

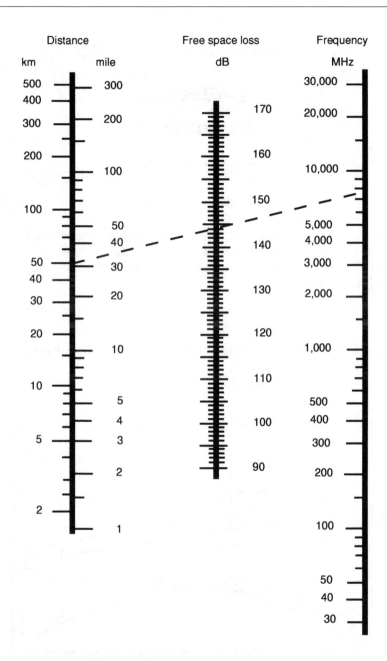

Figure 5.4 Free-space attenuation diagram.

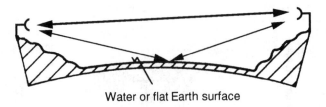

Water or flat Earth surface

Figure 5.5 Radio-relay path with ground reflection.

and a resulting phase shift of 360 degrees + 180 degrees (reflection), the two rays cancel one another.

The reflection points of all (desired) rays with a constant path length of half a wavelength delay are described by an ellipse, as shown in Figure 5.6, called the *first Fresnel zone.* The radius r of the Fresnel ellipsoid at any point is defined by the formula

$$r = \sqrt{(d_1 d_2 \lambda)/(d_1 + d_2)}$$

This formula makes it clear that the longer the wavelength λ (that is, the lower the RF frequency), the larger r is.

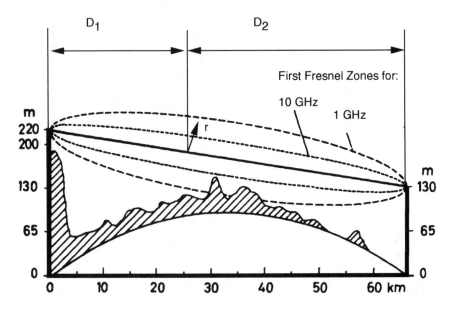

Figure 5.6 The first Fresnel zone.

A basic condition for optimum radio-relay transmission is that the path between two adjacent stations along the link should be completely free from obstructions, not only along the direct line of sight but also within the first Fresnel zone in order to take advantage of the ground reflection. This implies that, for a given link, the lower the frequency the higher the antenna tower must be. Consequently, using higher frequency bands would seem advantageous to reduce tower- and height-associated costs. On the other hand, as already shown, signal attenuation on the link increases substantially with frequency. As a compromise, therefore, the frequency bands between 4 and 7 GHz are preferred for high-capacity systems on long-distance (backbone) routes, with the frequency bands below 4 GHz tending to be reserved for low- and medium-capacity systems for rural and cellular mobile radio applications. Frequency bands beyond 10 GHz are mainly used for urban systems and backbone access systems in view of the rapidly decreasing hop lengths. Frequency bands around 20 GHz (with very short hops) are particularly useful for radio-relay access links from public to cellular networks and for interconnecting the cell base stations. In this case, the high path attenuation at these frequencies can be turned into an advantage since the same RF carrier frequency can be reused on more than one hop within the same area without the signals interfering with one another. This applies even more to frequencies around 60 GHz—where high oxygen absorption occurs—which are preferred for the emerging WLL systems.

5.2.3.2 Atmospheric Attenuation

Fortunately for us as human beings but unfortunately for the propagation of electromagnetic waves, the atmosphere is not really free space. Rather, it contains oxygen, nitrogen, and other gases as well as water (as rain, fog, and vapor), all of which hinder the propagation of electromagnetic waves especially at higher frequencies.

Table 5.4, which shows typical hop lengths for various frequency bands, clearly indicates the accumulated effects of frequency-dependent free-space attenuation and the atmospheric attenuation-related decrease of hop length at higher frequencies.

5.2.3.3 Fading

The nonhomogeneity of the atmosphere effects on the radio beam in such a way that the received RF signal normally is not constant but fades around a nominal value. The causes of fading are complex and still under investigation; with much simplification, however, they can be depicted as shown in Figure 5.7 and briefly described as in the following subsections.

Table 5.4
Typical Hop Lengths

Frequency Band (GHz)	Hop length (km)
2	60
4/5/6	50
7/8	45
11	35
13	25
15	20
18/20	10
30	5
60	0.5

Multipath Propagation Caused by Reflection From an Inversion Layer

High-altitude inversion layers may occur when masses of dry air descend into high-pressure areas or by advancing weather fronts, by trade winds above oceans, and by temperature inversions within limited altitudes of the atmosphere. Radio beams entering such inversion layers at a flat angle are reflected rather than passing through the layer and disappearing into outer space. Similarly to a ground-reflected wave, the inversion-reflected wave can interfere with the direct beam, thereby increasing or reducing the main signal, depending on its phase. As this type of fading can only happen at flat-angle reflection, long hops suffer more than shorter ones.

Multipath Propagation Caused by Reflection From Ground or Water

When a signal is transmitted over lakes, seas, or rice fields, as well as over very smooth deserts or large burned deforested areas, the direct beam and the reflected beam may be of equal strength. The goal of path planning will then be to plan the path in such a way that the reflected waves add rather than subtract.

Long-Term Fading

Low-altitude inversion layers may occur above flat ground in calm weather when there are higher temperatures at the higher altitudes (temperature inversion). Such conditions may prevail after sunset when the temperature of the lower layers drops.

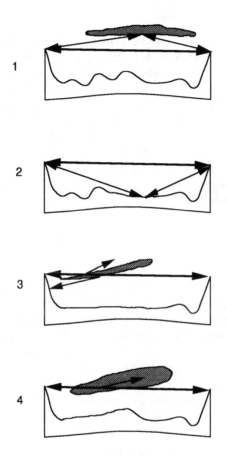

Figure 5.7 The major causes of fading.

Abnormal Diffraction Caused by Ducts

Specific climatic conditions near mountains or coastal lines as well as in tropical areas may create strong temperature inversion layers, well known for enabling extremely clear sight over hundreds of kilometers. Such inversion layers are most disturbing for radio-relay transmission because they favor so-called duct formation whereby, as in a waveguide, a substantial part of the direct beam stays in the inversion layer and disappears. Large fading margins need to be applied when planning radio-relay links in such areas.

5.2.3.4 Fading Correction

From the above discussion, it is clear that accurate path planning is crucial to limit fading. Even so, in many cases it will not be possible to fully prevent fading because of its complex origins. Consequently, there may be some random frequency-dependent (selective) and frequency-independent (flat) fading.

Overcoming these constraints to achieve transmission quality comparable to that offered by cable systems, however, is normal radio-relay engineering practice. In addition to *automatic gain control* (AGC), which is standard in radio-relay receivers to achieve an almost constant output signal from an RF input signal that is fading over a wide range, a number of optional corrective devices may be used to monitor and process the received signal continuously, making it possible to achieve real-time distortion compensation and thus to recover the original signal. The major corrective devices are briefly described in the following subsections.

Diversity Receiver

Fading caused by multipath propagation is a function of path length and frequency. The instantaneous distribution of the fading amplitudes therefore is not the same at different locations, and with two frequencies a different fading distribution is obtained at the same location. This means that if two different paths are used to transmit the same signal, only limited fading correlation is shown by the two signals provided the path or frequency difference is sufficiently large. Combining the two signals, or selecting the better of the two signals, therefore, can largely compensate for the fading effects. Depending on the actual conditions, one of the following diversity receivers will be used:

- A space diversity receiver, which combines two RF signals carrying the same baseband, received by two antennas that are vertically separated according to the frequency;
- A frequency diversity receiver, which combines the two RF carriers with the same baseband signal received by one antenna.

Both methods can also be combined to form a quadruple diversity system.

Adaptive Time-Domain Equalizer

This equalizer minimizes (time-related) intersymbol distortion. It is normally included in the demodulator, where it analyzes and reduces the distortion of the baseband of the demodulated signal by means of complex transversal filtering.

Adaptive Frequency-Domain Equalizer

This is optionally included in the receiver as part of the IF circuitry. It improves the spectrum of the received RF signal by correcting linear distortion introduced by propagation (for example, by selective fading).

Forward Error Correction Codec

This codec is used to correct any errors that remain after the equalizing and combining methods have been used. Limited to correcting isolated errors, the improvement it contributes is better the poorer the error rate before correction.

Adaptive RF Output Power Control

Instead of a constant high fade margin with a maximum transmitter RF output power, more efficient use of the radio frequency spectrum can be made by automatically adjusting the RF output power to the actual hop attenuation. This *automatic transmit power controller* (ATPC) saves energy and reduces interference.

Crosspolarization Interference Canceler

A *crosspolarization interference canceler* (XPIC) is used to suppress crosspolarization interference that would otherwise occur with the cochannel operation of two RF carriers on the same RF channel, one using horizontal and the other vertical polarization.

5.2.3.5 The Frequency Allocation

Radio-relay transmission takes place in free space using the frequency spectrum from 300 MHz to about 100 GHz. This free space, apart from not being free from absorbing water and gases, is neither freely available for everybody for uncoordinated simultaneous use nor is its use freely and infinitely expandable. Rather, it has to be shared with other services such as

- Mobile radio;
- Satellites;
- Broadcasting;
- Radio location;
- Radio navigation;

- Space research;
- Future transmission of electric power on radio frequencies from a spacecraft.

The international allocation of individual parts of the radio frequency spectrum to qualified users worldwide is the responsibility of the ITU. Within the ITU, two bodies execute this responsibility.

- The *World Administrative Radio Conference* (WARC), called *World Radio-communication Conference* (WRC) since 1995, holds regular meetings at which it allocates particular frequency bands to specific present and future services.
- *Radio Regulations Board* (RCB), founded in December 1992 as the successor to the *International Frequency Registration Board* (IFRB), settles the international rules for the assignment and registration of radio frequencies within the frequency bands defined by the WARC/WRC.

The WARC/WRC and RCB assign specific frequency ranges (called frequency bands) from the overall radio frequency spectrum to specific services. Until 1993, the definition of specific radio channels within a frequency band assigned for radio-relay transmission in so-called frequency plans was the responsibility of CCIR (*Comité Consultatif International de Radio*) which used to be an organ of the ITU. Since early 1993, it has become the responsibility of the newly constituted ITU Radiocommunications Sector in close cooperation with the new *Telecommunications Standardization Advisory Group* (TSAG).

For each frequency band, CCIR produced a frequency plan that was published in "Recommendations" giving details of bandwidth, channel spacing, number of channels, and center frequency with lower and upper bands. As an example, Figure 5.8 shows the frequency plan for the "lower" 6-GHz band as defined in CCIR Recommendation 383.

This plan was originally defined for 1,800-channel analog transmission and later extended to 140-Mbps and 155-Mbps digital operation. The 5.9- to 6.4-GHz band, which has a bandwidth of 0.5 GHz (500 MHz), is subdivided into two equal 250-MHz bands, known as the lower and upper bands. Within each 250-MHz band, eight channels are defined with a channel spacing of 29.65 MHz. In analog systems, the even-numbered channels are to be used crosspolarized relative to the odd-numbered channels in order to optimize channel decoupling. Thus, with channels 1, 3, 5, and 7 using horizontal polarization and channels 2, 4, 6, and 8 using vertical polarization, all eight RF channels can be operated in parallel using a single antenna.

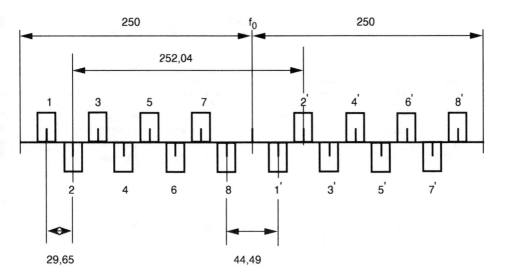

Figure 5.8 RF frequency plan for the lower 6-GHz band.

Similarly, the other bands below 10 GHz recommended for analog transmission have also been recommended for reuse by digital systems. This enables analog and digital systems to coexist on the same routes, thus supporting an easy long-term transition from analog to digital.

Specific RF channel arrangements have been defined by CCIR up to the 18-GHz band. For the next frequency band of 23 GHz, instead of a defined frequency plan with X channels arranged around a center frequency, only a homogeneous pattern has been recommended for 685 channels with a channel spacing of 3.5 MHz or, alternatively, 959 channels with a 2.5-MHz channel spacing throughout the 21.2- to 23.6-GHz band. This solution maximizes planning flexibility and allows a mix of high-, medium-, and small-capacity systems.

Table 5.5 summarizes the major RF frequency bands as assigned by WARC and defined by CCIR.

Radio-relay routes normally operate on a number of RF channels (similar to cable routes, which have a number of pairs in a cable), each with the same transmission capacity, which work in parallel within a specified frequency band in a so-called $N + 1$ (exceptionally $N + 2$) configuration. Here, the "1" and "2" refer to the number of standby channels. The number of channels N depends on the required transmission capacity of the route the maximum value of N depends on the chosen frequency band. An initial 1,800-channel 2 + 1 system operating in the 11-GHz band can, for example, be extended to 11 + 1

Table 5.5

Summary of Major Frequency Bands for Radio-Relay Transmission

Frequency Band (GHz)	Bandwidth (MHz)	CCIR Rec. No.	Channel Spacing (MHz)	Transmission Capacity		Number of RF Channels
				Analog Channels	Digital (Mbps)	
2	200	283	14	60–300	2, 8, 34	4–6
2/4	400	382	29	600–1,800	140/155	9
4	500	934	40	2,700	140/155	6
6.2	500	383	30	960/1,800	140/155	8
6.7	680	384	40	2,700	140/155	8
7	300	385	7	60–300	2, 8, 34	20
8	300	386	11.6	300–960	2, 8, 34	12
11	1,000	387	40	600–1,800	140/155	12
13	500	497	14	60–300	2, 8, 34	8
15		636	14		2, 8, 34	16
			28		140/155	8
18		595	5/7.5/27.5		2, 8, 34	188/125/35
			220		140/155	8
23		637	2.5		2, 8, 34	959
			3.5		140/155	685

without interrupting the installed system to cope with growth in the required transmission capacity, giving a total capacity of $11 \times 1,800 = 19,600$ channels.

5.3 RADIO-RELAY SYSTEMS

5.3.1 Definitions

Radio-relay transmission systems are classified by the CCIR into three categories; see Table 5.6.

In relation to the hop length, and consequently in relation to quality, radio-relay systems are either LOS systems, with optical or quasi-optical sight between the stations, or transhorizon (troposcatter) systems, whereby forward

Table 5.6

Radio-Relay Capacity Classification

Category	Analog (Number of Channels)	Digital Capacity (Mbps)
Small capacity or narrowband	up to 120	up to 10
Medium capacity	120–300	10–100
High capacity	above 300	above 100

scattering in the tropospheric layers enables the radiated RF energy to cover large distances beyond the horizon.

With regard to the mode of radiation, and consequently the number of stations receiving the same signal, radio-relay systems are either PP for stations in cascaded hops or P-MP where one central station communicates simultaneously with a number of surrounding stations.

With regard to the application, radio-relay systems are either public systems or subscriber systems. The public systems are used as transmission systems within the PSTN, and the subscriber systems connect single (groups of) subscriber(s) with the public network.

Subscriber systems can be divided into three groups:

- Urban broadband systems;
- P-MP systems;
- WLL systems.

With regard to installation, the radio-relay systems are either for *fixed operation* from a permanent site or *transportable* for operation from temporary sites. Transportable radio-relay systems are used for reporting and relief applications, which have in common that the equipment needs to be accommodated in easily transportable (usually weatherproof and shockproof) boxes, using small antennas (thus relatively high transmitter RF outputs), with convenient RF channel setting and simple hop alignment so that the system can rapidly go into operation.

As a performance yardstick, LOS radio-relay systems must comply with the *hypothetical reference connection* (HRX), with a length of 27,500 km (Recommendation G.801 for analog transmission and Recommendation G.821 for ISDN transmission), which CCITT has defined as the longest reference for worldwide transmission systems. In the case of transhorizon systems, CCIR has defined a shorter reference connection of 2,500 km (Recommendation 396). Figure 5.9 shows the 27,500-km HRX according to CCITT Recommendation G.821.

This HRX is divided into three classes of transmission quality; see Table 5.7.

The various bands of the frequency spectrum mentioned in Section 5.2 are globally allocated to radio-relay systems of the three quality classes as indicated in Table 5.8.

The quality of radio-relay routes depends on the equipment, system engineering, and route length. For analog systems, the noise level measured in a baseband telephone channel is taken as the quality criterion. According to CCIR Recommendation 395, on a route of length L km, used for telephony, the noise measured as a one-minute mean power in picowatts (pW), should not exceed

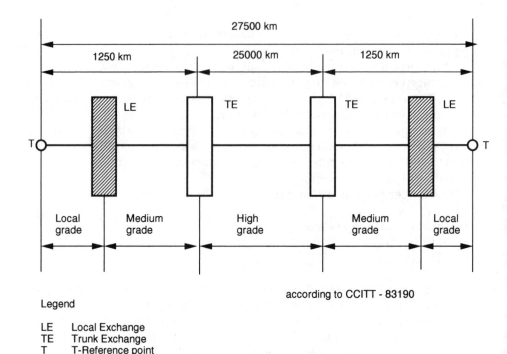

Figure 5.9 Hypothetical reference circuit for ISDN networks.

Table 5.7
Radio-Relay Quality Classification

Transmission Quality	Main Application
Local grade	Between subscribers and local exchange (including P-MP subscriber systems)
Medium grade	National networks
High grade	International networks

in any month 3 pW/km for more than 20% of the time or 47,500 pW for more than $L/2,500 \times 0.1\%$ of the time.

For analog transhorizon radio-relay routes, CCIR Recommendation 593 states that noise levels, which are considerably less stringent than for LOS systems, should not exceed 10 pW/km for more than 20% of the time or 63,000 pW for more than $0.5\ L/2,500\%$ of the time.

For radio-relay routes carrying television, CCIR Recommendation 555 recommends that the permissible noise, expressed as the ratio of the nominal

Table 5.8
Radio-Relay Frequency Band Allocation

Transmission Quality	Frequency Band (GHz)
Local grade	0.3–3 and 10–60
Medium grade	1–3 and 7–15
High grade	4–6

amplitude of the luminance signal to the *root-mean-square* (rms) amplitude of the weighted noise, should not exceed 57 dB for more than 20% of a month or 45 dB for more than 0.1% of a month.

In the case of digital systems, the noise performance is of lesser importance, so the quality criteria are defined in relation to the BER in a 64-Kbps channel. CCIR follows the quality objectives defined by CCITT in Recommendation G.821 for digital routes that may form part of an ISDN network with the following three criteria:

- *Degraded minutes* (DM) with a BER 1×10^{-6};
- *Severely errored seconds* (SES) with BER 1×10^{-3};
- *Total errored seconds* (ES) time interval of 1 sec during which a given signal is received with one or more errors.

For these three radio-relay performance quality grades, CCIR has recommended that in any month the quality—defined in DM, SES and ES—should not exceed the percentages of time outlined in Table 5.9.

An additional quality criteria defined by CCIR in Recommendation 557 is the *availability* of a radio-relay route or the *unavailability,* which is defined as the time without a signal or when the signal is of unacceptable quality. In the case of analog radio-relay routes, unavailability is defined as the percentage of time during which, for at least 10 consecutive seconds, the level of the baseband falls by 10 dB or more from reference level or, for any telephone

Table 5.9
Minimum-Quality Criteria for Digital Radio-Relay Routes

Quality Grade	CCIR Rec.	DM (%)	SES (%)	ES (%)
Low grade	697	1.5	0.015	1.2
Medium grade	696	1.5	0.04	1.2
High grade	594	0.4	0.054	0.32

channel, the unweighted noise power with an integrating time of 5 ms is greater than 10^6 pW.

The *unavailability* of digital radio-relay routes is also defined in Recommendation 557 as the percentage of time in which, for 10 consecutive seconds, the digital signal is interrupted or the BER in each second is worse than 1×10^{-3}.

Radio-relay systems are also classified into two groups, depending on their application. Applications such as transport, long-distance systems with numerous repeaters, or short routes within a long-distance cable or satellite transmission system consist the long-haul group. Urban access systems, with no or few repeaters make up a short-haul group.

5.3.2 Long-Haul Systems

Long-haul radio-relay systems are used for regional, national, and international networks. High quality is always essential for international networks and usually also for national networks. However, medium quality might be acceptable for some regional networks. Long-haul radio-relay systems are mostly LOS systems, although occasionally transhorizon radio-relay systems are used. Depending on the traffic volume, most long-haul LOS systems will have a high capacity, or at least a medium capacity. However, low-capacity long-haul systems are also used, for example, in less densely populated or less developed areas. The system configuration is almost always $N + 1$, and occasionally $N + 2$. In less developed countries, TV will quite often be carried on the protection channel.

5.3.3 Short-Haul Systems

Short-haul radio-relay systems are used in regional and urban public networks as well as in most private networks. They can be subdivided into five groups according to their application:

- Public access systems;
- Subscriber systems;
- TV and Broadcast P-MP distribution systems;
- Private systems;
- Transportable systems.

Short-haul systems are normally local grade quality, while the transmission capacity includes small-, medium-, and high-capacity systems.

In contrast to long-haul systems, which typically need detailed planning, special housings and power supply, large antennas, site-specific feeder runs, and high towers, short-haul systems are generally small, easy to install and maintain, and have a low power consumption. In particular, systems operating at 15 GHz and above usually consist of an indoor unit (rack-type or, for easy installation, already mounted in a cabinet) with the baseband circuitry and an outdoor unit consisting of an integrated antenna and the transmitter/receiver. The indoor and outdoor units are connected by a multipair cable carrying the IF signal (or, in the case of direct modulation, the baseband signal), supply voltages, and supervisory signals. Typically, the outdoor unit is housed in a compact lightweight weatherproof cabinet suitable for roof, wall, or mast mounting and designed for environmental conditions ranging from −30°C to +60°C.

Single-channel units can be used on one-hop terminal-to-terminal routes without repeater stations, thus eliminating the loss introduced by RF branching. Standby equipment, if justified at all, will operate in a hot-standby mode with a second single-channel unit or dual-channel units in 1 + 1 operation with dual polarization on a common antenna. The sizes of the integrated antennas range from 15 to 60 cm. In a few cases where longer hops or exceptionally high interference protection are required, external antennas, connected via a short feeder with the outdoor unit, are used.

Compact highly integrated short-haul radio-relay equipment can be connected directly to the mains power supply. A single-channel both-way unit with an RF output power of 0.1W will typically consume 20W to 40W. Because it is designed for installation and operation by nonspecialist users, it requires virtually no maintenance and can easily be relocated.

5.3.4 P-MP Systems

P-MP systems are mainly used to connect scattered populations to the public network, although they are occasionally also used to distribute TV and audio broadcast programs as mentioned under short-haul systems.

Exceptionally, for reasons of low cost and long range for small networks, P-MP subscriber systems for telephony, data, and video transmission use frequency bands below 1 GHz. However, most of the more sophisticated systems operate (in compliance with CCIR Recommendation 701) in the 1.5-, 1.8-, 2.0-, 2.2-, 2.4-, and 2.6-GHz bands, particularly in rural areas without frequency congestion problems. In urban areas, because of frequency congestion, the 10.5-, 12-, 19-, 23-, and 26-GHz bands are increasingly being used. The frequencies below 20 GHz are preferred in rainy areas, while the frequencies above 20 GHz are used in industrialized areas to further escape from the congested frequency spectrum.

In P-MP systems (see Figure 5.10), a central station serves a number of outstations located in LOS around the central station. If longer distances need to be covered, as is generally the case in rural areas, repeater stations can be added at approximately 40-km intervals up to a maximum distance of 500 km or, with some systems, even 1,000 km. The central station uses an omnidirectional antenna, or one or more sectorial or directional antennas directed toward the outstations. A repeater station operates as an outstation in the "upstream" direction (toward the central station) and as a central station in the "downstream" direction (toward the outstations and more distant repeater stations). Repeater stations use a directional antenna directed toward the central station together with an omnidirectional or sectorial antenna(s) covering the subscribers living in the surrounding area and or a further directional antenna toward a next repeater station. Outstations use directional antennas directed toward either the central station or their repeater station.

P-MP systems are a much more economical method of providing communication to scattered populations than cable or individual HF radio systems because a central station radiates a small number of channels that can be used by the subscribers on a shared *multiple access* (MA) basis, and outstations serve a number of subscribers living in the vicinity, so the investment and operating cost of the outstation can be shared by these subscribers. Cable or open wire only needs to be used for the short distance between the outstation and the premises of each subscriber.

The first P-MP systems introduced in the early 1970s were analog and operated in *frequency division multiple access* (FDMA) mode, meaning that each central station required a separate transmitter/receiver for each telephone channel (for example, eight radio transmitter/receivers for 64 subscribers). This was obviously still an expensive solution that found only a few applications.

Digitalization represented a real breakthrough for P-MP subscriber systems. Instead of using FDMA, it became possible to use *time division multiple access* (TDMA), which meant that the central station could use a single transmitter/receiver with only one RF channel to transmit a time frame with separate time slots for a number of telephone channels, generally 32 time slots in a 2-Mbps frame for some 400 subscribers or 64 time slots in a 4-Mbps frame for up to 2,000 subscribers. Some of the time slots can be assigned to 64-Kbps (or less) data lines and for common system use like signaling and supervision at the expense of a lower number of telephone subscribers or a higher blocking probability.

The distance between an outstation and the corresponding central or repeater station will range between 20 and 70 km, depending on topographic conditions. An outstation will be located in the middle of a group of subscribers so that the required two-wire subscriber connections can be kept as short as possible. An outstation can also serve coin boxes and low-speed data users.

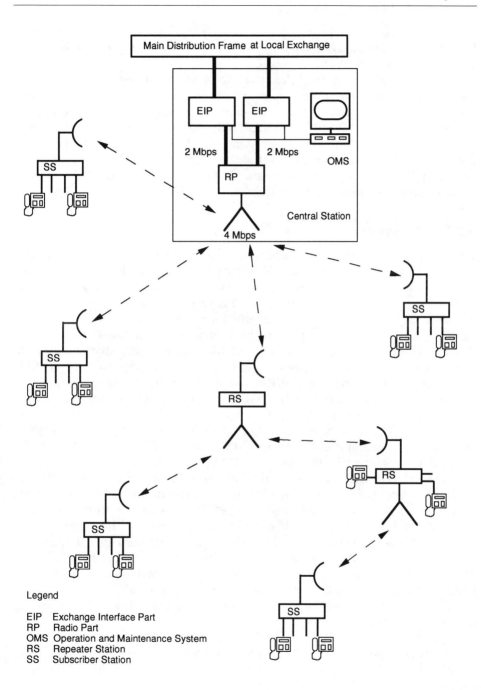

Figure 5.10 Typical P-MP network.

Specific lines can be allocated permanently to priority users like doctors, police, and the fire brigade.

Typical radio-relay facilities, like space diversity and 1 + 1 protection switching between central and repeater stations, can be implemented in P-MP systems, as can the use of a service channel on all stations and automatic network supervision. P-MP systems are radio-relay systems with a concentration function for access to a public exchange, but (apart from an intracall facility enabling communication between subscribers connected to the same outstation) they do not include any telephone switching functions. Consequently, they can be connected transparently to most existing public telephone exchanges.

5.3.5 WLL Systems

Wireless Local Loop (WLL) systems (also called RITL and RLL) are mostly also P-MP systems. Although an exact boundary is difficult to draw, it is probably best to say that WLL systems are short-distance systems connecting subscribers by radio with a *nearby* exchange (up to 10 km away) or *outside line plant* (OLP) access cable termination point (typically 100m away), whereas P-MP systems connect subscribers *far away* from the exchange (10 to 1,000 km) by radio directly with the exchange. In the case of this relatively long-distance application it is not usual to talk about a "loop," so this is not the domain of WLL. Moreover, most subscribers in a P-MP system still need a (local loop) cable access to their outstation. In other words, a WLL system is part of an access network OLP, whereas a P-MP system replaces an OLP.

Another distinction is that whereas WLL is (or will shortly become) B-ISDN capable, rural P-MP systems will remain N-ISDN (if ISDN at all) while urban P-MP systems will integrate into WLL.

Present WLL systems are mainly based on cellular radio systems—the so-called fixed-cellular systems—and thus operate in the 450-, 800-, 900-, 1,800-, and 1,900-MHz bands or on rural P-MP systems operating in the 1.5-, 2.4-, 2.7-, and 3.5-GHz bands. These adapted systems, still incorporating the frequency reuse concepts but without the costly technology for handing over moving subscribers from cell to cell, are being used to provide large numbers of potential subscribers with a fixed telephone service within a short period of time. Omnidirectional and sectorial antennas, typically 60 degrees, are installed at the central stations, whereas the subscriber stations can use simple directional antennas.

Apart from providing an excellent transmission quality at the earliest convenience those systems still provide a local terminal mobility since communication is possible within the confines of the cell in which the subscriber is located.

In addition to providing quick cost-effective connections to the PSTN, the future for WLL lies in the rapid introduction of B-ISDN local access by radio covering the last few hundred meters in an otherwise optical-fiber network. Ideally, this equipment will operate in the 60-GHz oxygen absorption band with (typically) 100-m hops without overreach, allowing almost unlimited frequency reuse with the minimum of coordination.

To prevent having to dig up roads and the access to buildings, which easily represent up to 90% of the cost of providing a cable connection between a subscriber and the local exchange, *radio distribution points* (RDP, also known as base stations in line with cellular practice) can provide a cost-effective solution for the last 100m from an optical-fiber cable termination to the subscriber. Such a WLL application is shown in Figures 4.19 and 4.20. The RDP can be equipped with a 60-GHz P-MP terminal operating in broadband TDM mode; each building or individual subscriber will have its own outstation operating in TDMA mode, as described for the P-MP systems. Outstations can be equipped with small, say 15-cm, antennas integrated with the miniature transmitter/receiver, built using MMICs connected to the subscriber's video-phone or multimedia terminal. Small passive repeaters, which are very efficient at higher frequencies, can then be used to connect subscribers who do not have a direct LOS with their RDP (possibly located around the corner).

Another application of WLL is the *multichannel multipoint distribution service* (MMDS), also called *local multipoint communications system* (LMCS), as wireless and "cableless" CATV used by over four million subscribers worldwide. MMDS, initially conceived for wireless TV distribution to subscribers located around a central radio base station that is connected to a CATV head end (see also Figure 4.22), is emerging as an integrated video, voice, and data distribution system to be used in the access network in competition with vested telecommunication operators. Present systems are mainly operating in the 2.5- to 2.7-GHz band as well as in the 12- and 40-GHz bands.

5.4 SPECIFIC APPLICATIONS

Radio-relay transmission provides the best solution for applications where the terrain or other conditions limit the use of optical fiber as well as in situations where traffic is too dense and the distance not long enough to justify the use of satellite transmission. The major applications for radio-relay transmission can be summarized as follows:

- Long-haul routes for national and international networks in difficult terrain;
- Dual-routed national networks with a radio-relay network parallel or meshed with an optical-fiber cable network;
- Digital backbone routes in countries where the infrastructures of the existing analog network can be used;
- Urban access routes between interurban optical fiber cable routes and the in-town terminal station (normally the telephone exchange);

- Emergency situations where the station locations have to be changed or replaced at short notice;
- Short-term project implementation following a national disaster, or in the event of war, or reconstruction after a war;
- Moderate density, distance, and growth rates;
- Semistationary operation in which the location may need to be changed from time to time;
- Links for pipelines, electricity and water distribution networks, railways, security networks, and other nonpublic users, especially where a transmission network is also needed during the construction phase of the distribution or security network;
- Access links from public to cellular radio networks and interconnection of cellular cells;
- WLL wherever road digging will be prohibitive for the use of cable;
- P-MP operation, for rural areas, TV and broadcast distribution, data collection and distribution in urban areas, and low-capacity corporate networks;
- MMDS operation for local TV distribution as well as integrated video, voice, and data distribution system in the access network.

Generally radio-relay transmission will be the preferred solution under the conditions outlined in Table 5.10.

Table 5.10
Typical Conditions for Radio-Relay Transmission

Subject	*Typical Condition*
Transmission capacity	Low, medium, high, and not very high
Distance	Short and medium, not very long
Geology	Mountainous, jungle, and earthquake areas. Not over lakes and seas or marshy land unless a cable solution would be even more expensive
Infrastructure	Hardly existing
Geography	Rural and urban continental, not transocean
Project implementation	Short time
Special circumstances	International events, emergency use after natural disasters, reconstruction during/after war or occupation
Operation begin	Earliest, thus justifying higher initial investment
Right-of-way	None or difficult to obtain
Vandalism and terrorism	Prevailing, therefore easier to protect a few radio-relay sites rather than cable routes
Network availability	Highest, justifying a radio-relay network as back-up for a cable network

Mobile Radio **6**

6.1 INTRODUCTION

After Heinrich Hertz proved in 1873 the existence of electromagnetic radiation, the base for all radio applications, it was the Italian physicist Guglielmo Marchese Marconi (1874–1937) who invented the radio. As a 21-year-old student at the University of Bologne, he constructed his transmitting and receiving equipment and took a first significant step by successfully transmitting radio signals over a few kilometers in 1896. Due to a lack of interest in his invention by the Italian Ministry of Post, Marconi left for the United Kingdom, where he demonstrated his radio between two Post Office buildings in London on July 2, 1896, and where he could file his first patent in the same year. This gave birth to wireless telegraphy, which could use long, medium, and short waves but not yet the very short waves required for radio-relay and satellite transmission. Marconi with his radiotelegraph successively succeeded to transmit signals over 16 km across the Bristol Channel in 1897, to bridge the English Channel in 1899, and to transmit the Morse signal "S" from St. Johns in Newfoundland on the American continent to Poldhu in Cornwall, England (at a distance of 3,500 km) in 1901.

HF-radiotelephony and radiotelegraphy were the major means of international long-distance communication in the 1920s to 1950s. Regular intercontinental HF-radiotelephone service was established in 1929 between New York and London and between Paris and Buenos Aires. HF-radio, also called short-wave radio, operates in the 3- to 30-MHz frequency band.

Short waves reflected in the upper atmosphere at an altitude of 80 to 100 km in the ionosphere, also called Heaviside layers, have the advantage that they can cover very long distances, even traveling around the world, without using any repeater stations. The quality of the transmission, however, is extremely variable, depending on time (day/night as well as season) and other ionospheric conditions, which vary, for example, as a function of sunspot activity. Furthermore, short-wave radio has a transmission capacity of just a few channels and can handle low-speed data only, which, therefore, is their typical domain of application.

Modern short-wave radio equipment has significantly improved the manner in which inconveniences of the unstable multiple reflection in the ionosphere are handled. The application of microprocessors supports the use of memory circuits to store frequency mode and gain control settings for a number of channels. Synthesizers operating in steps as low as 1 Hz, combined with extremely high receiver sensitivity, and steep filters reject disturbing adjacent signals. Thus, relatively inexpensive equipment can be manufactured for worldwide communication with no other infrastructure than a 12-V accumulator and a support for an aerial. Marconi's original invention of direct (unrepeated) radio contact between two distant points anywhere on the globe realized with his HF radio, in spite of satellites, has survived. HF-radio is still used the world over on ships, remote jungle outposts, oil rigs, and by Embassies and military representations that do not trust their communications to the networks of their host countries.

Beyond marine and intercontinental radio, Marconi developed his first mobile car radio in 1901. A steam-driven wagon was equipped with a transmitter, a receiver, and an approximately 5-m high cylindrical antenna mounted on the roof.

The first mobile radiotelephone service on land was set up by the Detroit Police Department in the United States in the early 1920s. By the late 1920s large and expensive car radios came on the market for private use in the United States. Paul Galvin, owner of a small company that made devices enabling the use of mains electricity by battery-powered radios, challenged his engineers to design a simpler car radio for the mass market. They succeeded in 1931; to celebrate their success, Galvin linked motion and radio, re-christening his firm "Motorola," still a major manufacturer of mobile radio systems. Soon corporate American cars were equipped with two-way radios; however, commercial service was not offered to the general public until 1946.

In the meantime, mobile radio found worldwide application in various independent *private mobile radio* (PMR) networks operating in the 80-, 160-, and 400-MHz bands mainly for

- Police, ambulance, fire brigade, and other security services;
- Taxi and other transport organizations;
- Service networks for utilities like gas, water, and electricity production and distribution as well as for railways and highways.

All these networks include mobile, handheld, and fixed transceivers and special PABXs, some of them with access to the PSTN. Thanks to the substantial technical progress with digitalization and computerization, those networks are becoming increasingly sophisticated and offer direct dialing, data transmission, and intelligent network features.

A special application of mobile radio communication, operationwise ranking between private mobile radio networks and cellular radio, is mobile radio trunking. With trunking it is understood that a small number of radio paths is shared by a large number of users of various private companies. Trunking became available in the late 1970s mainly to reduce the parallel operation of networks within the same crowded frequency bands. Trunking offers the advantage of infrastructure sharing (fixed radio stations and network control) and less sophistication than cellular networks (for example, no roaming and cell-handover).

The ultimate in trunking is presently being introduced as a digital international system standardized by ETSI under the name *Trans-European Trunked Radio* (TETRA), which, just as the European digital cellular system GSM (Global System for Mobile Communications), might evolve to a world standard.

A major development in mobile radio that is causing an enormous boost in telecommunications is currently happening with the emerging worldwide coverage of digital cellular radio. Cellular radio, even in its first analog version, is the most effective way of mobile radio. The radio coverage of large geographical areas is provided by seamless overlapping cells in a honeycomb configuration. Cellular radio networks provide the same, and even more versatile, services to mobile users as obtained at fixed locations from the conventional PSTN. Cellular radio is the fastest growing public telephone service. In the industrial world, cellular radio offers the highly mobile user continuous accessibility whether on foot or traveling by various modes of transport. In the developing world cellular radio offers a quick low-cost replacement for poor-quality fixed telephone networks, thus solving the problem of long waiting times for network connection.

By the time this book is published the first mobile cellular satellite systems (MSS, a subject of the next chapter) will be put into operation, thus extending cellular radio coverage to virtually any place on this globe [1–3].

6.2 MOBILE RADIO TECHNOLOGY

Mobile-radio transmission is the technology of generating, modulating, amplifying, and multidirectionally transmitting a modulated very high frequency through the atmosphere from a usually fixed central station to a number of mobile stations that selectively receive, amplify, and demodulate the desired signal, operating similarly the other way round from mobile to central station. Operation from the central or base station to a number of mobile stations thus is in P-MP mode, whereas operation from the mobile stations to the central station is in P-P mode. The waves used in mobile-radio transmission have a frequency ranging from 30 MHz to about 3 GHz. Basically, the same applies for mobile radio as for radio relay (as explained in Chapter 5) when considering radiation propagation and atmospheric attenuation of electromagnetic waves. Similarly, specific radio-relay techniques such as diversity, equalization, interference cancellation, error correction, and automatic RF output power adaptation are also applied for mobile-radio systems. Specific mobile radio technology, developed to provide orderly communication with and between numerous moving stations, mainly concerns:

- Service modes;
- Network types;
- Antennas;
- Power supply;
- Network trunking;
- Cellular radio, with its specific technologies concerning cell configuration, multiple access, time advance control, handover, roaming, frequency hopping, transport infrastructure, data transmission, and cellular radio networks;
- Cordless telephony;
- PCN;
- Paging.

In addition to these technologies, substantial technical and administrative efforts are required to protect mobile radio transmission against fraud. Transmission on optical fiber is practically immune against tapping. Radio-relay transmission can only be tapped at the high expense of radio-relay and multiplex equipment connected to an antenna located in the radio beam on a difficult disguisable tower. Mobile radio transmission can easily be tapped, however, so fraud protection devices are a necessity in mobile radio operation.

The frequency bands for mobile radio transmission have been allocated by WARC and are defined by CCIR as summarized in Table 6.1.

6.2.1 Service Modes

Communication between two stations of a mobile radio network requires forward (P_1) and reverse (P_2) paths, as shown in Figure 6.1. The radio path from a base station to a mobile station is usually referred to as downlink or forward link, while the path from a mobile station to a base station is referred to as uplink or reverse link.

In the "simplex" service mode the same frequency is used for uplink and downlink. An RF switch, activated by a "push-to-talk" key (integrated in the

Table 6.1
Frequency Bands Allocated to Mobile Radio

Frequency Band (MHz)	Major Application
(26.965–27.275	Citizen band)
29.7–40.7	PMR
68–87.5	PMR, Paging
87.5–108	Broadcasting, partially for paging
138–174	PMR, ERMES, POCSAC
174–230	Broadcasting Band III, partially free for PMR (such as 175.5 to 207.5 MHz)
380–400	TETRA public security (Europe)
410–430	PMR, TETRA, RadioCom 2000, POCSAC
450–470	PMR, TETRA, NMT 450, C 450, POCSAC
820–900	(D-)AMPS, CT2
890–980	NMT-900, TACS, GSM
1,600–1,800	TFTS
1,700–1,900	DECT, DCS 1800, PHS
1,900–2,200	PCS 1900, IMT–2000, UMTS

telephone handset), connects the antenna to the transmitter during speaking and to the receiver during listening at that station. With this sequential on/off communication, only one station can transmit at a given time but can be heard by all other stations of the network. Disciplined talking, thus, is a must for the simplex mode.

In the "duplex" service mode separate frequencies are used for uplink and downlink. The RF switch is replaced by an antenna-branching filter, the push-to-talk key can be eliminated, and two-way conversation is possible (similar to the conventional PSTN telephone) whereby the user can simultaneously speak and listen.

Semiduplex operation differs from duplex operation in such a way that one station, usually the central station, is equipped with an antenna-branching filter whereas the second station, usually a mobile station, is equipped with the less-expensive RF switch. Both simplex and semiduplex require the same speaking discipline. Sometimes an additional distinction is made between duplex and full-duplex in such a way that duplex refers to the use of an RF-antenna switch, two adjacent frequencies, and the on/off method of transmission. The term "full-duplex" is then used for the method described for "duplex" with two well-separated frequencies enabling the use of an antenna branching filter for simultaneous transmission and reception (usually one frequency in the upper band and the other in the lower band of the allocated frequency spectrum).

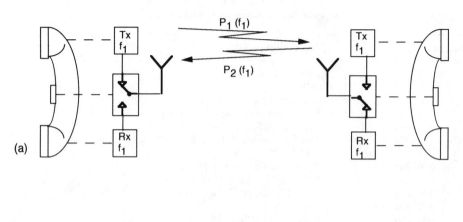

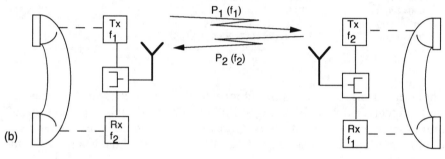

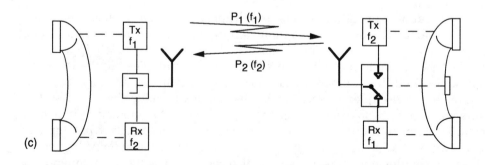

Figure 6.1 The three service modes: (a) simplex, (b) duplex, and (c) semiduplex.

The simplex and semiduplex service modes are still widely used in PMR networks. The (full-) duplex service mode is used for all mobile radio networks that are interconnected with the PSTN (for example, trunk radio and cellular radio). With digital systems, another version of duplex, called *time division duplex* (TDD), can be applied whereby uplink and downlink signals of each channel are on one single radio carrier (thus one frequency only per two-way channel) in different time slots.

6.2.2 Network Types

Mobile radio stations can communicate between themselves and with common base stations in mobile radio networks that are either

- Linear;
- Star-shaped;
- Meshed.

Linear mobile radio networks are used for maintenance troops for pipelines, railways, and HT-lines, for example. A number of fixed stations are located in such a way that radio coverage for the mobile stations is secured all the way alongside the pipeline or any other linear object. Figure 6.2(a) shows a linear network operating in the simplex service mode, which is one of the simplest network configurations whereby only one frequency is needed for traffic between the fixed station, between the fixed and mobile stations, and between mobile stations. The fixed base stations of linear networks normally are equipped with directional antennas; the mobile stations in any type of network are always equipped with omnidirectional antennas.

A linear network operating in duplex service mode is shown in Figure 6.2(b). Frequency reuse is normally applied using the same frequency pair on each over-next fixed station. The mobile stations are equipped with multichannel units so that they can be switched over to the different frequency pairs of the fixed stations along the network. The base stations are connected to a mobile network exchange via a telephone circuit usually leased from the public network. This mobile network exchange might be interconnected with the PSTN, in which case the mobile stations can be connected with the PSTN through their next base station. Duplex service between the mobile stations is possible with the assistance of the mobile network exchange; for example, M1 communicates with M3 through B1, the exchange, and B3. Direct communication between two nearby mobile stations is possible in the duplex mode too, as shown for M3 and M4 when the calling mobile station reverses its transmitter and receiver frequencies.

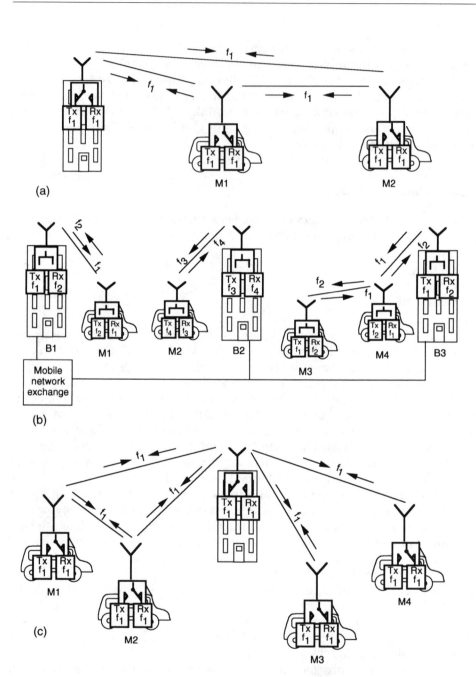

Figure 6.2 Typical network configurations: (a) linear network simplex, (b) linear network duplex, and (c) star network simplex.

A star network is commonly used if an area around a base station must be served. Figure 6.2(c) shows such a network operating in the simplex mode using one frequency only. Meshed mobile radio networks are used if large geographical areas must be served and communication between the mobile users throughout the network is required. An example of an intermeshed mobile radio network operating in the duplex mode is shown in Figure 6.3. The base stations, in addition to a transmitter/receiver serving their own area, are equipped with a second transmitter/receiver for communication between all base stations of the network on one common frequency pair (f_9/f_{10}). Frequency reuse will be applied wherever the geographical conditions allow, as, for example, for the base stations B1 and B3, assuming a different position for the two antennas of the same fixed station so that direct communication between the two base stations is possible but B1 cannot interfere with the mobile stations operating around B3 and vice versa. To enable simplex operation between mobile stations, those stations (such as M1 to M4) are equipped with a second receiver on the same frequency as the transmitter. Whereas the mobile stations M1 to M4 can communicate with each other and, if equipped as a multichannel unit, with all base stations, mobile station M5, for example, for restricted operation, is equipped for communication with its base station only.

To facilitate orderly communication with a minimum of mutual interference and a good degree of secrecy, in large networks (also in large linear and star networks) usually a selective calling procedure is implemented whereby only a called station can use the relevant operating frequency pair and a mobile station can initiate a call only if the network is free [4].

6.2.3 Antennas

The mobile stations mainly use quarterwave or (for better performance) 5/8-λ vertical polarized dipole rod antennas. Shorter-than-quarterwave whip antennas are obtained by either using a loading coil or bending the antenna conductor itself in a coil-like helical shape supported by a usually *glass-reinforced plastic* (GRP) or polyethylene insulator. Longer, higher gain antennas are the collinear rod antennas consisting of two interconnected quarterwave radiators one above the other. Multiband antennas enable the simultaneous operation of both the cellular and PMR services. Typically a 55-cm multiband whip antenna can be electrically tuned to the 450- and 900-MHz RF bands of cellular systems and the 80- and 160-MHz bands of PMR systems. Band filters then separate the frequency bands and reduce RF interference.

The base stations may use several antenna constructions:

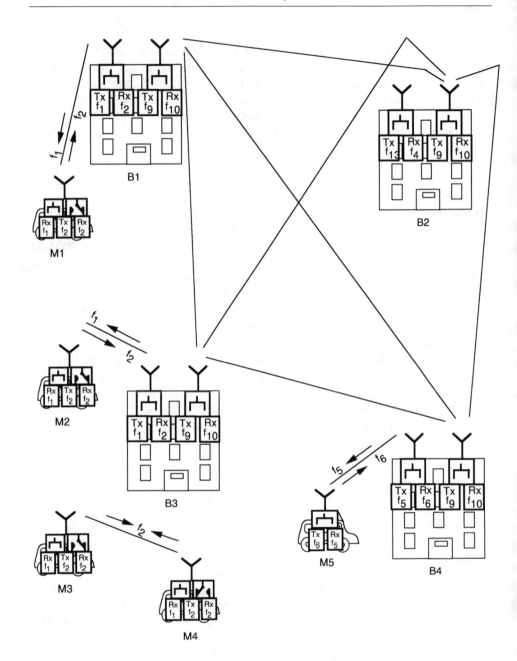

Figure 6.3 Meshed network.

- Omnidirectional vertical polarized rod antennas;
- Omnidirectional vertical polarized collinear antennas, using a number of quarterwave radiators above each other;
- Yagi antennas, with directing and reflecting rods;
- Stacked dipoles, with a number of dipoles above each other, facing in the same direction for directional antennas or circularly placed for omnidirectional antennas;
- Corner reflector antennas, used as simplified Yagi antenna;
- Reflector plate antennas;
- Helical antennas, with one or more coil-like radiator(s) in parallel;
- Log-period or end-fire antennas, with a number of reflector elements placed in front of each other resulting in a strong directional radiation.

Modern base station antennas widely use, and handsets too appear on the market using as an antenna, a *printed circuit board* (PCB) with arrays of stripline dipoles etched onto a board. The PCB is then housed in a suitable RF transparent weatherproof housing; or in the case of a handset a flat patch antenna is integrated into the housing, the handset itself assuming part of the antenna functions.

To improve antenna performance beyond their inherent physical limits, various antenna enhancement techniques are applied, such as

- Masthead electronics;
- Polarization diversity;
- Field component diversity;
- Pattern diversity;
- Smart antennas.

6.2.3.1 Masthead Electronics

Masthead electronics, as the name implies, consists of electronic circuitry located nearest to the antenna base preferably integrated in the antenna housing. This circuitry uses wideband low-noise receive amplifiers to improve the performance of the uplink from mobile to base stations, and power boosters to improve the downlink performance.

6.2.3.2 Polarization Diversity

In addition to space diversity requiring two physically sufficiently separated antennas and frequency diversity requiring two sufficiently spaced frequencies for the transmission of the same signal (as described in Chapter 5), similar improvements can be obtained with polarization diversity. The polarization of the received primary and the reflected signals normally will be different, so they can be received on two antennas: one horizontally polarized and the other vertically polarized. The two antennas do not need to be physically separated and thus both can be accommodated in the same antenna housing. These so-called "di-polar" antennas thus need much less space than conventional space diversity antennas. The di-polar antenna can especially be used in urban environments with heavy reflection from large buildings.

6.2.3.3 Field Component Diversity

Electromagnetic waves, as the name indicates, consist of electrical waves and magnetic waves. The two waves propagate in two different fields: an electric field and a perpendicular magnetic field. Field component diversity, also called energy density diversity, uses two colocated antennas—one to receive the electric field and the other to receive the magnetic field of incoming signals.

6.2.3.4 Smart Antennas

A smart antenna usually comprises an array of a relatively large number of individual small antennas, each of which can operate independently of the others. By electronically adjusting the relative phase of the signals applied to each array, the antenna as a whole can be made to point electrically in a particular direction without physically moving the array. With a static smart antenna, for example, the main output power and the maximum receive sensitivity can be directed in the main traffic direction, whereas a dynamic smart antenna dynamically points maximum power and sensitivity in the direction of the actual communicating mobile station. Even more sophisticated smart antennas such as the *spatial division multiple access* (SDMA) antennas allow the simultaneous use of any conventional (frequency, time slot, or code) channel by multiple users, none of which occupy the same directional location. An SDMA antenna performs an azimuthally selective processing, tracking the position of mobiles and determining, for example, frequency or burst period, to enable the multiple use of a single channel in different directions. In both supporting multiple conversations on a single conventional channel and allowing increased spatial reuse of channels, smart antennas with spatially directive transmission can significantly improve performance in terms of coverage, frequency reuse, reduced output power, reduced interference, and increased receive sensitivity.

Whereas such smart antennas can be used for base stations, compact planar polarization diversity antennas can be used as smart antennas for handsets, too. This smart antenna type is based on the principle that signal fades can be polarization-dependent and signals of pure vertical polarization at the base station can rapidly depolarize in dense built-up areas. A mobile station receiving a vertically polarized signal that is subject to a fade may receive a horizontally polarized component without fade. By dynamically changing the polarization characteristics then the probability of loosing a signal due to fading can be significantly reduced.

Another important advantage of the smart antenna for handsets is the higher personal protection against radiation. A rod antenna radiates its maximum of energy just in front of the most radiation-sensitive parts of the human body: the eyes and the brain. The human eye is particularly subject to thermal effects of radio wave absorption. The human brain is particularly subject to the internal currents generated by the electromagnetic field of the antenna. Handsets were introduced, therefore, with a flat planar antenna fitted in the rear of the handset to significantly reduce the radiation exposure toward eyes and brain. An even better solution—also protecting the hand from radiation—might be to accommodate the planar antenna not in the rear of the handset but inside a loop firmly attached to the rear of the handset leaving space for the hand between the handset and the loop.

The complex subjects of electrosmog and the detrimental effects of electromagnetic radiation go beyond the competence of this book and, therefore, are not further explained here [5–8].

6.2.4 Power Supply

For mobile radio transmission, even more than for the other transmission media, the power supply is a weak point, since batteries must be used not just for standby but for regular operation at variable load and, moreover, must be carried around. The battery should be small, light-weight, have a high *energy density factor* (EDF) combined with a long life and a low price. Those requirements have led to the exclusive use of rechargeable batteries in mobile applications in the following versions:

- *Sealed Lead-Acid* (SLA);
- *Nickel-Cadmium* (NiCd);
- *Nickel Metal Hybrid* (NiMH);
- *Lithium-Ion* (Li-Ion).

6.2.4.1 Sealed Lead-Acid (SLA) Batteries

Conventional SLA batteries are widely used for standby at base stations and for regular operation of vehicle-mounted mobile equipment. Their heavy weight prohibits portable use.

6.2.4.2 Nickel-Cadmium (NiCd) Batteries

The NiCd battery has been the dominant rechargeable battery for many years both for consumer and professional applications. The NiCd battery combines a high EDF with a constant discharge voltage (see Figure 6.4), a good cycle life between 500 to 700 cycles, and the ability to withstand storage in a fully discharged state. NiCd batteries, however, can suffer from a "memory effect," which refers to the feature of crystalline formation (with subsequent capacity reduction) of cells that are recharged before being sufficiently discharged. In order to retain their capacity, NiCd batteries should be fully discharged before recharging or otherwise be subjected to a regular "reconditioning" (slow, deep discharge to 1V/cell). The nominal voltage is 1.2V, which remains constant shortly before complete discharge. A battery pack usually contains 3 to 5 cells.

6.2.4.3 Nickel Metal Hybrid (NiMH) Batteries

NiMH batteries have similar characteristics as the NiCd batteries apart from an approximately 100% higher EDF and a slightly lower life of about 500 cycles after which the capacity falls to around 80% of its original value.

6.2.4.4 Lithium-Ion (Li-Ion) Batteries

Li-Ion batteries are a recent development and still have to prove that, compared with conventional NiCd batteries, an approximately 300% higher EDF, approximately double cycle life, and significantly lower weight, justify a three to five times higher price—especially because Li-Ion batteries have a steep discharge

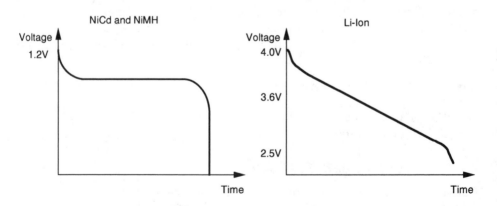

Figure 6.4 Typical voltage discharge curves.

voltage and require a more complex charging technique that has a direct impact on the battery performance. A constant charging current needs to be applied until the battery is charged to about two-thirds of its capacity; then the charger has to switch to constant voltage charging thereby steadily reducing the current until almost zero at full charge. The voltage at full charge is 4V but inconveniently reduces during discharge down to 2.5V.

6.2.4.5 Battery Choice and Management

From an environmental point of view, the choice of which battery to prefer is no easy matter. The NiCd battery contains the heavy toxic material cadmium; the NiMH battery contains oxides and hydroxides, which are suspected to be potential carcinogens; and even the newcomer Li-Ion contains organic solvents that must be handled carefully when the cells are disposed of or recycled.

To optimize the use of batteries, so-called battery management or "smart power" devices are being incorporated into mobile radio units. The smart or intelligent facilities range from simple charge control to sophisticated circuitry to provide fuel gauging (coulomb counting, constantly measuring input and output including self-discharge and leakage, and calculating the charge rate), charging control, and battery-to-host communication functions [9,10].

6.2.5 Trunked Radio

Conventional PMR networks require their own organization for network operation and either occupy part of the valuable frequency spectrum for their exclusive use or share a common radio channel with other networks and compete with one another for air time. In trunked networks, which usually include internal switching as well as interconnection with the fixed telephone and data networks, the subscribers share a common pool of channels that are allocated automatically for each call. When the call ends, the channel is returned to the pool and becomes available for others. Trunking thus offers a more efficient use of the frequency spectrum and wide-area coverage to subscribers who, moreover, do not need to care about investment, operation, and maintenance of the network. Trunked networks can be private networks set up by two or more companies for their common use or public trunked networks also called *public access mobile radio* (PAMR) networks for common use by public safety and emergency services, municipalities, utilities, transport companies, and service and maintenance organizations, for example.

Trunked radio networks are complementary to cellular systems. Cellular systems are aimed primarily at the individual user. Trunked radio networks offer corporate users specific mobile communication services like group calling, priority calling, direct mobile to mobile communication without infrastructure, limitation of access to PSTN or geographical coverage, fast call set-up, conference calls, call transfer, and status reporting. Trunked radio networks provide coverage mainly along roads, highways, waterways, and inside industrial and transport areas; while cellular radio provides coverage in large geographical areas by seamless overlapping cells combined with automatic handover between cells. In trunked networks charging is usually applied as a monthly fee based upon the subscribed services without an additional charge per call. A dynamic call duration limitation can adapt the permitted speech time to the actual traffic density.

A national open analog de facto standard for trunked networks developed by Motorola and Philips was issued by the U.K. *Department of Trade and Industry* (DTI) in 1990 under the term MPT 1327. Other de facto trunking standards are "Mobitex" for mobile data trunking developed in Sweden by Ericcson and SMARTNET developed in the United States by Motorola. A digital standard called TETRA was issued by ETSI in 1995. In the United States, where trunked radio is widely applied, it is called *specialized mobile radio* (SMR). The *Association of Public-Safety Communications Officials* (APCO) in the United States has issued their APCO-25 standard for digital trunking, especially for public safety communications.

6.2.6 Cellular Radio

Cellular radio, as the name indicates, provides radio coverage to *mobile stations* (MS) in large geographical areas by seamless overlapping cells (see Figure 6.5). Each cell has its own radio base station also called *base transceiver station* (BTS). A number of base stations are connected with a common *base station controller* (BSC). All BSCs of a coverage area are connected with a *mobile service switching center* (MSC), which performs the switching between the mobile subscribers and constitutes the interface to the PSTN/ISDN. The mobile subscribers traveling from cell to cell are automatically switched over to the base station of the next cell by a procedure called "handover" so that uninterrupted conversation is granted and no calls will be missed. The mobile subscriber automatically reports to the BTS in which cell he/she has arrived, and the MSC registers this in special registers so that calls to network subscribers can be routed to the appropriate cell where the called subscriber has reported its presence.

The first cellular mobile units were used in Japan in 1979; they still had a volume of 6,600 cm^3. The first portable unit introduced in 1987 was already in the 500-cm^3-volume range; currently, a cellular handy typically is in the

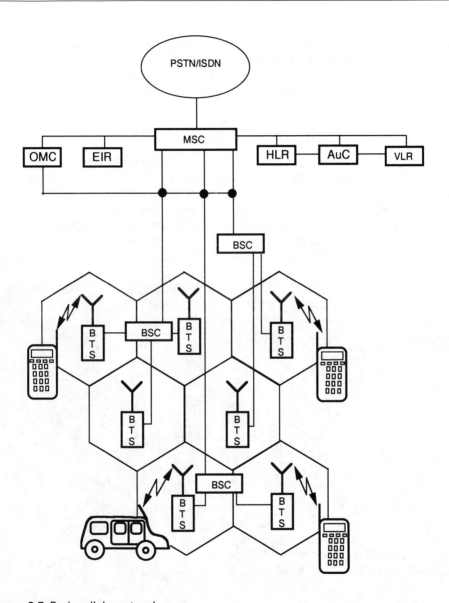

Figure 6.5 Basic cellular network.

150-cm^3-volume and 150-gram-weight range and has a battery capacity for 15 to 20 hr standby and 1 to 2 hr transmitting time.

The basic functions of the cellular network components are briefly explained in the following subsections.

6.2.6.1 Base Transceiver Station (BTS)

The BTS basically provides the air interface with the mobile terminals. The base stations in digital cellular networks emerged from indoor equipment mounted in a shelter on a specially prepared station site to a compact station integrating electronic equipment, antenna, and emergency power back-up into a small suitcase size weatherproof cabinet for simple wall mounting. Figure 6.6 demonstrates the compactness showing two complete masthead base stations (each with one transceiver) located at a corner of two cells, requiring an "infrastruc-

Figure 6.6 Compact cellular base stations (courtesy of Nokia).

ture" of a power outlet; a E1/T1 transmission line to the BSC; and a suitably located wall, pole, or other simple supporting structure [11].

6.2.6.2 Base Station Controller (BSC)

The BSC performs the radio resource management such as establishment and release of radio channels in response to MS or MSC request, switching over to a better radio channel if required during a call, and RF output power control of both the BTS and the MS, including voice activity detection and *discontinuous transmission* (DTX), to limit RF interference and increase the MS autonomy by saving battery power. The BSC, moreover, acts as a data server to their base stations.

6.2.6.3 Mobile Service Switching Center (MSC)

The MSC provides the switching and registration functionalities to connect the mobile subscribers to the PSTN/ISDN or to each other. Fundamental elements of a MSC are as follows.

Home location register (HLR) is the component in which each mobile subscriber of the coverage area of the MSC is registered with their significant static data such as access capabilities, subscribed and supplementary services, and dynamic data concerning the actual location of the MS. This information enables the MSC to route incoming calls immediately to the called mobile subscriber.

Visitors location register (VLR) stores all the information about nonregistered mobile subscribers who have entered the coverage area of the MSC. The VLR thus is a dynamic subscriber database, which performs intensive data exchange with the "visiting" MS as well as with the HLR of that MS. The data thus stored within the VLR "follows" the MS upon entry into an adjacent MSC area. This VLR data allows the MSC to set up incoming and outgoing calls with the visiting MS.

Authentication center (AuC) provides the authentication of a subscriber attempting to use the network. The AuC stores the authentication information and ciphering keys necessary to protect the operator and the mobile subscribers against unauthorized use of the network. Corresponding information is stored in the *subscriber identity module* (SIM) card, which the subscriber has to insert in a MS before it can be used. Instead of a SIM card, some cellular systems use a *user identification module* (UIM) or a *subscriber identity security* (SIS) card.

Equipment identity register (EIR) checks out the status of the MS's identity number. Each MS has its own identity information called *international mobile station identity* (IMEI). This number can be checked against "white" lists of authorized and "black" lists of unauthorized (such as stolen, cloned, and numerous other types of fraud) and "grey" lists of suspected mobile stations. Such color lists on a global scale, for example, are maintained for GSM terminals by the *Central Equipment Identity Register* (CIER) in Dublin.

Operation and maintenance center (OMC) takes care of the control of the system elements, upgrading of services, security management, and billing and accounting.

6.2.7 Cell Configuration

Seamless radio coverage in cellular networks can be obtained with a side-by-side arrangement of cells in a honeycomb configuration as shown in Figure 6.5. The cell configuration, apart from offering a seamless coverage, is determined by two fundamental requirements: minimum interference and maximum traffic capacity. The minimum interference requirement has led to special cellular cell cluster arrangements. The maximum coverage has led to a hierarchical cell structure.

6.2.7.1 Cellular Cell Clusters

A cellular base station equipped with an omnidirectional antenna is usually located at a central point of the cell or a BTS equipped with a sectorial or corner antenna is located in a corner of a cell. With the "sectorial illumination" three base stations are normally located at the common corner of three cells, thus reducing the number of required sites. To prevent interference from adjacent cells a group of frequencies is allocated to a specific cluster of cells. This cluster is repeated throughout the network, thus reusing each frequency on a nonadjacent cell. Clusters with 3, 4, and 7 cells are normal for omnidirectionally illuminated cells, whereas sectorially illuminated cells use 3-site/9-cell, 4-site/12-cell, or 7-site/21-cell clusters as shown in Figure 6.7.

A different approach is made with the concentric-cell cluster shown down-left in Figure 6.7. This 60-degree concentric sectorized cluster provides a very high capacity in the center, for example, in the central part of a city where the traffic density is highest. The capacity of the cells then gradually decreases outward from the densest traffic in the center. Base stations equipped with sectorial antennas are located on concentric circles. All frequency groups can be repeated on each circle but not in the same order so that an angle of 120 degrees is observed between cells using the same set of frequencies (for example, as indicated with the three black arrows in Figure 6.7).

Given the fixed number of radio channels in a cellular system, the choice of cluster type depends on the required traffic capacity and the permissible

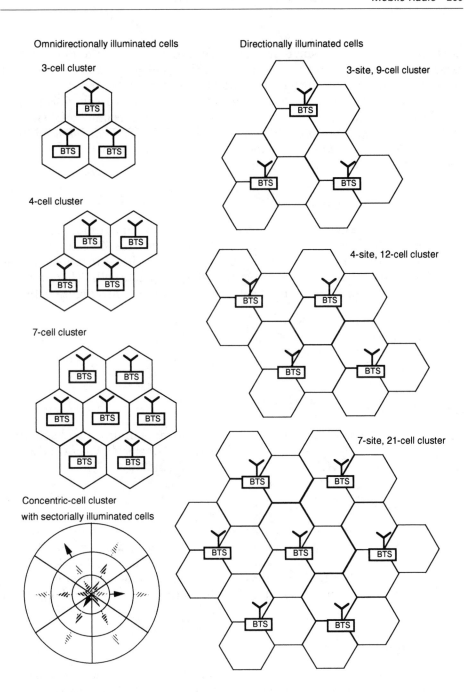

Figure 6.7 Typical cell clusters.

interference. Basically, a small pattern will be preferred for metropolitan application and a large pattern for rural areas. Most analog systems use a 21-cell cluster, whereas 9-cell clusters are widely used for digital cellular systems. Further efforts are currently directed toward developing a 1-cell cluster, thus a cell-to-cell frequency reuse, by combining complex technologies such as wideband frequency hopping, fractional loading, DTX, *code division multiple access* (CDMA), and *adaptive channel allocation* (ACA, automatically assigning channels based on the instantaneous interference situation).

6.2.7.2 Hierarchical Cell Structure

Cellular radio initially was developed to serve vehicles quickly moving through large geographical areas divided into relatively large cells. In the meantime, however, more and more pedestrians moving relatively slowly in crowded areas are using cellular radio, thus requiring small high-capacity cells. Therefore, the following hierarchical structure of cells has evolved (see also Figure 6.8):

- Standard, or macrocells;
- Microcells;
- Picocells;
- Extended cells.

6.2.7.3 Macrocells

Macrocells are the standard cells for cellular networks in urban and suburban areas serving fast-moving users. Macrocells have a diameter between 1 and 35 km and an RF output power of the BTS between 2W and 50W.

6.2.7.4 Microcells

Microcells are used in city centers, shopping malls, sports grounds, airports, railway stations, and campuses, for example, to serve slow-moving users. Microcells typically have a diameter between 100m and 1,000m and an RF output power below 1W. The BTS antennas are positioned below rooftop level. The radio propagation occurs at LOS, through diffraction, and by scattering around buildings.

6.2.7.5 Picocells

Picocells are applied in office complexes, underground parking, and subways, for example, to serve slow-moving users over distances of 10m to 100m. The BTS output power is typically 100 mW.

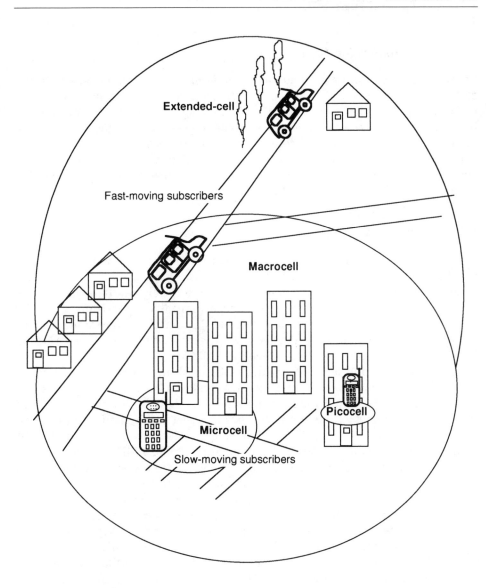

Figure 6.8 Hierarchical cell structure.

6.2.7.6 Extended Cells

Extended cells are used to provide coverage for fast-moving users over distances up to 70 km in rural areas, along railways and highways, and in coastal areas including off-shore locations. The RF output power of the BTS ranges between 30W and 100W.

6.2.7.7 Cell Application Modes

These cells types can be applied side-by-side or as stand-alone cells within a coverage area when additional capacity is required. In situations where there is sufficient traffic-handling capacity but coverage is inadequate, a macrocell BTS can operate as host for microcells and picocells to be located in those undercovered areas. Extended cells likewise can be used as stand-alone alongside macrocells, or in another concept one BTS can operate two groups of channels—one group for calls in the macrocell and the other group for the calls in the surrounding extended cell [11–13].

6.2.8 Multiple Access

To enable simultaneous conversations of several subscribers within a cell, in analog cellular radio systems FDMA (as explained in Subsection 5.3.4) is used whereby the BTS requires a separate transceiver for each simultaneous conversation within its cell. The number of transceivers thus depends on the traffic load and the grade of service to be provided. In digital cellular radio systems two different approaches are currently followed: TDMA/FDMA and CDMA. Figure 6.9 depicts both versions as well as FDMA.

The combination of TDMA with FDMA as shown in Figure 6.9 is based on the currently most used digital cellular system GSM. Each radio channel carries a TDM signal with eight time slots for eight traffic channels. Eight simultaneous conversations, however, hardly meet traffic load requirements; so several transceivers normally need to be operated in parallel, thus in FDMA mode [also called *multicarrier* (MC) mode]. For GSM a 25-MHz frequency band is available with a total of 124 RF channels each 200-kHz wide, resulting in a maximum of 124 FDMA × 8 TDMA = 992 (full-rate, 13-Kbps) channels, respectively, 1,984 half-rate (6-Kbps) channels per network. Each BTS can be equipped with up to eight transceivers, thus providing 64 channels. For the new DECT system the access method is defined with the term MC/TDMA/TDD, which stands for multicarrier-TDMA with *time division duplex* (TDD).

CDMA is a spread-spectrum technology derived from military application, where it is used to reduce the effects of deliberate jamming or other interference resulting from battle conditions. *Spread spectrum* (SS) increases the bandwidth of a digital signal to make it less vulnerable to interference. There are three spectrum spreading procedures, namely, *time hopping* (TH-SS), *frequency hopping* (FH-SS), and *direct sequence* (DS-SS). In DS-SS the information spectrum is spread into a bandwidth (also called chip rate) many times wider than the bandwidth of the user signal using a pseudo-random sequence of microbits. Each data bit is encoded with a binary sequence of up to several thousand microbits. The ratio of the chip rate to bandwidth or bit rate of the user signal

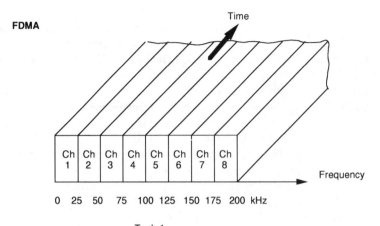

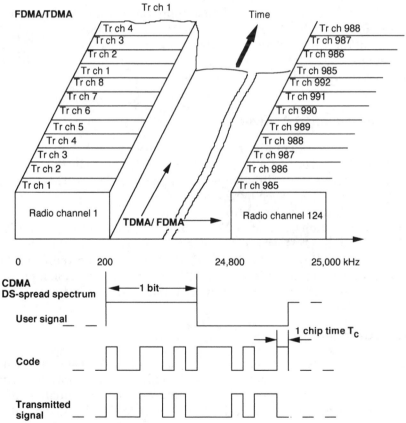

Figure 6.9 Cellular multiple-access methods.

is referred to as the processing gain. For each user a different code is applied so that several users can operate on the same frequency. The DS-SS access method thus is based on code division multiplexing and, therefore is commonly referred to as CDMA. A receiver using the allocated coding can recover the original information using the same binary sequence to decode the encoded data bits. All the other channels using another code are perceived by the receiver as noise. The capacity of a CDMA system, therefore, is determined by the signal-to-interference ratio; with CDMA there is no fixed capacity of a cell, but as the number of connections in a cell grows, the background noise increases and thus the quality for all users degrades. CDMA, due to DS-SS, tolerates a higher path loss, thus permitting larger cells than TDMA systems operating at the same RF power level. CDMA also facilitates frequency planning since cell-to-cell frequency reuse is possible in most applications.

The first cellular radio network operating CDMA was put into operation in Hong Kong on September 28, 1995. After one year of operation the manufacturer, Qualcomm of the United States, reports an operational system capacity of 283 subscribers per megahertz per base station compared with 85 on GSM and 11 on AMPS networks [14].

6.2.9 Time Advance Control

The mobile stations of cellular networks are at varying distances from their base station. The bursts, in digital cellular systems, received by the base station, therefore, are spread due to different radio propagation delays. This time spread requires a guard time between bursts, resulting in a lower spectrum efficiency. In order to minimize this guard time a time adjustment of the bursts, coming from the mobile stations, is made at the mobile station depending on its distance from the BTS. In the GSM network, for example, this adjustment covers a range of 233 ms, which allows correction for cells of a maximum radius of 35 km. The time advance of each MS is monitored by its BSC as a criterion for handover and also to help a MS to preset its time advance prior to handover.

6.2.10 Handover

A user of a cellular radio network while moving within a cell and between cells will be subject to strong varying reception conditions. The BSC permanently monitors the link conditions of each MS within its area, accumulating information on BER, transmitted power level, level of received signal, and time advance. The required quality measurements are made both by the BTS and the MS, whereby the MS measures parameters of its instantaneously used traffic channel as well as of those received from neighboring cells and from within the cell.

Based on this information, the BSC instructs the BTS and the MS to "handover" the ongoing call to a better traffic channel either within the cell or to a neighboring cell.

6.2.11 Roaming

Roaming is the name given to the feature that cellular radio network users can be called via unique telephone numbers wherever they "roam" around through the coverage area of their network. A user of a specific cellular radio system, such as the GSM system, can use a MS equipped with his or her SIM card in every GSM network worldwide, provided the operator of that network adheres to the international GSM roaming agreement. Under that international roaming agreement, bills for calls made in any of the associated networks will only be received from the home operator.

In order to establish roaming, some basic administrative, legal, and technical requirements need to be fulfilled. The technical requirements mainly address the following:

- The air interface between MS and BTS should be uniform.
- The exchange of signaling information between operators (e.g., information about roaming subscribers) must be standardized.
- The subscriber number series must be coordinated to enable identification of each visiting user and its home network.
- Definitions of functions and services are to be harmonized.

An extended definition of roaming is that the SIM card can be used physically in other networks even when a different system is used, for example, an Asian (900-MHz) GSM subscriber using his SIM card for communication on the American (1,900-MHz) PCS-1900 network with a handset rented in the United States.

6.2.12 Frequency Hopping

For radio-relay transmission the station sites and antenna heights are carefully planned to obtain optimum conditions for LOS transmission. For mobile communication only the location of the base stations can be carefully planned, while the mobile user moves around and receives reflected signals from varying directions rather than exclusively in a direct LOS from the base station. For PMR varying reception with the occasional necessity for the user to move to a nearby location with better reception is usually accepted. For cellular radio, however, the user expects good reception wherever he moves within the coverage area. At the BTS space diversity can combat uplink propagation problems. Therefore, taking advantage of the sophisticated signal processing that is possible with digital transmission, a special form of diversity is applied to improve the downlink reception for digital cellular radio, called *frequency hopping* (FH). As previously explained, a set of frequencies is assigned to each base station; with FH each user, while communicating with its base station, hops at every TDMA time slot pseudorandomly between one of the frequencies of the assigned set of frequencies. The hopping sequences of course are arranged such that no collisions occur with other mobile stations of the same base station area. To obtain a maximum diversity effect, the frequencies should preferably be spread over the whole available band. Still, time slots allocated to a fading frequency might get lost. To facilitate the retrieval of such lost information, the information bits can be coded and interleaved over several time slots.

6.2.13 Transport Infrastructure

A cellular radio network, in order to serve mobile users, needs a fixed infrastructure connecting the various base stations and switching centers of the network. A national cellular radio network may consist of some thousand radio sites that need to be connected with their BSCs and MSCs. This infrastructure or transport network can be implemented making use of leased lines of the transport transmission systems existing for PSTN and ISDN. Tariffs for leased lines in most countries are high (reportedly varying between 20% and 50% of total network operation cost); so more and more cellular network operators, subject to national regulations, see the opportunity to save cost and improve network performance by either using alternative infrastructures (such as using transmission networks of utilities) or building their own radio-relay-based infrastructure. Figure 6.10, which is a variation of Figure 6.5, depicts the transport infrastructure of a cellular network. A group of macro- and microcell base stations, as an example, is interconnected with their BSC via an optical fiber ring. Another group of microcell base stations is interconnected via a linear (for example, radio-relay) link with their BSC. The BSCs are interconnected with each other and their MSC by radio-relay links by optical fiber cable or leased lines. To obtain a maximum of flexibility, the transport network interfaces of the fixed cellular radio equipment shall allow connection to any available transmission media: twisted copper pair, coaxial cable, radio relay, or optical fiber. Cross-connect transmission systems tailored to cellular networks exist that groom, switch, and multiplex the traffic between the infrastructure nodes to obtain a more cost-effective use of the infrastructure lines and to facilitate future expansions [15].

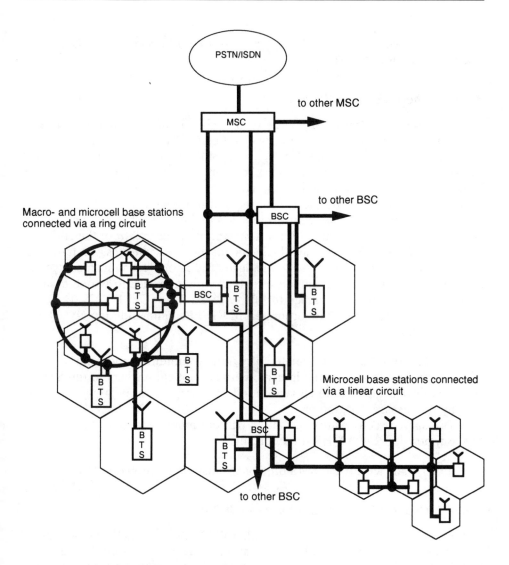

Figure 6.10 Transport infrastructure of a cellular network.

6.2.14 Data Transmission

Although cellular networks were primarily implemented for speech transmission, the digital signal transmission presents the prerequisites for the introduction of advanced data services such as *short message service* (SMS), *general packet radio service* (GDRS), and *high-speed circuit switched data* (HSCSD) on the GSM networks and SMS and *cellular digital packet data* (CPDP) on D-AMPS networks.

6.2.14.1 Short Message Service

SMS is based upon an ETSI standard to ensure international user interface commonality. It is designed to improve call completion rates by forwarding a call that otherwise would fail through to a SMS *service center* (SC). When a caller (from in- or outside the network) cannot reach the required MS because it is switched off, busy, or out of the coverage area, a messages of up to 160 characters can be left at the SC for subsequent retrieval by the called MS. The GSM network keeps track of waiting messages in the relevant HLR. The message is delivered by the SC to the MS upon receipt of an "alert" from the HLR that the destination MS is again available on the network. Success or failure of message transfer is reported to the message originator. The short message can be either purely numerical; standard text; or alphanumeric, in which case the user can edit a standard message in line with CCITT or CEPT T/CS 34-15.

Special SMS features incorporated in ETSI Recommendation GSM 03.40 version 4.6.0 are

- The ability for a user to delete a previously submitted short message;
- The facility to allow mobile users who are not subscribers to the SMS to submit one reply to a message, in effect a prepaid reply service paid for by the message originator;
- Transparent conveyance of data to or from mobile stations;
- Support for X.400 MHS interworking.

Voice mail within the SMS on GSM is also being prepared [16].

6.2.14.2 General Packet Radio Service (GPRS)

GPRS is a wireless version of LAN or rather WAN introducing the advantages of packet switching into the GSM network at times of spare transmission capacity. GPRS interworks with public data networks using Internet protocol, OSI connectionless network protocol, and X.25. Special GPRS support nodes interconnect GSM networks with public data networks. These support nodes keep track of the location of mobile stations to enable forwarding packets even if a mobile station is not in its home area.

6.2.14.3 High-Speed Circuit-Switched Data (HSCSD)

HSCSD is an enhancement of the current 9.8-Kbps circuit-switched GSM service based upon the use of multiple time slots [in a technique called *variable rate reservation access* (VRRA)], thus achieving multiples of 9.8 Kbps without compression, for example, seven time slots for the full ISDN 64-Kbps rate. Combining multiple time slots and video compression (for example, according to ITU-T standard H.261 mentioned in Chapter 2) supports video transmission via GSM with acceptable picture quality and a user rate around 24 Kbps.

Figure 6.11 indicates typical applications for GPRS and HSCSD data transmission [17,18].

6.2.14.4 Cellular Digital Packet Data (CDPD)

CDPD is being developed by a U.S. consortium as an overlay system on the AMPS voice network for data transmission on analog and digital AMPS net-

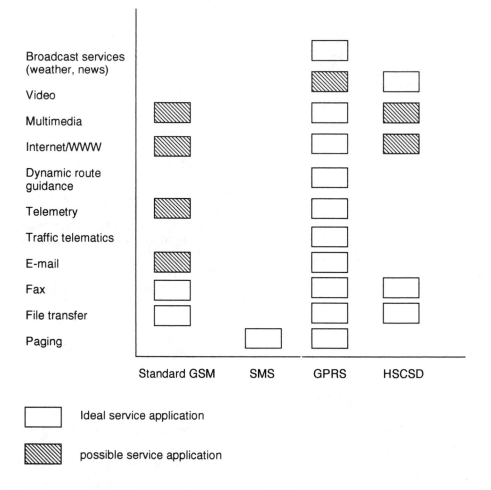

Figure 6.11 Typical data transmission services on GSM.

works. CDPD sends 19.2-Kbps data packets whenever idle times are detected on the cellular voice network. CDPD is supported by Internet TCP/IP and OSI *connectionless network protocol* (CLNP), X.400, and X.500 protocols. CPDP supports automatic encryption of data and includes precautions to prevent data from reaching the wrong destination [12,19].

6.2.15 Cellular Radio Networks

Both analog and digital cellular radio networks are operated in parallel in most countries. Most analog networks have the (rapidly diminishing) advantages of larger coverage, lower infrastructure cost, and lower subscription and call charges. The advantages of the digital networks can be summarized as follows:

- Better voice quality;
- Higher system capacity (due to more voice channels per radio carrier, reduced cell size, and improved frequency reuse);
- More efficient use of the scarce frequency spectrum;
- Higher security against eavesdropping and fraud;
- Easier integration of computer technology (a must to handle such complex tasks of handover, subscriber authentication, transborder roaming and billing);
- Better coexistence with emerging IN services.

The following analog cellular radio systems are currently still in operation:

- Japanese *Mobile Control Station System,* MCS-L2, third-generation system; the first generation MCS was introduced in 1979 as the world's first commercial cellular system;
- The Scandinavian *Nordic Mobile Telephone,* NMT 450 and 900;
- The North American *Advanced Mobile Phone System* (AMPS) and *Narrowband*-AMPS (NAMPS);
- The British *Total Access Communication System* (TACS) and *extended* TACS (ETACS), with the derived Japanese systems *Japan*-TACS (JTACS) and *narrowband*-TACS (NTACS); TACS originally was a modified version of AMPS;
- The German system C 450;
- The French system RadioCom 2000.

The major data of those analog systems are summarized, in order of system introduction, in Table 6.2.

The digital cellular radio systems presently in operation include the following.

Table 6.2
Summary of Major Data of Analog Cellular Radio Systems

System	Year of Introduction	Frequency Bands		Access Mode/ Modulation	RF Channels	
		BTS→MS (MHz)	MS→BTS (MHz)		Qty	Bandwidth (kHz)
MCS-L2	1988(1979)	870–885	925–940	FDMA/PM	2,400	12.5
NMT 450	1981	463–467.5	453–457.5	FDMA/FM	180	25
					225	12.5
NMT 900	1986	935–960	890–915	FDMA/FM	1,000	25
					1,999	12.5
AMPS	1983	869–894	824–849	FDMA/FM	832	30
NAMPS	1991	869–894	824–849	FDMA/FM	2,580	10
TACS	1985	935–950	890–905	FDMA/FM	1,000	25
ETACS	1988	917–950	872–905	FDMA/FM	1,640	25
JTACS	1989	870–885	925–940	FDMA/FM	600	25
NTACS	1991	870–885	925–940	FDMA/FM	1,200	12.5
C450	1985	461.3–465.74	451.3–455.74	FDMA/FM	222	20 (25)
RadioCom	1985	424.5–427.9	414.8–417.9	FDMA/FM	170	20

The European *Global System for Mobile Communications* (GSM) introduced in 1992, operates 992 channels in FDMA/TDMA access mode in the frequency bands 935- to 960-MHz downlink and 890- to 915-MHz uplink.

The *North American Digital AMPS* (D-AMPS), introduced in 1994, operates in the same frequency band as AMPS and is implemented in two versions— one version according to EIA/TIA standard IS-54 operating in FDMA/TDMA access mode with three time slots in each 30-kHz RF channel (thus increasing the capacity by a factor three compared with AMPS) and a second version according to standard IS-95 operating in CDMA access mode.

The *Japanese Digital Cellular* (JDC) system, introduced in 1992, is also specified in two versions, namely, as PDC 800 (Personal Digital Cellular) system operating in the 800-MHz band and as PDC 1500 operating in the 1.5-GHz band.

An aeronautical digital cellular radio system is the European *Terrestrial Flight Telecommunications System* (TFTS), which provides *aeronautical public correspondence* (APC) communication between airborne equipment and a network of terrestrial radio base stations connected with the fixed telephone and data networks. TFTS is intended for high-quality communication under dense air traffic conditions for short- and medium-haul continental flights. TFTS is thus complementary to truly global satellite APC systems such as Skyphone (introduced in 1991 using the INMARSAT satellite network) serving aircrafts on long-haul flights even over remote areas and oceans. TFTS, contrary to the nonstandardized aeronautical analog cellular systems operating in the 800-MHz band since 1984 in the United States (known as "Airfone") and since

1986 in Japan, is an ETSI-standardized system for which WARC-92 allocated and the ITU recommended in 1993 the frequency bands 1,670 to 1,675 MHz (ground-to-air) and 1,800 to 1,805 MHz (air-to-ground).

TFTS consists of (see Figure 6.12) *airborne stations* (AS), *ground stations* (GS), *ground switching centers* (GSC), and *service centers* (SC). The airborne station includes radio equipment, telephone handsets with credit card readers, facsimile machine, and data terminals. The ground stations provide the radio coverage for communication with the airborne stations. Three types of ground stations are used:

- Low-power GS, located within the airport area for communication with aircrafts at the airport and at take-off and landing within a range of a few kilometers;
- Medium-power GS, with an RF output power of typically 2W, located near an airport to cover the airport approach area within a range of about 50 km and an altitude of about 0.5 to 2.5 km.
- High-power GS, with an RF output power of typically 20W, located "en route" for communication with aircraft in the air corridors within a range of about 350 km from the HP-GS and an altitude of about 2.5 to 13 km.

In total 45 terrestrial radio base stations are planned to be in operation by the end of 1996 to provide pan-European coverage including the Middle East and North Africa. Deployment in the United States and South-East Asia is under study [20].

6.2.16 Cordless Telephony

Cordless telephony (CT) started as and basically still is a low-cost (mainly one-cell) wireless telephone access. CT does not provide handover, so continuity of conversation can only by assured by means of a single cell or by arranging sufficient overlapping coverage of a few adjacent cells. To enable frequency coordination-free application, a *dynamic channel selection* (DCS) technology is applied whereby an interference-free channel is automatically selected for each call. The major applications of CT in their order of introduction are as follows:

- Wireless extension of the residential telephone;
- Wireless extension or replacement of the desktop telephone in a corporate PABX network;
- Wireless access to the PSTN in pedestrian areas.

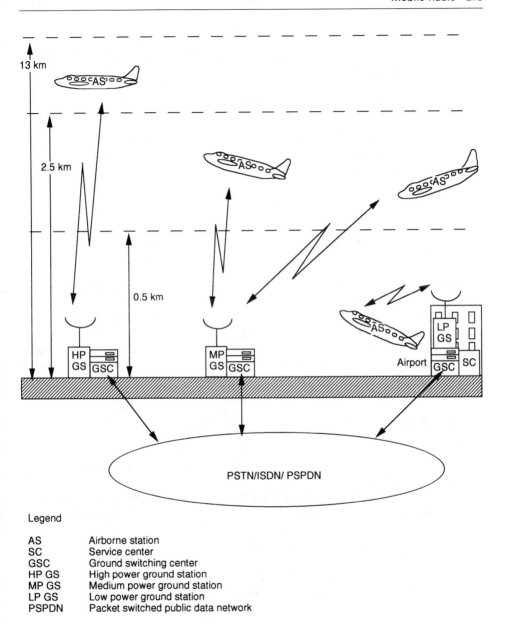

Legend

AS	Airborne station
SC	Service center
GSC	Ground switching center
HP GS	High power ground station
MP GS	Medium power ground station
LP GS	Low power ground station
PSPDN	Packet switched public data network

Figure 6.12 TFTS system architecture.

Figure 6.13, based upon the Japanese *personal handyphone system* (PHS), depicts all three applications.

6.2.16.1 Wireless Extension of the Residential Telephone

Cordless telephones provide telephone mobility at home with a miniature base station in the fixed telephone unit. Moreover, with a base station located in a residential area, additional telephone subscribers can immediately be connected to the PSTN without extending the underground cable network (as described in Subsection 5.3.5), thereby also providing terminal mobility.

6.2.16.2 Wireless Extension or Replacement of the Desktop Telephone in a Corporate PABX Network

Cordless operation in corporate networks is possible with a base station attached to a so-called "wireless PABX." Cordless handsets can replace or supplement the desktop extension. Both the desktop and the cordless telephones can be called simultaneously, thus increasing the mobility of employees and significantly reducing unsuccessful call attempts. Cordless telephony can quickly provide extra capacity without going to the expense of installing additional wiring or modifying existing wiring.

6.2.16.3 Wireless Access to the PSTN in Pedestrian Zones

Cordless telephony originally was a one-way portable radio access to suitably placed base stations (called "Telepoints," respectively, "Pointel" in France and "Greenpoint" in the Netherlands). Such base stations are located in shopping zones, airports, and sports and leisure grounds. Cordless telephones employ a low transmit power serving very short distances typically up to approximately 200m. Subscribers using their cordless telephone initially could call into the PSTN but could not be called. In the meantime, cordless telephone users can register with a particular base station and then be called by that base station. Paging simultaneously over a few adjacent cells is also possible, whereby a called person can respond using a CT for setting up a call via the nearest base station.

6.2.16.4 Cordless Systems

The major cordless systems are

- The British Cordless Telephone version 2 (CT2), an analog system introduced in 1988 (after CT0 in 1980 and CT1 in 1985) mainly in the United

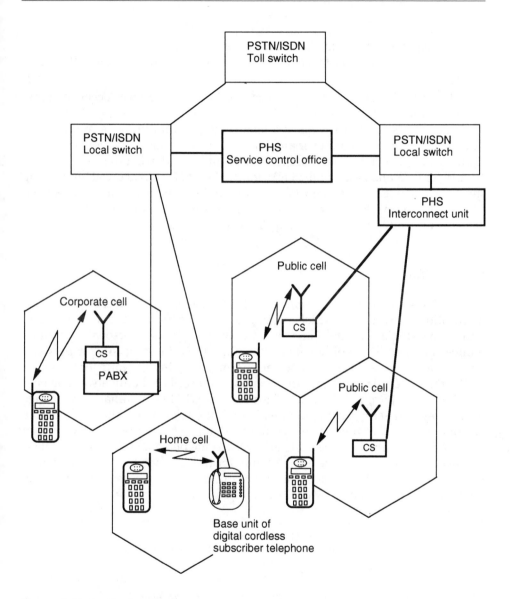

Figure 6.13 Cordless applications.

Kingdom and in the Asian Pacific region, lately loosing ground to the rapidly growing competing digital cellular networks;

- Digital European Cordless Telecommunications standard (DECT)—in view of growing worldwide interest, newly baptized as Digitally Enhanced Cordless Telecommunications System;

- The Japanese Personal Handyphone System (PHS);
- The North American Personal Access Communications System (PACS).

The DECT, PACS, and PHS systems have a 32-Kbps speech channel ADPCM coded in line with ITU-T Recommendation G.721, supporting voice, high-speed data (up to 64 Kbps at a later stage), and multimedia services. All three systems include handover and roaming. The RF output power of the base stations ranges between 10 and 100 mW and for the handies is typically 10 mW for PHS and DECT and 25 mW for PACS. DECT is conceived for walking speed, whereas both PHS and PACT can handle downtown motorcar speed. The major data of those cordless telephone systems are summarized in order of system introduction in Table 6.3 [21–23].

6.2.17 PCN

Personal communications network (PCN) is the name for the emerging telecommunication networks in which person-to-person communication is possible under one single national—and eventually worldwide—personal telephone number independently of home or office. In a PCN, in order to offer *personal communications services* (PCS, generic term prevailing in North America for digital micro-cellular services), the telephone subscriber should always be near a switched-on terminal. Instead of a personal CT, the terminal may change during day and season using the residential phone at home, the office phone at work, the cellular phone while commuting, and a CT while being away from those terminals, always accessible under the same personal telephone number. To ensure full network coverage, not only for traffic roads but for office buildings, hotels, shopping centers, sporting grounds, and residential areas, too, numerous microcells and picocells are required in the network.

Table 6.3
Summary of Major Data of Cordless Telephone Systems

System	Year of Introduction	Frequency Band (MHz)	Access Mode	Channels RF	Traffic
CT2	1988	864–868	FDMA/TDD	40	40
PHS	1995	Indoor 1,895–1,906.1	TDMA/TDD	77	308
		Outdoor 1,895–1,918.1			
DECT	1996	1,881.792–1,897.344	TDMA/TDD	10	120
PACS	1997	Uplink 1,850–1,910	TDMA/FDD	200	1,600
		Downlink 1,930–1,990			

Whereas the aforementioned analog and digital cellular networks were optimized for quick-moving vehicle-mounted terminals, the emphasis in PCN is on low-cost lightweight handsets in a slow-moving densely populated environment. Basically a digital cellular radio network can evolve into a PCN by enhancing the capacity (for example, by using 1.8- or 1.9-GHz bands that provide three to four times more RF channels than the 800- and 900-MHz bands) and by reducing the cell dimensions by adding micro- and picocells.

The emerging personal communication systems are the Digital Cellular System operating in the 1800-MHz band (DCS 1800) derived from GSM 900, and the Personal Communication System operating in the 1900-MHz band (PCS 1900) derived from GSM 900 for operation on the American continent.

6.2.18 Paging

Paging is the most economical and consequently most widely used form of mobile radio communication. Over 90 million subscribers worldwide—mainly in North America and South-East Asia—have prescribed to a public paging service. Paging started in the 1950s, mainly as on-site paging, as a one-way selective-calling communication alerting a person about a waiting message. The early pagers sent a tone-only signal, the bleep, alerting the paged person to callback from the nearest telephone. Modern pocket or wrist-watch pagers receive small messages of 20 to 240 characters—either numerical or alphanumerical—on a small LCD screen, and the paged person can immediately respond in an *interactive voice response* (IVR) via his two-way pager unit. Sophisticated pagers can be connected to a printer and receive up to 10,000 characters as e-mail. The different types of pagers can be summarized as follows:

- Tone-only pager, providing an alert only, where a few different tones may indicate a different caller, a degree of urgency, or an otherwise agreed upon information;
- Numeric pager, with a small display for short messages comprising digits and sometimes other characters as space, hyphen, and the letter "U" for urgent messages;
- Alphanumeric pager, with a larger display and storage capability, accommodating the full set of letters and digits and some other characters.

Paging systems operate in the VHF and UHF bands and use high-power transmitters (typically 100W to 250W) covering large cells or in the FM-subcarrier technique using existing commercial FM and TV broadcast transmitters. The frequency bands for FM (87.5 to 108 MHz) broadcast transmission are divided in wideband channels, of which approximately 50 kHz remains unused for the broadcast signal and thus can be used to accommodate "subcarriers" modulated with additional information, for example, to implement paging "environmentally friendly" without using additional frequency spectrum and station sites.

Pagers are identified by a *receiver identification code* (RIC) for message routing and subscription to different paging services. A pager thus can have multiple RICs depending on the quantity of subscribed services. A RIC typically consists of five sections:

- Section 1: World zone;
- Section 2: Country;
- Section 3: Operator;
- Section 4: Pager address;
- Section 5: Number of batch that contains the message.

The world zones are based upon CCITT Recommendation E.212:

- Zone 2: Europe and Mediterranean region;
- Zone 3: North America and Carribean region;
- Zone 4: Middle East, North Asia, Indian subcontinent, Indo China, and greater China region;
- Zone 5: South-East Asia, Australia, Oceania and Pacific Islands region;
- Zone 6: African continental region;
- Zone 7: Central and South American region.

Thus far, international roaming has been limited to neighboring countries, such as *Euromessage* introduced in Germany in 1990 based on POCSAG standard operating in the 466-MHz band with roaming facilities mainly in cities in Switzerland, the United Kingdom (using *Europage*), France (using *Alphapage*), and Italy (using *Teledrin*). ERMES paging will allow worldwide roaming.

Current paging systems are mainly based upon the CCIR No. 1 Radiopaging code developed by Philips and introduced in 1981 as a national standard by the *British Post Office Code Standardization Advisory Group* (POCSAG). Those systems have a capacity of 100,000 subscribers per RF channel, a speed of 1,200 bps (later extended to 2,400 bps), and a maximum of two million paging addresses. POCSAG operates in the 138- to 174-MHz and 420- to 470-MHz bands.

A further paging system likewise developed by Philips is called APOC (for "advanced paging operators code"). APOC, with a bit rate of 2,400 bps, supports a message compression of alphanumeric text that substantially increases channel capacity enabling long e-mail reception.

A widely distributed digital paging system based upon a proprietary protocol is Motorola's FLEX paging system released in 1994. FLEX is a paging system serving up to 150,000 subscribers per RF channel, with a transmission speed of 1.6, 3.2, and 6.4 Kbps and a limit of one billion paging addresses. ReFLEX is an advanced version of FLEX offering two-way paging, whereas InFLEXion includes voice paging.

The latest developments in paging technology are launched under the acronyms: ERMES, TDP, and CPP [22,24].

6.2.18.1 European Radio Messaging Service (ERMES)

ERMES is a digital European Radio Messaging Service operating at a speed of 6.4 Kbps. ERMES was approved by the ITU for worldwide use in October 1994. Within the 169.4- to 169.8-MHz band ERMES operates 16 RF channels, each with four time slots. A 35-bit RIC permits a total of over 34 billion individual identification codes. The ERMES standard specifies six interfaces as shown in Figure 6.14 and summarized as follows.

- I1: Air interface between BS and paging terminal (pocket pager);
- I2: Management and alarm interface between BS and the *paging area controller* (PAC);
- I3: Information and control interface between PAC and the *paging network controller* (PNC);
- I4: Interface between PNCs of different networks to facilitate roaming;
- I5: Physical interface between PNC and the national access network;
- I6: Dialogue interface for telephony and nontelephony access.

ERMES transmits its data in hourly sequences with 24 sequences per day. Within each sequence there are 60 cycles, each with five subsequences. The subsequences are divided in 16 batches, in which the messages and addressing information are transmitted to the pocket pagers. The pager is only required to turn on (wake up) and listen for messages within its nominated batch; during the other 15 batches the pager is turned off (asleep), thus saving battery life. Depending on the number of messages received, battery replacement is required once every seven to nine months.

Worldwide roaming between ERMES networks is being implemented, and deferred message delivery even into another country will then be possible [25].

6.2.18.2 Telocator Data Protocol (TDP)

The recently established "Telocator Data Protocol" issued by the *Personal Communications Association Industry* (PCIA), the leading international trade associ-

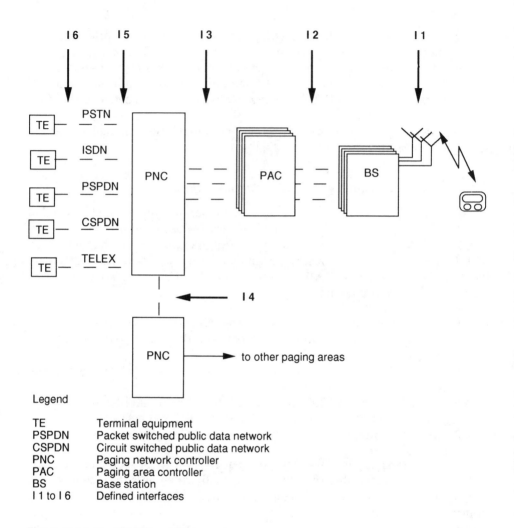

Figure 6.14 The ERMES architecture.

ation representing the wireless communications industry, allows transmission of 64-kb messages in 8-bit segments from PC-based files and documents both in one-way and two-way modes [26].

6.2.18.3 Calling Party Pays (CPP)

The operating mode "calling party pays," developed by Motorola, basically implies that paging operators derive their income with value-added services from the messages sent, using premium rate telephone call charging for message retrieval rather than from pager rental and monthly rates [27].

6.2.18.4 The Paging Outlook

Recent paging improvements will ensure that paging, despite worldwide cellular radio penetration, will survive as low-cost, small-sized, easy-to-use *narrowband personal communication services* (NPCS). It is even estimated that by the end of this century the number of pagers will double to 120 million worldwide.

6.3 SPECIFIC APPLICATIONS

Mobile radio is the choice transmission medium for all applications where an individual communication is to be established, not with a fixed location, but with a certain person irrespective of its location, at a lower price than would otherwise be possible using satellite communication. Mobile radio communication generally will be less expensive than satellite communication wherever population density and existing public infrastructure—roads, electricity supply, and fixed telephone network—economically justify the deployment of the relevant fixed part (base stations, control, and switching equipment) of a mobile network. Mobile radio transmission increases the personal mobility up to the ultimate goal that people are able to communicate at all times at all places, no matter where they may be.

Figure 6.15 indicates the application domains of the various mobile radio systems mentioned in this chapter. The future mobile systems shown in this figure are the *mobile satellite system* (MSS, which is a subject of the next chapter) and *Universal Mobile Telecommunication System and the International Mobile Telecommunications* (UMTS/IMT-2000, described in Chapter 10).

The major applications for mobile radio can be summarized as follows:

- Paging through pocket pagers;
- Cordless communication at residential sites, at corporate on- and off-site locations, and at places of pedestrian agglomeration;
- Vehicle bound simplex and semiduplex push-to-talk communication within PMR, PAMR, and trunked networks with limited through connection with fixed networks;
- Mobile full-duplex communication (with vehicle, portable, and handheld terminals) within wide-area cellular networks fully interconnected with fixed networks;

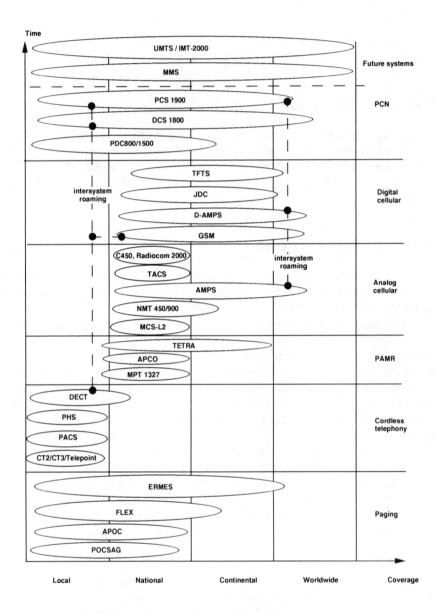

Figure 6.15 Application domains of various mobile radio systems.

- Mobile full-duplex communication (mainly with handheld terminals) in PCN and PCS networks;
- Semifixed application as WLL with additional on site terminal mobility;
- Airborne communication in *aeronautical public correspondence* (APC) networks through terrestrial infrastructure connected with fixed networks.

References

[1] Huurdeman, Anton A., *Radio-Relay Systems,* Norwood, MA: Artech House, 1995.

[2] Williams, Ken, "HF Still Has Much to Offer," *Communications International,* Vol. 13, No. 5, 1987, pp. 17–20.

[3] Robertson, David, "Spark of Genius," *Mobile Communications International,* Issue 25, Sept./ Oct. 1995, pp. 113–128.

[4] Giese, Werner, *Funksprechen Möglichkeiten und Anwendungen Land- und Seefunk,* Stuttgart: Verlag Berliner Union, 1971.

[5] Fletcher, Peter, "Masters of Disguise," *Communications International,* Vol. 22, No. 8, 1995, pp. 61–62.

[6] Roy, Richard, and Martyn Warwick, "Journey into space," *Communications International,* Vol. 22, No. 8, 1995, pp. 49–52.

[7] Shelley, Martin, "Intelligent Antennas and Urban Mobiles," *Mobile Europe,* Vol. 5, No. 10, 1995, pp. 99–100.

[8] Hubermark, Sven H., "Could Phoning Seriously Damage Your Health?" *Mobile Europe,* Vol. 4, No. 7, 1994, pp. 36–38.

[9] Kemp, Peter, "Battery Impact on Cellular Electronics," *Mobile Europe,* Vol. 5, No. 10, 1995, pp. 83–90.

[10] Robinson, Heather, "Charge of the Light Brigade," *Mobile Europe,* Vol. 6, No. 6, 1996, pp. 31–32.

[11] PrimeSite, "Simply an Extraordinary Base Station," *Discovery,* Nokia Telecommunications' Customer Magazine, Vol. 40, 1st quarter 1996, pp. 22–25.

[12] Balston, D. M., and R. V. Macario, editors, *Cellular Radio Systems,* Norwood, MA: Artech House, 1993.

[13] Feldmann, M., and J. P. Rissen, "GSM Network Systems and Overall System Integration," *Electrical Communication,* 2nd quarter 1993, pp. 141–154.

[14] Jacobs, Dr. Irwin, "In Defense of CDMA," *Communications International,* Issue 23, Nov. 1996, p. 95.

[15] Ganev, Tsviatko, et al., "The Ericsson DXX Cross-Connect System in Mobile Networks," *Ericsson Review,* Vol. 73, No. 2, 1996, pp. 75–88.

[16] "Focus on GSM, Getting The Message," *Mobile Europe,* Vol. 5, No. 9, 1994, pp. 23–26.

[17] Gilchrist, Philip, "One Stop Data Shop From GSM's GPRS," *Mobile Communications International,* Issue 29, Mar. 1996, pp. 62–64.

[18] Hämäläinen, Jari, "GSM's Support of High Speed Data," *Mobile Communications International,* Issue 27, Dec. 1995/Jan. 1996, pp. 72–78.

[19] Wong, Peter, "It's Useful, and It Comes in a Packet," *Mobile Asia-Pacific,* Vol. 2, No. 6, Dec. 1994/Jan. 1995, pp. 20–22.

[20] Campet, G., "The TFTS Air-to-Ground Telephone System," *Electrical Communication,* 4th quarter 1994, pp. 353–358.

[21] Fletcher, Peter, "The Office Goes Cordless," *The Mobile Revolution,* a supplement to *TE&M and Communications International,* 1994, pp. 24–27.

[22] Kandiyoor, Suresh, and Peter van den Berg, "DECTS's Potential for Public Network Operators," *Mobile Communications International,* Issue 27, Dec. 1995/Jan. 1996, pp. 87–90 and 96.

[23] Hamano, Takayoshi. "PHS: The Technology and Its Prospects Outside Japan," *Mobile Communications International,* Issue 26, Nov. 1995, pp. 54–56.

[24] Brown, Mike, "Paging and the Mass Market," *Telecommunications,* Vol. 30, No. 3, 1996, pp 69–73.

[25] Cooper, Charles, "ERMES: A World Standard Contender for Paging," *TELECOM Asia,* Vol. 6, No. 10, 1995, pp. 68–74.

[26] Kaneshige, Thomas, "Two-Way Pager Wager," *Communications International,* Vol. 22, No. 12, 1995, pp. 28–29.

[27] Chow, David, "Technology Choices are Key To Paging's Prospects," *Mobile Communications International,* Issue 28, Feb. 1996, pp. 42–45.

Satellites 7

7.1 INTRODUCTION

Satellite transmission presents the youngest version of radio transmission. The bright, but initially almost unnoticed, idea of using an in-space Earth-orbiting satellite as a relay for communication between two distant stations on Earth came from the visionary British science fiction writer Arthur C. Clarke. In the technical magazine *Wireless World,* Volume 51 (October 1945), Clarke wrote a prophetic article entitled "Extra-Terrestrial Relays," with the subtitle "Can Rocket Stations Give World-wide Radio Coverage?" In that article Clarke proposed the peaceful use of the World War II weapon V-2, in a combination of rocketry and radio-relay engineering to operate solar-powered relay stations in space. He conceived the idea of placing satellites in orbit at a distance of slightly less than 42,000 km from the center of the Earth (corresponding to about 35,860 km from the surface of the Earth) so that the satellite would move synchronous with the Earth and thus have a *geostationary* position relative to their Earth stations. Clarke likewise proposed arranging three satellites in this orbit equidistantly around the Earth to provide worldwide coverage.

After various experiments in the United States and the Union of Soviet Socialist Republics, the first satellite in space was *Sputnik 1,* a 58-cm ball, transmitting space telemetry information on 20 and 40 MHz beginning October 4, 1957 for 21 days. The second satellite, *Explorer I,* was launched on January 31, 1958 by the U.S. Army. *Explorer I,* a 2-m-long 14-kg-heavy cylinder, orbiting on a 360/2500-km ellipse was a big success. Telemetry information transmitted for nearly five months led to the most important discovery of the 1957/58 International Geophysical Year: the "Van Allen radiation belts" around the Earth.

Still in the same year the U.S. Air Force succeeded in launching the first communication satellite *Score* (standing for signal communication orbit repeater experiment). *Score* was a delayed-repeater satellite, which received signals from Earth stations at 150 MHz, stored them on tape, and later retransmitted them. The 68-kg *Score* was placed in a *low Earth orbit* (LEO), similarly to what will be applied for new global personal mobile satellite services within a few years, on an ellipse with 182-km perigee and 1,048-km apogee. *Score* was still battery powered, so that transmission finished after 12 days.

A further experimental communication satellite still used as a passive repeater, however, was *Echo I*—an aluminized plastic balloon with a 30-m diameter and a weight of 75.3 kg. *Echo I* was launched on August 12, 1960 and placed into orbit at about 1,500-km altitude circling the Earth in 118.2 min.

Direct-repeated satellite transmission started with *Telstar I*, which was successfully launched by the NASA (the U.S. civil National Aeronautics and Space Administration, founded in 1958) on July 10, 1962. *Telstar I*, a 87-cm sphere weighing 80 kg and equipped with solar cells for generating electricity, belonged to AT&T and was developed in the Bell Laboratories. One Earth station was located near the west coast of the Atlantic Ocean in Andover, Maine, USA and two nearest to the east coast of the Atlantic Ocean in Goonhilly Downs, at Land's End in Cornwall, U.K., and in Pleumeur Bodou, Brittany, France. The U.S. and French Earth stations used large horn-reflector antennas with a horn diameter of 20m and a reflector length of 54m (see Figure 7.1). The antenna with a 340-ton weight could be moved in azimuth and elevation to track the satellite at an accuracy within one-twentieth of a degree. The whole antenna, including the electronic equipment, was protected by a huge Dacron and synthetic rubber radome with a diameter of 64.5m to protect the sensitive equipment from humidity, wind, and rapid temperature changes—a technology that after 35 years is planned to be reapplied for the highly sensitive tracking of the future IRIDIUM personal communication mobile satellite system.

Telstar I, in an elliptical LEO between 960 and 6,140 km with an orbital period of 158 min, had less then half an hour per day common visibility in the United States and France, respectively, in the United States and the United Kingdom. *Telstar* was used for experimental telephone, image, and TV transmission including color TV transmission of a surgical operation. Unanticipated radiation damage resulting from the newly discovered Van Allen belts unfortunately interrupted the transmission in early 1963.

In 1963 NASA started launching their *synchronous communication* (SYNCOM) satellites in an orbit at an altitude of 35,786 km, as calculated by Arthur C. Clarke. The SYNCOM satellites were produced by Hughes, who, contrary to AT&T, accepted the long signal delay as a challenge. SYNCOM I, launched on February 14, 1963, unfortunately went silent 20 sec after ignition of its apogee motor for synchronous orbit injection. SYNCOM II, launched July 26, 1964, was successfully placed in synchronous orbit and worked perfectly for telephone, teletype, and facsimile transmission between Africa, Europe, and the United

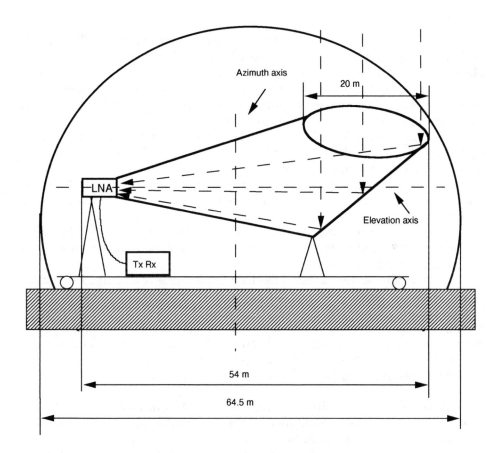

Figure 7.1 Basic construction of the first generation of Earth stations.

States. Thus, SYNCOM II was the world's first operating synchronous satellite, realizing Clarke's vision within 18 years.

SYNCOM III was launched August 19, 1964, and transmitted TV pictures from the Olympic Games in Japan.

The positive results with *Telstar* and SYNCOM encouraged the U.S. government to operate communication satellites on a commercial bases. The *Communication Satellite Corporation* (COMSAT) was founded in 1963 to further the commercial development of communication satellites. Satellite operation, however, should not be a national but a global affair. Therefore, in accordance with Resolution 1721 of the United Nations, the *International Telecommunications by Satellite* (INTELSAT) consortium was founded in July 1964, COMSAT being one of the major consortium members. INTELSAT's task is to design, develop, construct, establish, and maintain the operation of a global commercial communications satellite system on a nondiscriminatory basis. INTELSAT, with its headquarters in Washington, DC, began with a charter membership of eleven nations. Currently INTELSAT is an international commercial cooperative of governments and telecommunications organizations (the *Signatories*) from 140 member nations, which together own the space segment and operate the global communications satellite system.

The implementation of the INTELSAT network started with the launch of INTELSAT I, also called *Early Bird* on April 6, 1965. INTELSAT I was located over the Atlantic Ocean, so commercial international telephone (240 channels) and television transmission between the United States and Europe—thus, the era of commercial space telecommunications—could and did, indeed, start on June 28, 1965.

A second global satellite system was established by INTERSPUTNIK, founded in 1971 and at the time serving 14 socialist countries who had not joined the INTELSAT system.

A third global network called INMARSAT (originally standing for International Maritime Satellite Organization, in 1995 changed to the International Mobile Satellite Organization) was created in 1979. INMARSAT evolved to become the only provider of global mobile satellite communications for commercial and distress and safety applications at sea, in the air, and on land.

In addition to these *global* operating satellite networks, quite a number of *regional* operating networks have been implemented:

- ARABSAT, Arab Satellite Communications Organization;
- AUSSAT, Australian Satellite Corporation serving since 1985 the Australian continent and Papua New Guinea;
- EUTELSAT, European Telecommunication Satellite Organization. Beyond telecommunication services, EUTELSAT offers radio-navigation (EUTELTRACS), meteorological, and space research services;
- AsiaSat, Asian Satellite Organization serving South-East Asian countries;
- PAS, PanAmSAT, the first private global satellite network operator.

Next to regional satellite networks a large number of national satellite networks, also called Domsat networks, evolved, including as the first national satellite system ANIK (Eskimo term for "little brother") in Canada (starting operation in 1972), COMSTAR WESTSTAR Galaxy and SBS (Satellite Business Systems) in the United States, Molnija Raduga and Gorizont in the Union of Soviet Socialist Republics, and PALAPA in Indonesia. Beyond being a domestic system, PALAPA also serves the neighboring countries.

In addition to the aforementioned international, regional, and national satellite network operators, numerous other organizations use the satellites of these networks to provide satellite transmission services for radio and TV broadcasting, news and information gathering, and business telecommunication in so-called *very small aperture terminal* (VSAT) networks and in *direct-to-home* (DTH) networks. The latest version of a DTH network called "DirecPC" was launched in September 1996 as a hybrid solution allowing PC users to access the Internet with a standard dial-up connection through the PSTN and receive data in a 400-Kbps satellite link directly via a 60-cm home-mounted dish.

In the first decade, telecommunication satellites were mainly used for long-distance continental and intercontinental broadband and TV transmission and for narrowband long-distance transmission as a replacement for the poor-quality HF communication. The advent of broadband optical fiber transmission shifted the application of satellites in the second and third decades toward thin-route P-MP systems, TV and data-distribution systems, and the aforementioned VSAT systems. Now in its fourth decade, satellite transmission will experience a further significant growth with the introduction of *mobile satellite systems* for personal communication and *fixed satellite systems* for broadband data transmission [1,2].

7.2 SATELLITE TECHNOLOGY

A satellite network in its simplest form consists of two Earth stations communicating with each other via a satellite as shown in Figure 7.2. To minimize the power requirements for the satellite, the frequency for the signal from the satellite to the Earth station, called downlink, is always lower (and thus the attenuation is lower) than the frequency for the uplink from the Earth station to the satellite.

Satellite technology basically is radio-relay technology. In fact, many elements of satellite technology have been taken from well-proven radio-relay technology in order to reduce the problems that were expected in the development of the new satellite systems. To master the extremely long distances (causing high attenuation and long delays) and to provide mainly (multi)point-to-multipoint communication rather than point-to-point (requiring highly effective antennas), specific satellite technology concerns

- Satellite services;
- Satellite orbits;
- Satellite launching;
- Antennas;
- Satellites;
- Earth stations;
- Multiple access [3].

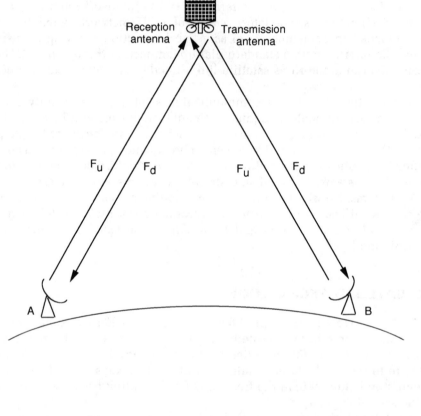

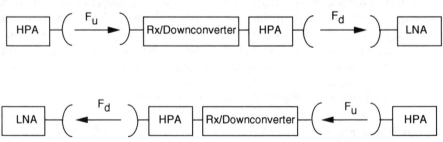

Figure 7.2 Basic satellite network.

7.2.1 Satellite Services

Communications by satellite can cover the whole range of voice, data, facsimile, and video transmission. The telecommunication satellite services are classified into three groups (see also Figure 7.3):

- *Fixed satellite services* (FSS);
- *Direct-broadcast satellite services* (DBS);
- *Mobile satellite services* (MSS).

7.2.1.1 Fixed Satellite Services

In FSS networks the signals are relayed between fixed Earth stations, which are relatively large, complex, and expensive. The Earth stations are connected to the conventional terrestrial telecommunications network, and the service is mainly intended for long-distance operation. A special version of FSS is the VSAT service using small antennas up to 2.4 m in diameter and terminals that are directly located at the users' premises. A VSAT terminal in a star network typically communicates via a satellite directly with other VSAT terminals or in a double-hop twice via satellite through a common larger Earth station, usually called a *hub station.*

7.2.1.2 Direct-Broadcast Satellite Services

In DBS networks broadcast and television signals are transmitted from a central large Earth station via a satellite to a "receive-only" Earth station (such as TV-RO and HiFi-RO), which are either at CATV head-end stations or directly located at individual homes in so-called DTH service.

7.2.1.3 Mobile Satellite Services

In MSS networks the signals are relayed via satellite between a large fixed Earth station (normally connected with the terrestrial telecommunications network) and small mobile stations fitted to a ship, aircraft, vehicle, or portable handheld terminals. Currently, this refers to all INMARSAT and some EUTELSAT services. At least three new operators, however, are well on the way to implementing satellite networks for global personal communication systems scheduled to operate before the end of this century under the names Globalstar, Iridium, and Odyssey (see Section 10.6).

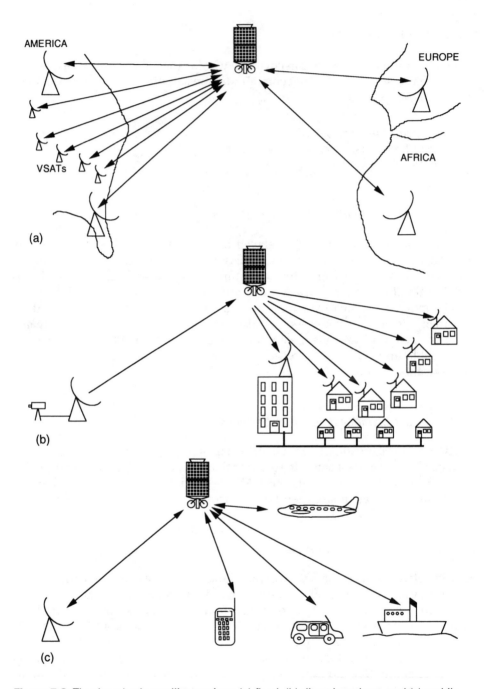

Figure 7.3 The three basic satellite services: (a) fixed, (b) direct-broadcast, and (c) mobile.

7.2.1.4 Satellite Frequency Bands

Specific frequency bands (as shown in Table 7.1) are reserved for satellite transmission, similarly as explained for radio-relay systems in Chapter 5, and numerous subbands have been allocated for FSS, DBS, and MSS services within those frequency bands.

7.2.2 Satellite Orbits

A satellite orbit is the lifetime location of a satellite in space. A satellite remains in that orbit as long as its centrifugal force is in balance with the gravitational attraction of the Earth and other cosmic influences. The rotation time or orbit period T (hr) for which a satellite for a given height h remains balanced in orbit is given by the formula

$$T = 2\pi\sqrt{(h + R)^3/GM}$$

where R = Earth's radius = 6,378.155 km, G = Gravitational constant = $6.67 \cdot 10^{-11}$ Nm^2/kg^2, and M = Earth's mass = $5.95 \cdot 10^{24}$ kg. The relation between the distance from the Earth's center and the circular orbit period T is indicated in Figure 7.4 on the horizontal axis.

A satellite orbit, in accordance with Keppler's first law, is an ellipse with the center of the Earth at one focus. The point at which the satellite is closest to the Earth is called the "perigee," while the point at which the satellite is furthest away from the Earth is called the "apogee." The most frequently applied circular orbit is a special case of the elliptical orbit with both foci coinciding with the center of the Earth. Keppler's second law says that the line joining the satellite with the center of the Earth sweeps over equal areas in equal time

Table 7.1
Frequency Bands Allocated to Satellite Transmission

Band Nomination	Frequency (GHz)
L	0.4–0.46
	0.6–0.8
S	1.5–2.7
C	3.4–8.4
Ku	12.4–18
K	18–26.5
Ka	26.5–40
Q	33–50

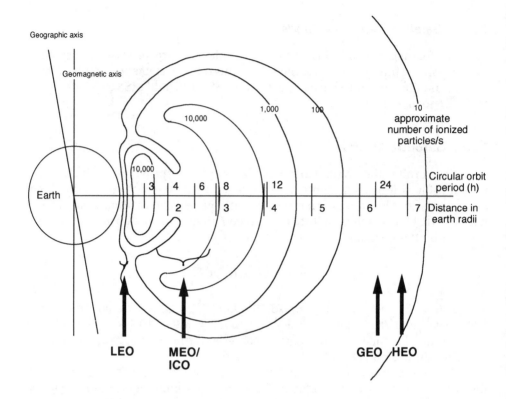

Figure 7.4 Satellite orbits and the Van Allen radiation belts.

intervals (thus achieving its highest speed in apogee). The point where this line meets the equator is called the "subsatellite point." The distance between this subsatellite point and the satellite, thus the altitude above the mean equatorial radius of the Earth, is a major characteristic of the satellite—determining the coverage area, the orbit period and thus the time of satellite visibility, the signal propagation delay, the path attenuation, and the influence of the two Van Allen radiation belts. Four distinct altitude ranges are used for telecommunications satellites, moving in the same direction as the Earth's rotation, in orbits as follows (see also Figure 7.4):

- *Low Earth orbit* (LEO) between 500 and 2,000 km;
- *Medium Earth orbit* (MEO) between 5,000 and 15,000 km and also called *intermediate circular orbit* (ICO);
- *Geostationary Earth orbit* (GEO), also called Clarke orbit, at 35,786 km;
- *Highly elliptical orbit* (HEO), with an apogee that may be beyond GEO.

7.2.2.1 Low Earth Orbit

LEO, located below the Van Allen belts, will be used for some of the future MSS for personal communications and FSS networks for broadband data transmission. The major advantages of LEO are as follows:

- The ability to use handheld "Earth stations" with an omnidirectional antenna and a relative low RF output power;
- Lowest signal delay; for example, for a two-way conversation via a satellite at an altitude of 1,000 km the delay is only about 13 ms (total for uplink and downlink);
- Satellite path diversity eliminates signal interruption due to path obstruction (Figure 7.5 demonstrates handover from satellite A to satellite B and path diversity between satellites B and C);
- Low satellite launching cost with direct injection into the orbit of several (for example, seven for the Iridium system) satellites in one launch.

The disadvantages of LEO are as follows:

- The orbit period at 1,000-km altitude is in the order of 100 min and the visibility at a point on the Earth some 10 minutes only, requiring 40 to 80 satellites in six to seven planes for global coverage.
- Frequent handover is necessary for uninterrupted communication.
- During times of the year that the orbital plane is in parallel to the direction to the sun, a satellite in LEO is eclipsed (in the shade of the Earth) for almost one-third of the orbit period. Consequently, there is a significant demand on battery power with up to 5,000 charge/discharge cycles per year, which with present NiCd batteries reduces satellite lifetime to 5 to 7.5 years.

7.2.2.2 Medium Earth Orbit

MEO, also called ICO, located between the Van Allen belts, will also be used for future mobile satellite systems. The major advantages and disadvantages mentioned for LEO apply at a slightly lesser degree for MEO too. The relevant parameter for satellite operation in MEO at an altitude of 10,000 km compare with LEO as follows:

- Signal delay is about 70 ms;
- Orbit period is six to eight hours providing slightly over one hour local visibility; around 10 satellites in two planes, each plane inclining 45 degrees to the equator, are required for global coverage;

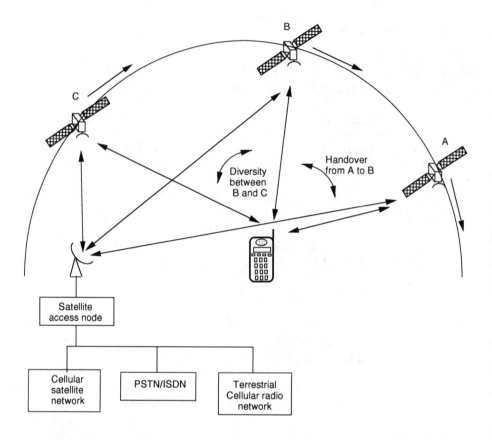

Figure 7.5 Mobile satellite system architecture (*After:* ICO).

- Fewer eclipse cycles so that battery lifetime will be more than 7 years;
- Lower cosmic radiation with subsequent longer life expectancy for the complete satellite system;
- Higher average elevation angle from user to satellite, minimizing probability of LOS blockage;
- Higher RF output power required for both the satellites and the handheld terminals.

7.2.2.3 Geostationary Earth Orbit

GEO, with an orbit period equal to the rotation of the Earth (23.94 hr), is essentially used for commercial telecommunications satellites with stationary Earth stations for the following reasons.

- The satellite remains stationary with respect to one point on Earth. Therefore, the Earth station antenna can be beamed exactly toward the satellite without periodical tracking.
- Three satellites in GEO can cover the entire surface of the Earth with some overlapping, except for the polar regions beyond 76°N and 76°S latitude. The three satellites are normally spaced at 120-degree intervals and located over the equator, above the Atlantic Ocean, the Pacific Ocean, and the Indian Ocean, respectively serving the *Atlantic Ocean Region* (AOR), the *Indian Ocean Region* (IOR), and the *Pacific Ocean Region* (POR).
- The Doppler shift (affecting synchronous digital systems) caused by a satellite drifting in orbit (because of the permanently changing gravitational attraction of the moon and to a lesser extent that of the sun) is small for all Earth stations within the satellite coverage.

Figure 7.6 shows the world coverage obtained with three satellites at GEO positions.

The disadvantages of GEO operation, compared with LEO and MEO, are

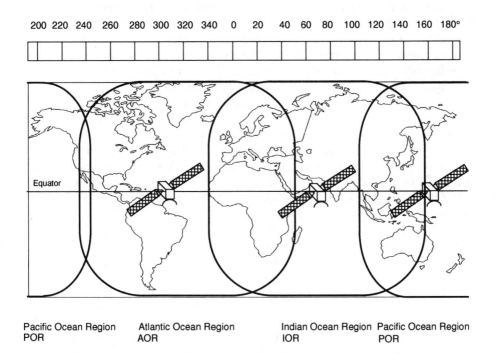

Figure 7.6 World coverage with three satellites at GEO positions.

- The long signal delay due to the large distance between satellite and Earth stations. This distance is 35,800 km if the satellite is in zenith for an Earth station and 41,000 km at the minimum elevation angle of 5 degrees. For the electromagnetic waves traveling at the speed of light this causes a round-trip signal delay of a minimum 240 ms and maximum 270 ms and a full-duplex delay of 480 ms to 540 ms. For telephony via satellite this used to be disturbing; however, echo cancellation devices developed in the 1980s reduced the problem. For data transmission, especially when using error-correcting protocols that require retransmission of blocks with detected errors, complex circuitry with high-capacity buffer devices are required to overcome delay problems.
- The required high RF output power and the use of directional antennas aggravate GEO operation for mobile handheld use.

To achieve the most efficient use of the geostationary orbit, preventing mutual interference and assuring an orderly operation of the many satellites moving in the same plane, a strict adherence to internationally agreed upon conditions is required. Severe requirements on keeping satellites within their assigned position have been set up and codified by the *International Radio Regulations* (IRR). These requirements are regularly reaffirmed and updated at the WRCs and specified in detail in relevant CCIR Recommendations.

A variety of perturbing forces, however, cause the satellite to drift out of its assigned position toward a so-called *inclined orbit*. By far the most important pertubations are the lunar, and at a lesser degree the solar gravitational forces, which cause the satellite to drift in latitude or in a north-south direction. The longitudinal drift, or east-west drift, is caused by fluctuations in the gravitational forces from the Earth due to its nonspherical shape and by fluctuations in solar radiation pressure. To counteract these pertubations the satellite needs station-keeping devices as described in Subsection 7.2.4. Figure 7.7 depicts these drifts of a satellite in an orbital slot limited to a variation of ±0.1 degree as specified at WARC-79 for fixed satellite services, and ±1 degree for inclined orbit operation to reduce fuel consumption. (See also Subsection 7.2.5.7.)

After reaching the end of its operational life a satellite has to be removed from its orbital slot into a "graveyard" orbit some 200 km above GEO.

7.2.2.4 Highly Elliptical Orbit

HEO, contrary to GEO, also covers the polar regions. HEO was chosen, therefore, by the Soviet Russian Space Authorities for their Molniya system in order to facilitate launching from their territory and ensure coverage throughout the Union of Soviet Socialist Republics. Molniya has an orbit period of 12 hours,

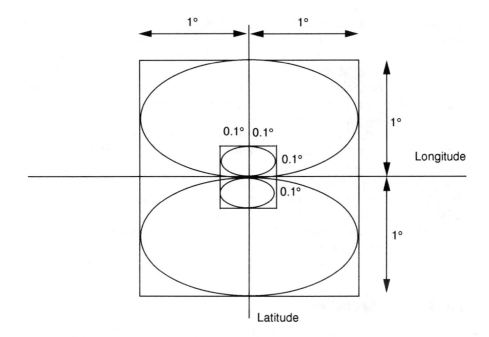

Figure 7.7 Satellite movement within an orbital slot.

eight of which can be used for coverage over the operational region (previous Union of Soviet Socialist Republics, Cuba, and China). For continuous regional coverage three satellites are required. HEO, compared with GEO, provides the advantage of lower launching cost and a higher elevation angle for ground stations, which reduces atmospheric losses. The disadvantages, however, are significant, such as the necessity of constant satellite tracking at the Earth stations, compensation of signal loss variation, extensive eclipse periods, and more complex control of the satellites and the Earth stations.

7.2.2.5 The Choice of Satellite Orbit for Mobile Satellite Systems

Satellite communication with mobile terminals require a special type of signal propagation concerning beam elevation and RF power. A handset terminal communicating in a MSS with a satellite can only have a limited RF output power and a small nondirectional antenna, thus a very low EIRP, which aggravates the possibility of communication with a satellite in GEO. For the future MSS for global personal communication including the extensive use of handsets, the satellites need to be in LEO or at least in MEO. This, however, implies that geosynchronism is lost and that the satellite's footprint shrinks. Thus more satellites are required to achieve permanent global coverage, the number of satellites depending on the altitude. The choice of altitude among others depends on the Van Allen belts, the reason why altitudes are selected either below 1,500 km or slightly beyond 9,000 km. An altitude below approximately 400 km is also uneconomical in view of the still significant Earth gravity requiring high station-keeping efforts.

Although with MSS there are few multiple paths, the Earth-satellite path can be obstructed by buildings and trees depending on the elevation of the satellite. From 0-degree to 20-degree elevation, the signal propagation, with frequent multipath propagation, and significant signal blockage from obstacles, is similar to the propagation of terrestrial systems. From 20 to 40 degrees there are practically no multiple paths other than in the form of diffuse reflections and there is less signal blockage. Above 40 degrees the propagation improves, and over 70 degrees elevation the propagation is always of excellent quality. Table 7.2 summarizes the MSS characteristics as a function of orbit [3–7].

7.2.3 Satellite Launching

Satellite launching is a crucial action in the creation of a satellite telecommunication system. Launching into LEO or MEO can be done in one step. To launch a satellite into GEO, however, three successive orbits are required as shown in Figure 7.8. The satellite is first transported into a circular parking orbit. Ignition of the rockets in the third stage brings the satellite into an elliptical *geostationary transfer orbit* (GTO). When the satellite is at the apogee of the GTO, an apogee

Table 7.2
Comparison of MSS Characteristics

Characteristics	LEO	MEO	GEO
Altitude (km)	600–1,500	9,000-11,000	35,800
Number of satellites	40–80	8–20	2–4
Orbital period (hr)	1–2	6–8	24
Two-way signal delay (ms)	10–15	150–250	480–540
Satellite lifetime (years)	3–8	8–12	10–15
Elevation angle	medium	best	good
Call handover	frequent	infrequent	none
Handheld operation possible	yes	yes	restricted
Space segment cost	highest	lowest	medium
Gateway cost	highest	medium	lowest
Operation	complex	medium	simplest
RF output power	lowest	medium	highest

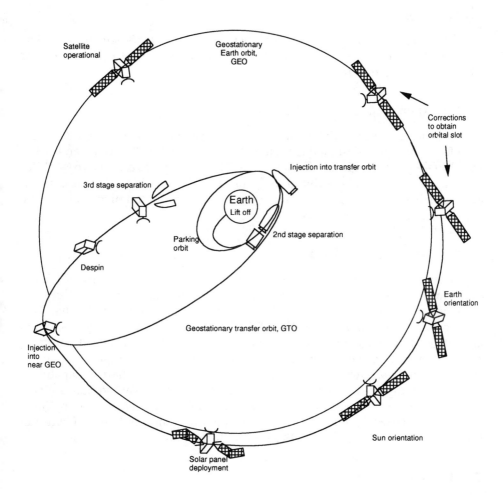

Figure 7.8 Launch and orbit insertion of a GEO satellite.

kick motor is fired to bring the satellite through "near GEO" (an elliptical orbit with an apogee near to geostationary height) into the geostationary orbit.

The vehicles used for satellite launching are divided into two categories: reusable launchers and *expendable launcher vehicles* (ELVs), which are destroyed in space during the course of their mission.

7.2.3.1 Reusable Launchers

Reusable launchers are the U.S. Space Shuttle and the Russian Energia. These *Space Transport Systems* (STSs) launch one or, depending on size, several satellites into a low Earth parking orbit between 300- and 500-km altitude. The satellite then must use its own propulsion for injection into the GTO, near GEO, and finally into its orbital slot.

7.2.3.2 Expendable Launcher Vehicles

Expendable launcher vehicles such as the European Ariane (see Figure 7.9, the only ELV especially designed for satellite launching), the U.S. Delta and Atlas, the Chinese Long March, and the Japanese H-2 rockets place one or two satellites successively into parking orbit, transfer orbit, and near GEO; whereas expendable launchers such as the U.S. Titan and the Russian Proton place the satellite into its orbital GEO slot.

7.2.3.3 Satellite Launching Sites

Satellites usually are designed and manufactured specifically for a certain launcher. Presently more than 50% of the telecommunication satellites are launched with an Ariane ELV at Kourou spaceport in French Guiana. This spaceport is conveniently located at 5.2°N, which allows taking maximum advantage of the Earth's rotation to impart velocity to the satellite and minimize required energy for launching and maneuvering into orbit. Figure 7.9 shows Ariane 5, the latest version of the Ariane series with a dual-launch capability. Other launching sites are Cape Canaveral at 28.5°N in the United States, Kagoshima in southern Japan at 32°N, Taiyuan satellite launch center in China 640-km southwest Beijing, and the Baikonur Cosmodrome launch facility of the Krunichev Space Center in Kazakstan (48°N) [5,8].

7.2.4 Antennas

The function of an antenna for satellite transmission is to provide shaped uplink and downlink beams for the transmission and reception of the communication signals as well as for the satellite telemetry and tracking purposes. Various solutions are applied for onboard satellite and Earth station antennas such as parabolic single and dual polarized, dual-band multihorn feed transmit and receive, dual gridded (for horizontal and vertical polarization), as well as active (steerable and switchable) microstrip direct radiating antenna arrays. In addition to the characteristics summarized for radio-relay antennas in Subsection 5.2.2, satellite transmission antennas must meet specific requirements.

1. Common requirements for satellite and for Earth station antennas:
   - Low noise temperature, so that the effective noise temperature at the receive side of the Earth station, which is proportional to the antenna

Figure 7.9 Ariane 5 (courtesy of Arianespace).

temperature, can be kept low in order to reduce the noise power within the downlink carrier bandwidth;
- High antenna gain-to-noise temperature ratio *G/T*, the antenna performance *figure-of-merit* (FOM) defined as the ratio between the antenna receiving gain and the system noise temperature;
- High *effective isotropic radiated power* (EIRP), the measure of transmitted power generated by the high-power amplifier and focused by the antenna.

2. Requirements specific for satellite antennas:
   - Low weight and small dimensions to facilitate launching and orbital station keeping;
   - (Optionally) reflectors in orbit deployable.

3. Requirements specific for Earth station antennas:
   - High directive gain, to focus its radiated energy into a narrow beam toward a satellite without interference to the satellites in adjacent orbital slots;
   - Easily steerable, so that a tracking system (if required) can point the antenna beam accurately toward the satellite taking into account the satellite's drift.

4. Requirements specific for ship Earth station antennas:
   - Highly efficient azimuth and elevation control that is immediately responsive to the various ship's movements such as rolls, pitches, yaws, heaves, sways, and surges in response to sea and land wind, due to change of course, and speed acceleration and deceleration.

Satellite antennas in relation to their footprint (covered portion of the Earth's surface) are divided into the following four categories:

- Global antennas, with a beam covering one-third of the Earth's surface, which is typical for C-band operation;
- Hemi(spherical) antennas, covering a "hemisphere" of about 20% of the Earth's surface, thus serving a continent or other large geographical region, applied for C-band too;
- Zone antennas, covering major portions of continents, particularly those with heavy traffic;
- Spot antennas, beaming high power to high-traffic regions, hence enabling the use of Earth stations with smaller antennas and facilitating mobile use.

A spot antenna can use a single narrowbeam high-gain small reflector, or a single large dish can produce many spot beams using many small feed horns positioned such that their signals are reflected in separate narrowbeams. Different spot antennas or feed horns can be switched on and off in compliance with traffic requirements. Spot antennas produce a high EIRP and facilitate frequency reuse and the use of less sensitive Earth stations. An even more advanced application of spot beaming is the recently introduced "scanning spot beam system." This system, derived from radar application, essentially combines TDMA access with fast-moving electronically steered antenna beams from a phased-array antenna. At a sweep rate of typically 0.01 sec, various Earth stations are polled in an TDMA procedure. This system, for example, with a fixed spot beam for coverage of major cities and a scanned spot beam for the remainder of the country, saves energy and radio spectrum and reduces Earth station cost. Typically the beam will cover 1% of a nation at any instant as opposed to a beam continuously covering the entire country at a much lower power flux-density.

Earth station antennas are used with reflector diameters ranging from 18m (previously 32m) to 0.2m depending on the EIRP of the corresponding satellite and the required service. One antenna used to be required for each corresponding satellite. Recently introduced multibeam Earth station antennas support simultaneous communication with up to 20 satellites. Instead of reflecting the signals from a reflector into a single focus, a multibeam antenna reflects the signals from a reflector to a specially curved rectangular plate. This plate functions as a "focal region" (instead of a focal point). A number of detectors are located in this focal region, each communicating with a different satellite. If satellite tracking is required, the small detectors can move instead of moving the large reflector [5,9].

7.2.5 Satellites

A telecommunication satellite is a radio-relay station temporarily located (5 to 18 years) in space to receive radio signals from Earth stations, amplify the signals, and transmit them back to other Earth stations. To counteract the forces acting on the satellite movement, and to keep the antennas of a satellite constantly pointing to a specific region of the Earth, the satellite must be stabilized in its orbital slot. Figure 7.10 illustrates the three axes that need to be controlled:

- The yaw axis, with movement in the direction toward the Earth;
- The pitch axis, perpendicular to the orbital plane pointing south;
- The roll axis, in the direction of the orbital velocity.

In relation to the applied spacecraft orientation toward the Earth, satellites are divided into two categories: spin-stabilized satellites and three-axis stabilized satellites.

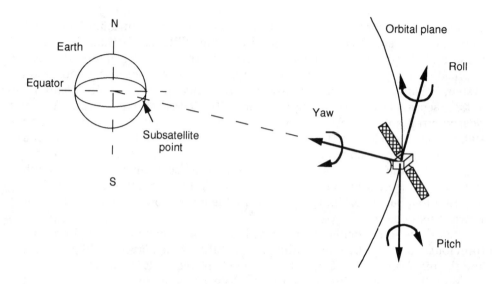

Figure 7.10 Three axes of satellite stabilization in orbit.

Both categories are shown in Figure 7.11 with the INTELSAT satellites series I to VII-A. INTELSAT series I, II, and III are of the complete satellite spin-stabilized construction. The series IV and VI are dual-spin satellites, whereas the series V and VII are three-axis stabilized satellites. Figure 7.12 is a photograph

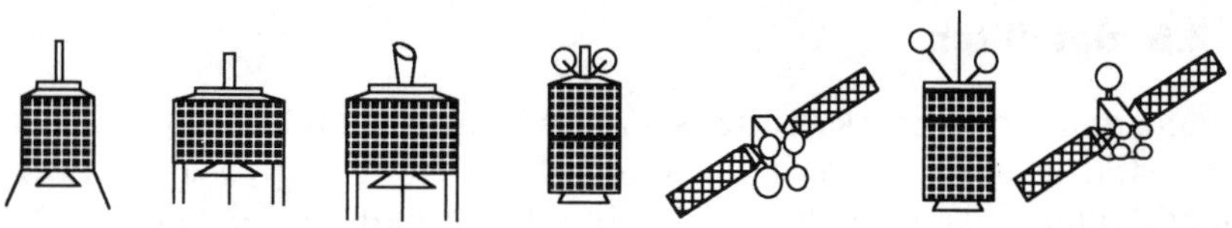

INTELSAT	I	II	III	IV	V	V-A	VI	VII	VII-A
Year	1965	1966	1968	1971	1980	1985	1989	1992	1994
Transmission capacity:									
Telephone channels	240	240	1200	6,000	12,000	15,000	24,000	18,000	22,500
with DCME							120,000	90,000	112,000
TV channels	1	1	1	12	2	2	3	3	3

Figure 7.11 INTELSAT satellites series I to VII-A.

Figure 7.12 INTELSAT 706 satellite in preparation for launching (courtesy of Arianespace).

of an INTELSAT VII satellite prior to placing in an Ariane 4 launcher at Kourou in May 1995.

A satellite basically consists of following subsystems:

- The payload;
- The structure;
- The telemetry tracking and command;
- The attitude and orbit control;
- The reaction control;
- The apogee boost motor;
- The electrical power.

7.2.5.1 Spin-Stabilized Satellites

The spin-stabilized satellite spins around its principal axis of inertia, whereby the gyroscopic stability is obtained by the rotation of the spacecraft's body. To lock the antenna on to the Earth, satellites with a single global antenna usually despin the antenna whereas in so-called "dual-spin satellites" the whole payload and its supporting structure are despun by counterrotation, thus permitting the accommodation of several antennas. Spin-stabilized satellites are invariably cylindrical in shape and symmetrical about the spin axis. They cannot have broad-span solar-cell arrays, relying instead on solar panels flush-mounted on their external surfaces.

7.2.5.2 Three-Axis Stabilized Satellites

The three-axis stabilized satellite obtains its gyroscopic stability by one or more momentum wheels or a set of reaction wheels. Three-axis stabilized satellites have less symmetry and power-generation restraints than the spin-stabilized satellites since the solar cells are mounted on external wing-type panels that can be unfolded in near-GEO and accommodate more solar cells than can be flush-mounted on the satellite surface.

7.2.5.3 The Payload

The payload or communication subsystem is the actual radio-relay station consisting of antennas, transponders, and high-power amplifiers. Figure 7.13 gives a simplified schematic diagram of a typical payload. The transmit/receive antenna array is connected to an antenna feed network that separates the (received) uplink from the (transmitted) downlink beams and distributes the downlink signals to the relevant antennas.

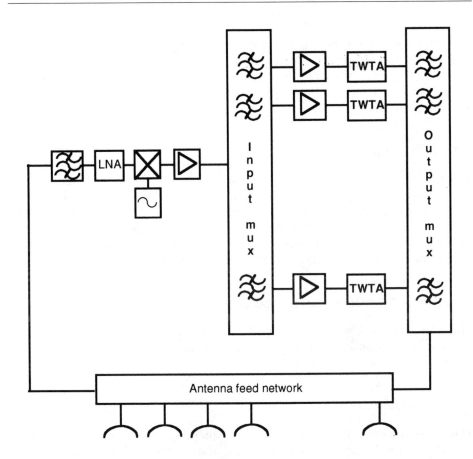

Figure 7.13 Payload schematic diagram.

The transponder consists of a bandpath filter, a *low-noise amplifier* (LNA), a wideband receiver/downconverter, an amplifier, and an input multiplexer.

The individual channels are separately amplified in a power amplifier. Solid-state amplifiers are used for output power up to about 10W. For GEO operation, however, 50W is typically required at C-band and 80W to 120W at Ku band, which with sufficient reliability and efficiency can only be produced in *traveling wave tube amplifiers* (TWTAs). To enable operation of the TWTA in a linear mode a driver amplifier is normally used in front of the TWTA.

An *output multiplexer* (OMUX) combines the powerful signals from the TWTAs and simultaneously surpresses the signal harmonics and spurious noise generated by the TWTAs. Variable power dividers may be used at the input of the OMUX to provide the necessary power split for the various (by ground command) selected antenna transmit coverages.

More advanced digital transponders can regenerate the received signal and switch them to different destinations.

The high reliability required for an uninterrupted operation throughout the lifetime of a satellite is achieved by duplicating and automatic standby switching for critical elements.

7.2.5.4 The Structure

The structure or platform is the mechanical support for the payload and all other satellite subsystems. The structure, beyond supporting the subsystems, carries the mechanical loads generated during the launch and provides micrometeorite and thermal protection for the internal subsystems. Therefore, the structure must be light in weight yet very stiff in order to minimize vibrations. *Carbon-fiber-reinforced plastics* (CFRP) are normally used.

7.2.5.5 The Telemetry Tracking and Command

The *telemetry tracking and command* (TT&C) subsystem monitors all satellite subsystems and continuously transmits to Earth sufficient information for determination of the satellite attitude, status, and performance.

7.2.5.6 The Attitude and Orbit Control

The major functions of this subsystem are to maintain accurate satellite position and antenna pointing for the operational life of the satellite and to control satellite maneuvers in the transfer and near GEO orbits. Special sensors, such as Earth censors and sun sensors, provide the required data for the satellite control.

7.2.5.7 The Reaction Control

The reaction control corrects orbit drift and changes the longitude location of the satellite upon commands given by the ground control via TT&C in response to the attitude and orbit control data. This subsystem executes keeping the station within the orbital slot. The control is executed by radial and axial thrusters, operated from gas under pressure in tanks or by small rocket motors.

Keeping the station in east-west (longitudinal) direction typically requires 5% of the annual fuel consumption; whereas for the north-south (latitudinal) direction, 95% is required. To extend the useful in-orbit life of an otherwise intact satellite, north-south station keeping is increasingly terminated. This results in a 95% reduction in fuel consumption but leads to inclined orbit operation (see Figure 7.7), which might require modifications at the Earth stations to track the satellite inclination as well as additional coordination in case of shared radio-relay frequency use.

7.2.5.8 The Apogee Boost Motor

The apogee boost motor is a solid rocket built into the satellite to propel it from transfer orbit to GEO. The motor is fired by a ground command, and after firing the spent motor remains as part of the in-orbit mass of the satellite.

7.2.5.9 The Electrical Power

Photovoltaic cells care for the electricity supply of satellites. At GEO the solar radiation is 1.45 kW/m^2, thus a solar array with an efficiency of 15% generates over 200 W/m^2. Micrometeorites, however, reduce the efficiency by 20% to 40% during the operational lifetime of the satellite.

On a spin-stabilized satellite the solar array consists of concentric cylindrical panels around the body of the satellite. The capacity thus is limited by the size of the satellite and ranges between 100W and 1,000W. Moreover, only one-third of the solar array is exposed to the sun at any time. Three-axis stabilized satellites use rigid panels hinged together in two wings that can be deployed once in GEO. Each wing is connected to a boom that rotates one revolution per day to keep the face of the solar cells pointing toward the sun; thus typical dc power rates of 1 to 2 kW can be obtained.

The output of the solar array is connected to a usually NiCd battery to obtain a more constant voltage and to bridge the regular eclipse periods. To safeguard the battery life, special efforts are required from the satellite's thermal control system to keep the battery temperature within confined limits. In addition to fuel consumption for the reaction control, solar array efficiency decrease, and battery aging are the major factors limiting the operational lifetime of a satellite.

Figure 7.14 shows a three-axis satellite (*Kopernikus,* a German domestic telecommunication satellite) with two solar panels connected to a rotating boom. To improve the efficiency of the solar panels, a fixed optical solar reflector is placed perpendicularly to each solar panel wing. The capacity of the solar generator (at life's end) is approximately 1,500W. Two redundant NiCd batteries, each with a capacity of 35 Ah, support operation during sun eclipse periods. The satellite provides TV, radio, and data transmission throughout Germany, operating in the Ku band, and was launched October 12, 1992.

Figure 7.14 *Kopernikus,* a German domestic telecommunication satellite.Photo: Bosch (courtesy of Bosch Telecom GMBH).

7.2.6 Earth Stations

Earth stations communicate with one another, usually in a P-MP network via a satellite transponder. An Earth station thus must be able to receive and amplify the weak signals received from the satellite and transmit strong signals to the satellite. A simplified functional block diagram of a digital Earth station is given in Figure 7.15. In order to transmit the digital baseband information over a satellite channel that is a bandpass channel, it is necessary to transfer the digital information to a carrier wave at the appropriate frequency by a technique called *digital carrier modulation.* In the transmit direction the encoded signal (with error-correction coding) is modulated onto an IF frequency and fed to an upconverter that transfers the IF to the uplink frequency. The IF is 70 MHz for a transponder with a 36-MHz bandwidth and 140 MHz for transponders with a higher bandwidth (for example, 44-, 54-, 72-, or 90-MHz). A *high-power amplifier* (HPA), usually with a TWT (in the range of 100W to 3 kW) or klystron (1 to 3 kW) for narrowband operation, or an Impatt diode amplifier or GaAs FET amplifier, is used in the case of low-power application, for example, below 400W. At the receive side the weak RF signal passes a low-noise amplifier

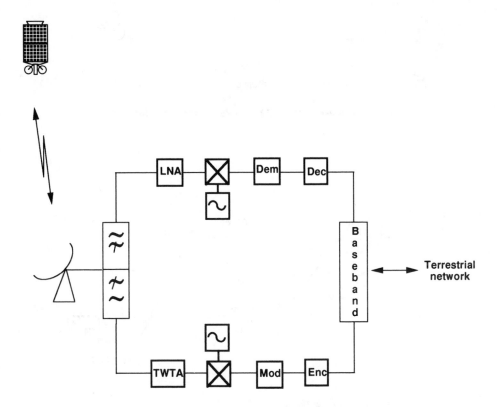

Figure 7.15 Functional block diagram of a digital Earth station.

before being downconverted, demodulated, decoded (to effect error correction), and fed to the BB equipment for delivery to the terrestrial network. The sequential stages of amplification and attenuation in a complete satellite system are shown in Figure 7.16.

Earth stations can be divided into four groups:

- Large Earth stations for international public service, also called "gateway" or hub stations, interfacing a satellite network with the terrestrial telephone and data networks;
- Middle-range Earth stations designed for direct corporate applications, for example, for *international business service* (IBS) networks;
- Small Earth stations usually for mono user stations with antenna diameters of 1.8m to 4.5m called VSAT and antenna diameters between 0.2m and 1.2m called USAT (for *ultra small aperture terminal*).

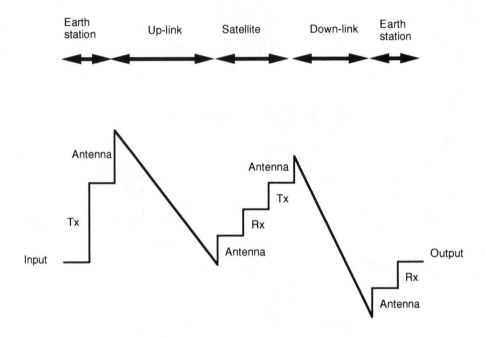

Figure 7.16 Satellite system level plan.

- Mobile Earth stations ranging from ship-, airplane-, and vehicle-mounted Earth stations, via transportable Earth stations, to handheld dual (cellular radio and satellite) terminals.

Earth stations must conform to international technical and operational standards and need to be approved by the satellite operator before taking up operation with a satellite. Specific Earth station categories with different minimum quality requirements for various applications have been defined by the satellite operators; for example, categories A through H and Z, ranging from international voice data and TV for category A and domestic leased service for category Z for operation via the INTELSAT network [3,6].

7.2.7 Multiple Access

In terrestrial cable and radio-relay networks, a substantial amount of the installed transmission capacity is unused and available to meet future traffic increases. In contrast, for transmission via expensive satellites with a still relatively short operating lifetime of 10 to 15 years, the luxury of unused capacity cannot be afforded. Sharing a satellite's transmission capacity by several users, therefore, is a necessity. Basically this is achieved by *multiple-access* (MA) techniques (as already explained in Subsection 6.2.8), which were originally developed for satellite application.

In the case of analog transmission, FDMA is applied in the following versions:

- *Single channel per carrier* (SCPC), where each telephone channel independently modulates a separate RF carrier and is transmitted in parallel (separated by a protective guard band) with $n - 1$ other carriers to the satellite, thus occupying $1/n$ of the transponder bandwidth. FDMA-SCPC is mainly used for analog small capacity systems.
- *Multi channel per carrier* (MCPC), where the transmitting Earth station frequency-division multiplexes several single-sideband suppressed carrier telephone channels into one carrier baseband that frequency modulates the RF carrier.
- *Companded FDM* (CFDM), applied on MCPC where the telephone channels are compressed in amplitude before modulation and expanded after demodulation in order to improve the SNR; typically a 36-MHz transponder can accommodate 2,100 companded voice channels instead of 1,100 uncompanded channels.

For digital transmission TDMA is applied so that several Earth stations in a satellite communication network use the same carrier for transmission with short bursts via a common transponder on a time division base within a periodic time frame called the TDMA-frame. The transponder receives one burst at a time, amplifies and upconverts it, and retransmits it back to all the Earth stations of the network. As shown in Figure 7.17(a) each Earth station can receive the entire retransmitted burst stream and extract the relevant bursts.

In a more advanced approach called *satellite-switched* TDMA (SS-TDMA) [see Figure 7.17(b)], a switch on board the satellite, controlled by a ground station, switches the received bursts to the desired Earth stations, thus, in this example, increasing the traffic capacity by a factor of three.

With CDMA each user employs a particular code address and spreads its signal over the whole available transponder bandwidth so that all Earth stations of a network can transmit simultaneously without frequency or time division and with low interference.

A capacity increase of satellite links is also obtained by means of *digital speech interpolation* (DSI) (see Section 2.5).

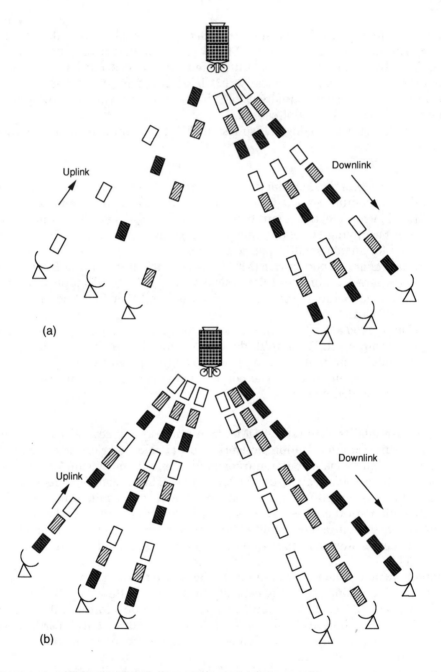

Figure 7.17 (a) TDMA and (b) SS-TDMA operation.

These FDMA and TDMA operation modes assume a fixed assignment of the channels, respectively, bursts to the various Earth stations of a network. This, however, implies that a channel or a burst remains vacant during nonsignal periods of an Earth station, and thus transponder capacity is wasted. To improve transponder efficiency, therefore, *demand assignment multiple access* (DAMA) has been introduced. DAMA, in a complex procedure, uses dynamic satellite circuit sharing from a common pool, allocating space segment resources (RF-carriers or RF-channels for analog transmission and bursts for digital transmission) on demand under control of a ground station. When a call is completed, the used resource is returned to the common pool for reassignment. DAMA thus adapts the transmission to traffic changes and reduces call blocking.

The first DAMA version introduced by INTELSAT in the early 1970s is the *Single-channel-per-carrier PCM multiple-Access Demand assignment Equipment* (abbreviated SPADE). SPADE applies distributed control, without needing a master station, using a computer-controlled *demand assignment signaling and switching* (DASS) system at each Earth station for self-assignment of circuits. A network idle/busy table, based upon continually updated circuit allocation status data exchange between the participating Earth stations, informs all Earth stations which channel can be assigned.

One of the simplest DAMA modes applied for packet-switched data signals is the ALOHA mode of access developed at the University of Hawaii, whereby an Earth station starts transmitting "randomly" and from the quality of the signal retransmitted by the satellite can deduct whether its signal was received at the destination without collision or overlapping with signals from other Earth stations of the same network or if retransmission needs to be attempted. The ALOHA mode is a tradeoff between low-cost user equipment and inefficient channel utilization with a maximum channel throughput of only 18%.

Slotted-ALOHA (S-ALOHA) synchronizes the start of transmission for all stations of a network and provides time slots of the same duration as a packet length. A station can start transmitting a packet only at the start of a time slot, thus eliminating partial overlapping and increasing the maximum channel throughput to 36%.

Reserved-ALOHA further improves the efficiency as users are given the opportunity to place a reservation for a slot over a given number of frames that then cannot be accessed by other users.

Modern DAMA systems, beyond providing 100% channel throughput by typically requesting service reservation from a central station through an order wire channel in a fully automatic computerized procedure, include advanced facilities such as guard band reduction and adaptive link quality and power control. With guard band reduction RF carrier inaccuracies of the other Earth stations of a network are evaluated and compensated, reducing both the adjacent carrier interference and the width of the guard band and increasing the bandwidth available for actual traffic. With adaptive link quality and power control for an optimized bandwidth/power balance, at favorable atmospheric conditions a significant amount of link margin can be used for additional traffic.

DAMA equipment has reached maturity and now provides the most flexible and cost-effective multiple access requiring modest up-front investment [3,10,11].

7.3 COEXISTENCE OF RADIO-RELAY AND SATELLITE SYSTEMS

Coexistence of radio-relay and satellite systems is necessary in many cases and is possible provided sufficient care is taken to minimize mutual interference. Figure 7.18, as an example, shows the various interference possibilities for a

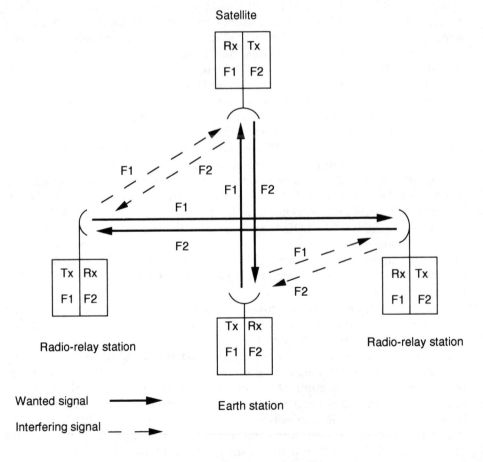

Figure 7.18 Interference paths between radio-relay and satellite systems.

radio-relay link operating on the same frequency pair as a nearby Earth station. The solid lines indicate the wanted signal and the dashed lines the interfering signal paths.

To limit the possible interference to an acceptable level, CCIR has defined *sharing conditions* mainly concerning the radiated power of the radio-relay equipment. The maximum permissible EIRP, for example, has been defined in Recommendation 406-6; whereas a coordination area has been defined (in Recommendation 359-5) around Earth stations in such a way that the possibility of mutual interference with terrestrial stations situated outside that area may be regarded as negligible. Normally, such coordination areas have a radius of 200 to 300 km around the Earth station.

The coexistence conditions as previously outlined do not apply for the operation of mobile Earth stations for which exclusive frequency bands are always required without frequency sharing with other services.

A special problem occurs when satellites, toward the end of their operational life, apply inclined orbit operation (see Subsection 7.2.2.3) to save fuel. From that moment the orbital inclination will change continually and an Earth station operating with such a satellite may have to track the inclined satellite with its antenna beyond the limit so far coordinated for the shared radio-relay operation, so new coordination might be required. This mainly concerns stations in countries at higher latitudes. In CCIR Report 1142 the regions that are liable to such interference from inclined orbit satellites are indicated [12,13].

7.4 SPECIFIC APPLICATIONS

Satellite transmission is the only possible choice where worldwide coverage is required. Moreover, it can also be economical for domestic and regional operation in the following circumstances:

- Scattered population in a large number of widely dispersed towns and villages (with less than about 5,000 inhabitants) where the terrain is hostile or difficult, such as deserts, mountains, and islands;
- National and/or regional TV distribution;
- Rapid implementation of national networks;
- Flexible network configuration with changing locations;
- Temporary use in, for example, the event of a natural disaster or war, or to support major exhibitions and sports events of national, or even more, international interest;
- Corporate networks connecting widely dispersed locations [2].

References

[1] Pritchard, Wilbur L., "The History and Future of Commercial Satellite Communications," *IEEE Communications Magazine,* Vol. 22, No. 5, 1984, pp. 22–37.

[2] Huurdeman, Anton A., *Transmission: A Choice of Options,* Paris: Alcatel Trade International, June 1991.

[3] Ha, Tri, T., *Digital Satellite Communications,* New York: McGraw-Hill Publishing Company, 1990.

[4] Nelson, Robert A., "Satellite Constellation Geometry," *Via Satellite,* Vol. 10, No. 3, 1995, pp. 110–122.

[5] Gallagher, Brendan, editor, *Never Beyond Reach: The World of Mobile Satellite Communications,* London: INMARSAT, 1998.

[6] Minoli, Daniel, *Telecommunications Technology Handbook,* Norwood, MA: Artech House, 1991.

[7] Sharrock, Stuart, "Four Weddings And A Regulatory Nightmare," *Communications International,* Vol. 22, No. 12, 1995, pp. 4–8.

[8] Floury, G. and J.-L. Cazaux, "Benefits of New Technologies in Satellite Transponders," *Electrical Communication,* 4th quarter 1994, pp. 371–377.

[9] "Multibeam Antenna," *Mobile Europe,* Vol. 3, No. 12, 1993, p. 22.

[10] Herter, E., and H. Rupp, *Nachrichtenübertragung über Satelliten,* Berlin, Heidelberg, New York, and Tokyo: Springer-Verlag, 1983.

[11] Dankberg, Mark, "Advanced Technology in DAMA Networking," *Telecommunications,* Vol. 30, No. 8, 1996, pp. 68–77.

[12] Huurdeman, Anton A., *Radio-Relay Systems,* Norwood, MA: Artech House, 1995.

[13] CCIR Report 1142, "Frequency sharing between the fixed service and the fixed satellite service using satellites in slightly inclined geostationary orbits," 1990.

The Choice of Solutions 8

8.1 GENERAL

The five transmission media described in this book—copper lines, optical fiber, radio relay, mobile radio, and satellites—are basically complementary. The choice between them depends primarily on a detailed evaluation of transmission requirements, the prevailing geographical and infrastructural conditions, and implementation and operation cost. Each media has its own domain of maximum cost effectiveness. This complementary nature is demonstrated in Figure 8.1, where for each media the optimum relation with transmission capacity and covered distance is roughly indicated. The figure has obviously been simplified to emphasize its message that the domains of maximum cost effectiveness are as follows:

- Copper lines: for small- and medium-capacity transmission over a short distance;
- Optical fiber: for medium- and high-capacity transmission over any distance;
- Radio relay: for medium- and low-capacity transmission over short and medium distances and for the local loop;
- Mobile radio: for small-capacity transmission and mobility over short and medium distances.
- Satellite: for low- and medium-capacity transmission over medium and long distances for dispersed area, and for mobility over any distance.

The typical domains of maximum cost effectiveness for each media are given in the "Specific Applications" sections at the end of Chapters 3 to 7. None of the media will optimally meet all the requirements. In the overlapping

The information contained in this chapter is almost entirely taken from Anton A. Huurdeman, *Transmission: A Choice of Options,* Paris: Alcatel Trade International, June 1991.

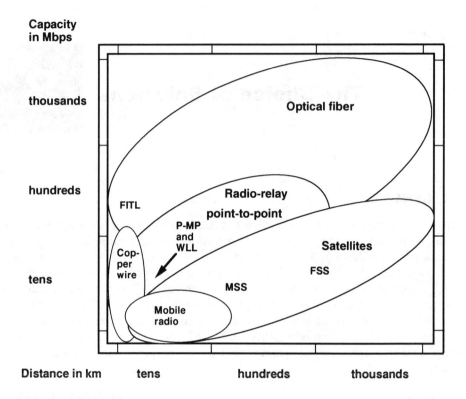

Figure 8.1 Complementary nature of the transmission media.

areas especially, to determine the most cost-effective solution, careful consideration has to be given to the future rate of traffic increase, the geographical conditions, and the operational requirements.

8.2 COMPARATIVE CONSIDERATIONS

The complementary nature of the transmission media as well as the combined (hybrid) application of two or more media for one route or network is illustrated with a few examples in the following subsections.

8.2.1 Access Network

A special situation is given in the local loop where FITL, WLL, and in the near future even satellite can provide cheaper solutions than traditional copper wire.

Figure 8.2 demonstrates this situation with an approximate indication of the relation between distance and cost of the four solutions.

8.2.2 Transocean Transmission

Another special case is transocean transmission. Despite the tremendous increase in capacity of the optical fiber submarine cable and the frequent prediction of the demise of transocean satellite transmission, satellite transmission still maintains a share of the traffic. The big advantages of satellite transmission are flexible P-MP connectivity direct to the centers of traffic at a very high reliability (typically 99.9%) and cost insensitivity to distance. In contrast, the essentially cost-effective circuits on submarine optical fiber still need to be routed through vulnerable terrestrial routes. Submarine cable restoration accounts for roughly 3% of Intelsat's annual revenue; satellite-restoring cable routes apparently do not appear.

8.2.3 Radio-Relay Back-up for Optical Fiber Routes

Radio relay is widely used as a back-up for terrestrial cable routes. Dual routing of cable and radio relay either in parallel or, even better, in a meshed network

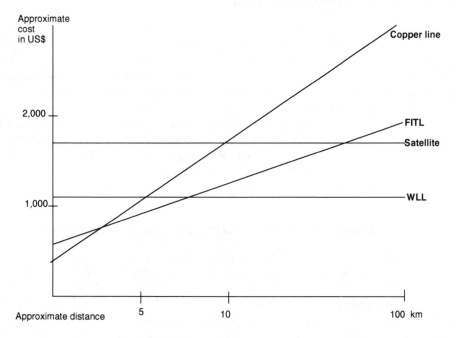

Figure 8.2 Approximate cost/distance relation for local loop solutions.

has already been adopted by many administrations for analog transmission. Now with digital transmission prevailing, dual routing is even more important to guarantee the uninterrupted transmission of data.

8.2.4 Nonstandard Applications

Optical fiber, no doubt, is the major transmission medium for all "standard" applications that do not require mobility, a quick implementation, nor the crossing of difficult geographical areas; otherwise satellite and radio-relay transmission are more cost effective. When earthquakes happen, volcanoes erupt, or war and revolutions create havoc with the infrastructure, satellite communication is often the only link with the outside world. Even in highly developed Japan, satellite and mobile radio were the only means of communication between the rescue workers, the authorities, and the dispersed citizens when the big earthquake flattened Kobe in 1995.

Satellites will continue to reach locations that are not accessible for fiber or radio-relay links, especially in the P-MP mode.

8.2.5 High Transmission Capacity

Optical fiber line transmission equipment is already operating with transmission capacities of 2.5 Gbps for terrestrial routes and 5 Gbps for submarine cable. Radio-relay equipment with transmission capacities of 155, 310, and 622 Mbps at first view does not appear to keep pace with the high transmission speeds of optical fiber. Radio-relay systems, however, usually operate a number of RF channels in parallel. A high-density radio-relay route can operate some 20 RF channels in parallel, thus providing a total transmission capacity of, for example, 20 × 622 Mbps equals 12.5 Gbps, well in line with the transmission capacity of optical fiber routes. An actual example is the trans-Canadian optical fiber route with 622-Mbps radio-relay equipment used partly in parallel and partly (in mountainous areas) instead of optical fiber cable.

8.2.6 New Network Free From Existing Equipment Constraints

The unification of West and East Germany provides a unique opportunity for comparing different transmission media and analyzing the cost structure of a fully digital telecommunications network. The recently completed construction of a new telecommunications network in former East Germany (in which by Western standards almost no telecommunications infrastructure existed) starting from a "green field" situation clearly demonstrates the complementary nature of the transmission media. Even in this network, which could be constructed using the most modern equipment and with few constraints about existing systems but where early implementation was of highest priority, the investment is almost equally divided between optical fiber and radio-relay systems. Table 8.1 illustrates the investment structure, indicating approximate percentages elaborated from initial project planning data.

8.2.7 Future Trends

The market development shown in the next chapter also predicts that radio-relay and satellite transmission will remain complementary to and competitive with optical fiber, each media comfortably keeping its market share.

Radio-relay in high-capacity transmission systems—thus far used primarily for national and international long-distance routes—will shift to low- and medium-capacity systems for access and interconnection links in cellular networks and as WLL.

Satellite systems, likewise, will shift from long-distance application to medium-distance application, increasingly with VSATs and USATs, and in global personal communication MSS networks.

Table 8.1
Approximate Cost Structure of the Digital Telecommunication Network
in the Former East Germany

Investment	Percentage
Total telecommunications network investment structure	
Outside line plant	40
Switching	25
Transmission	10
Cable	9
Sites and buildings	9
Terminal equipment	7
Total	100
Structure of the (10%) transmission	
Multiplexing	56.5
Optical fiber (including FITL)	18.5
Radio relay (including WLL)	15
Earth stations	4
Total	100
Structure for transmission media alone	
Optical fiber	50
Radio relay	40
Earth stations	10
Total	100

Mobile radio and the new MSS networks, rather than replacing the established media, will create significant additional traffic in areas not cost effective for optical fiber and radio relay and thus will indirectly require more investment on optical fiber and radio relay in the distribution networks to handle the additional traffic.

8.3 SELECTION CRITERIA

Normally the transmission medium selection process in any given case requires a detailed evaluation of the prevailing conditions. As an aid to this evaluation process, various relevant selection criteria are summarized in Table 8.2.

Table 8.2
Main Criteria for Selecting Specific Transmission Solutions

	Best Solution				
Criteria	*Copper Line*	*Optical Fiber*	*Radio Relay*	*Mobile Radio*	*Satellite*
Transmission Capacity					
Low and medium	*	*	*	*	*
High		*	*		*
Very high		*			
Distance					
Short	*	*	*	*	
Medium		*	*		*
Long		*	*		*
Very long		*			*
Geology					
Flat area, soft soil	*	*		*	
Mountainous			*	*	*
Jungle			*	*	*
Marshy and lakes			*	*	*
Oceans		*			*
Geography					
Industrialized area		*		*	
Urban	*	*	*	*	
Rural	*		*	*	*
Population Density					
Low, scattered				*	*
Medium	*	*	*	*	
High		*	*	*	
Infrastructure					
Electricity and roads					
Good	*	*		*	
Bad			*	*	*
Nonexistent				*	*
Right of way					
Easy to obtain	*	*			
Difficult to obtain			*	*	*
Existing cable ducts	*	*			
Existing radio-relay buildings and towers			*	*	

Table 8.2
(continued)

Criteria	Copper Line	Optical Fiber	Radio Relay	Mobile Radio	Satellite
Project Implementation					
Standard	*	*			
Short time			*	*	
Very short time					*
Environment					
Electromagnetic radiation	*				
Earthquake zone			*	*	*
Network					
Fixed stations	*	*	*		
Stations on flexible sites			*	*	*
Stations at short distances with frequent drop and insert	*	*			
Star topology	*			*	*
Ring topology		*	*		
Lower capacity spurs in high-capacity networks		*	*		
Access from public to cellular networks and interconnection of cells			*		
Minor extensions of copper line networks	*				
Private Customers					
Companies with various sites					
In urban areas		*	*	*	
In isolated areas			*	*	*
Pipelines, highways					
New		*		*	
Existing			*	*	
Communication required during construction			*	*	*
Special Circumstances					
International events (e.g., sport and festivals)			*	*	*
Emergency at natural disaster			*	*	*
Reconstruction after war or occupation			*	*	*
Flexible TV studio access			*		*
Operation Mode					
Mobile			*	*	
Dual routing					
Terrestrial		*	*		
Intercontinental		*			*

Transcription Market 9

9.1 GENERAL

The technical and legislative progress that is enabling the construction of the *Global Information Infrastructure* (GII)—as presented by U.S. Vice President Al Gore at the World Telecommunication Development Conference of the ITU held in Buenos Aires, Argentina, on March 21–23, 1994—is giving an enormous impetus to the telecommunication market. The telecommunication market (here defined as the total world market of telecommunications terminal equipment, cables, switching, and transmission equipment, excluding real estate and civil works (such as buildings, access roads, and cable laying) and without telecommunication operation services), roughly calculated at a value of US$150 billion in 1990 and US$200 billion in 1995, is estimated to further increase at an annual rate of 15% to about US$400 billion by the year 2000. This expansion is supported by annual growth rates of reportedly 25% to 40% in data communications, 25% in mobile communications, and still 8% to 9% in voice communications. By the beginning of the next century, the telecommunication market is expected to overtake the automotive industry in terms of economic importance. The geographical development of the telecommunication world market is roughly as shown in Figure 9.1.

The introduction of new telecommunication services require that ever greater volumes of information have to be circulated at ever higher speeds. To transport such high volumes of information, many new transmission networks will be required. Moreover, the computerization of the telecommunication equipment is causing a shift of control and routing functions from the switching to the transmission domain. The number of switching levels in national telecommunication networks is being reduced from typically five to three, supported by three transmission networks: a local transport network, a regional transport network, and a national transport network.

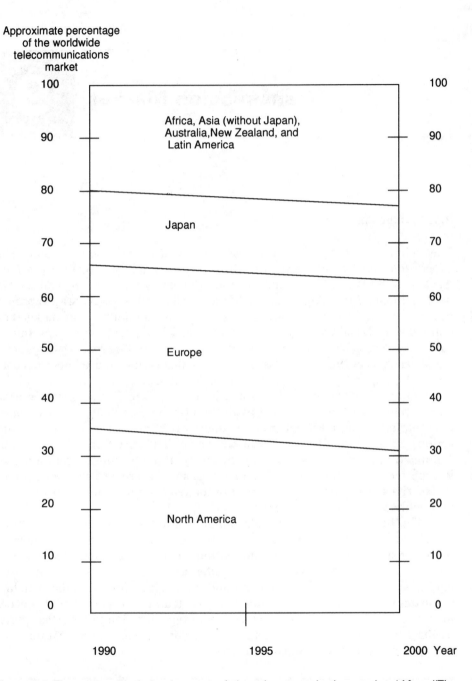

Figure 9.1 The geographical development of the telecommunication market (*After*: "The Sagatel Yearbook").

Access networks for N-ISDN and even more for B-ISDN require significant volumes of transmission equipment both for optical fiber transmission mainly on new lines as well as for broadband operation on existing copper lines.

Telecommunication networks thus are evolving from switch based toward transmission based. The transmission market, here defined as the total market for the transmission equipment and the telecommunication cable in the domains as described in this book, without real estate and civil works, will therefore contribute significantly to the telecommunication market growth.

9.2 TRANSMISSION MARKET DEVELOPMENT

An evaluation of various market studies and estimates indicates an approximate transmission market development as shown in Figure 9.2, in a simplified way linearly presented for the 10-year period from 1990 to 2000. The trend illustrated in Figure 9.2 is briefly commented on for each domain [1,2].

9.2.1 Multiplex

The increase of the multiplex market might be surprising at first because the large market of PCM equipment for the analog to digital conversion will completely shift to the switching domain for the existing analog circuits and to the terminals for the ISDN circuits. This loss, however, will be overcompensated by the expected equipment volumes for crossconnecting, digital data interfacing, voice and video compressing, and ISDN line terminating in addition to the continued large volumes of higher order multiplex and the still surviving modems.

9.2.2 Copper Lines

Copper cable is hardly used anymore in new distribution networks and will be replaced by FITL and WLL in the access networks. Symmetrical cable and coaxial line equipment thus will disappear. To extend the life of existing copper wire plant and delay the introduction of optical fiber cable, however, significant volumes of copper line enhancement systems will be required, such as pairgain and xDSL systems. Coaxial cable will still be used for CATV networks mainly in hybrid networks with optical fiber between head ends and distribution hubs. Furthermore, coaxial cable will increasingly be used as radiating cable to extend broadcasting and mobile radio services into tunnels, subways, and other confined areas where direct radio transmission does not meet the requirements [3].

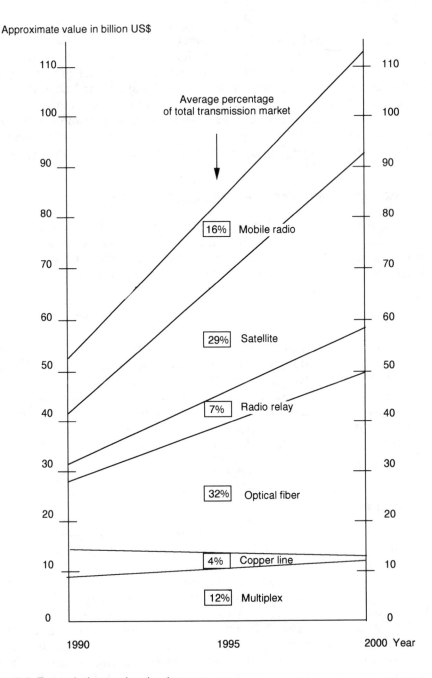

Figure 9.2 Transmission market development.

9.2.3 Optical Fiber

Optical fiber, because of its striking advantages, has become the dominant transmission medium and will cover the major part of additional required transmission capacity worldwide with new submarine cables and terrestrial links for the GII. FITL, still in FTTZ, FTTC, and FTTB versions but increasingly as FTTT, will become the rule rather than the exception for new access networks (see Figure 10.3).

9.2.4 Radio Relay

The radio-relay transmission market, many times forecasted to disappear, will not only survive but will even increase. It will shift from predominantly high-capacity long-distance transmission systems to highly integrated, easy-to-install, compact, low-capacity, short-distance systems as is illustrated in Figure 9.3.

More than 50% of the market increase will go to new WLL systems. Other even more optimistic market analysts forecast a US$10–13 billion market for WLL at the turn of this century [4,5].

9.2.5 Mobile Radio

The driving forces on the mobile radio market are well known under the acronyms GSM, D-AMPS, and JDC for the digital cellular radio systems; DECT, PACS, and PHS for the digital personal communication systems; TETRA for international trunked systems; and ERMES and FLEX for digital paging.

In the early 1990s optimistic market observers, including the author of this book, dared to forecast around 36 million cellular subscribers in 1995 and a maximum of 100 million at the beginning of the next century for cellular radio including PCN. Instead of the forecasted 36 million, almost 80 million analog and digital cellular subscribers worldwide were reported at the end of 1995. By the turn of this century more than 200 million cellular subscribers are now expected, roughly distributed as shown in Figure 9.4.

The enormous expansion of the GSM system illustrated in Figure 9.4 is supported by the operation of more than 175 GSM networks in over 90 countries at the end of 1996 and an expected 230 networks in over 100 countries by the end of this century.

The mobile radio market, in addition to the strong growing cellular radio systems, also includes the more constant paging and private mobile radio and private access mobile radio systems. Paging systems, currently with almost 100 million subscribers worldwide, are expected to grow to over 120 million by

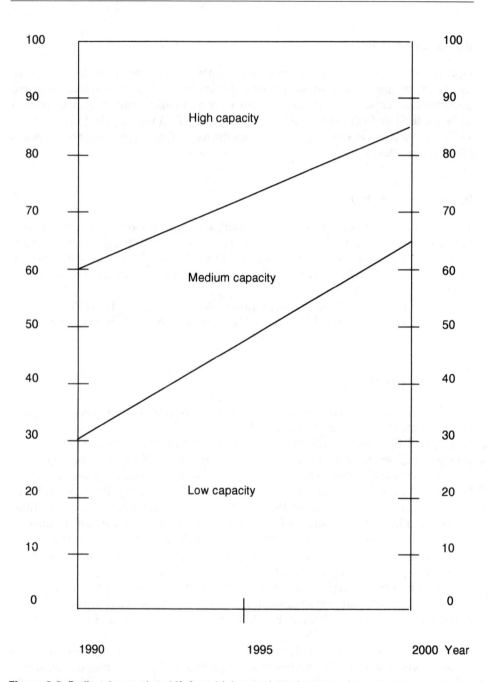

Figure 9.3 Radio-relay market shift from high-capacity to low-capacity systems.

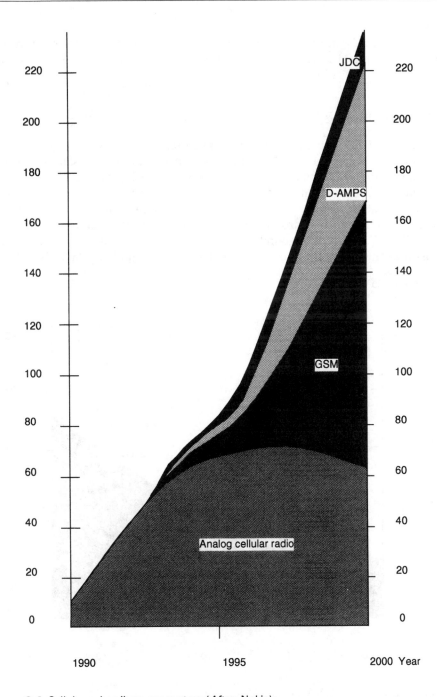

Figure 9.4 Cellular subscribers per system (*After*: Nokia).

the end of this century, contributing to the mobile radio market an average annual amount of roughly US$1.5 billion and PMR and PAMR contributing with roughly US$3 billion [6–8].

9.2.6 Satellites

Satellite transmission reportedly will experience the biggest expansion over the next few years. Some 155 commercial telecommunication satellites will be launched into GEO from 1996 to 2000, thus increasing the annual number of launchings from the current 15 to over 30. Furthermore, at least part of the numerous satellites for the announced future mobile satellites systems for personal communications (described in Chapter 10) will be launched into LEO and MEO. Figure 9.5, elaborated from information published in mid-1996, illustrates the distribution of the expected satellite market growth [9–10].

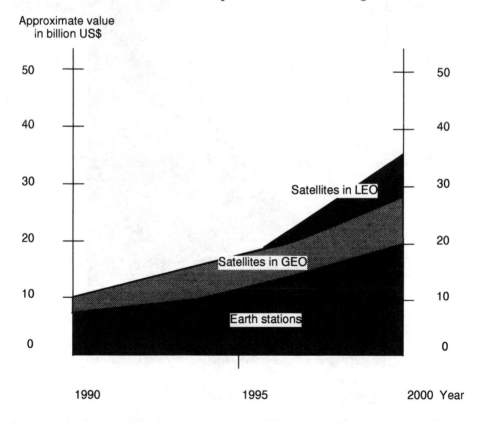

Figure 9.5 Expected growth of the satellite market.

References

[1] Shankar, Bhawani, "WTO and the Politics of Deregulation," *Telecommunications*, Vol. 30, No. 6, 1996. pp. 31–32.

[2] Masin, Michael T., "The Last Mile Telecomms Liberator," *Communications International*, Vol. 22, No. 10, 1996, p. 103.

[3] Järvinen, Hannu "Evolution of the Access Network," *Telecommunications*, Vol. 28, No. 9, 1994, pp. 21–24 and 134.

[4] Channing, Ian, "Where There's a WLL," *Communications International*, Vol. 22, No. 1, 1996, pp. 14–17.

[5] Huurdeman, Anton A., *Radio-Relay Systems*, Norwood, MA: Artech House, 1995.

[6] "World Cellular Subscribers 1988–1995," *Mobile Communications International*, Issue 28, Feb. 1996, p. 72.

[7] "Introduction of PCS 1900 Switch Gives NOKIA Turn-key Capability," *Discovery*, Nokia Telecommunications' Customer Magazine, Vol. 40, 1st quarter 1996, pp. 12–13.

[8] Hellström, Kurt, "GSM—the Technology for the 21st Century," *GSM World Focus 1996*, London, Mobile Communications International, p. 3.

[9] Katherine A. Schuerholz, "The Earth Station It's Not Just a Dish Anymore," *Via Satellite*, Vol. 11, No. 5, 1996, pp. 22–32.

[10] Boeke, Cynthia, and Fernandez Robustiano, "Via Satellites Global Satellite Survey," *Via Satellite*, Vol. 11, No. 7, 1996, pp. 16–26.

Future **10**

10.1 GENERAL

For almost 200 years, telecommunications developed as a profitable arm of national postal services, mainly providing for national and some international telegraphy and telephony communication. A major change from national analog telephone networks to global digital integrated voice data and video communication networks emerged in the last 30 years, driven by three major developments: digitalization, computerization, and deregulation.

Digitalization supported the integration of voice, data, and video communication from subscriber to subscriber via one single bearer at an improved quality.

Computerization supported the decentralization of network control with a shift of control and routing functions from the switching to the transmission and terminal domains. Computerization now supports a convergence of telecommunications and computer technologies, called *computer telephone integration* (CTI), combining the desktop telephone and desktop computer into an integrated personal and business multimedia communication terminal. Digitalization and computerization will bring an overall integration of information technologies such as computers, facsimile, videophone, and TV with the plain old telephone, thus integrating voice with full-motion video, text, and data into multimedia subscriber equipment.

Deregulation started in the United States with the "Carterfone case" in 1968—the FCC permitted noncarrier-provided terminal equipment to be attached to the telephone network—and resulted in the worldwide creation of competition in telecommunication through liberalization, deregulation, and privatization of the telecommunication services. Four milestones highlight deregulation: the divestiture of the monolithic AT&T into the *Regional Bell Operating Companies* (RBOCs) in 1982, the complete liberalization of telecommunications in the United Kingdom in 1984, the privatization of NTT in 1985, and the end of the voice service monopoly in the European Union on January 1, 1998.

Digitalization, computerization, and deregulation of the telecommunication networks are paving the way toward the information age, with a different world economy and different lifestyles that require ever greater volumes of information to be circulated at ever higher speeds. Figure 10.1 gives an impression of the bit rates for various telecommunication services.

To transport such high volumes of information, new networks will be required. These new networks will bring people access to interactive multimedia services in their workplace and their homes. Multimedia terminals will give access to data banks; to video-on-demand; and other service providers operating on high-speed, high-capacity telecommunication links in multimedia networks called information highways or even information superhighways. A typical multimedia network, as shown in Figure 10.2, mainly consists of

- Service providers giving access to stored programs;
- Broadband switching;
- Broadband transmission (the information highway);
- Broadband access lines (optical fiber, radio relay, and copper lines with compression equipment);
- Subscriber multimedia terminals.

Movies are digitized with the MPEG standard, typically at 2 Gbyte per 100-min movie; and some 100 to 200 movies can be requested from a 200- to

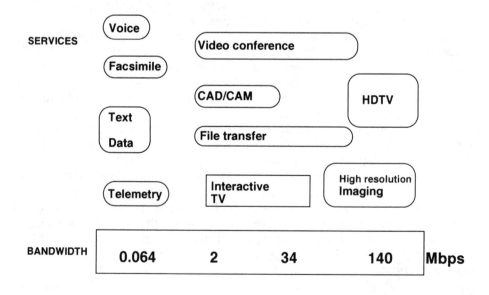

Figure 10.1 Telecommunication services versus bandwidth.

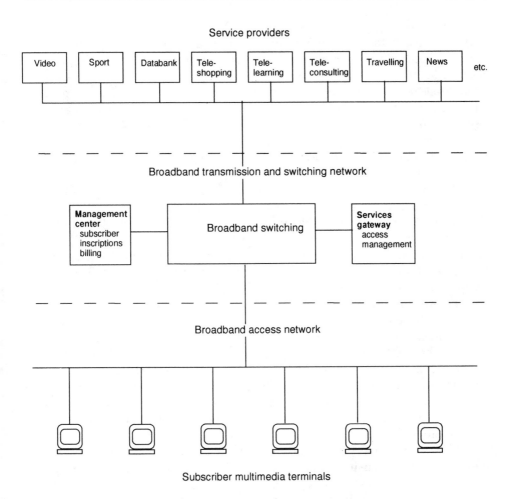

Figure 10.2 Typical multimedia network.

400-Gbyte store. In addition, TV can be selected from a few hundred stored programs.

Various national broadband information highways will be interconnected to the GII so that ultimately "everybody on this planet will be able to obtain the right answer to her/his question in a matter of seconds, at affordable cost." Plans for broadband information highways are most advanced in the United States and Western Europe. A dozen experimental information highways have recently been launched in the United States, United Kingdom, France, Germany, and Japan. In the United States, the Information Infrastructure Task Force coordinates the technological and legislative activities to construct a *national infor-*

mation infrastructure (NII). In Europe the European Union works out guidelines and priorities for a *European broadband infrastructure* (EBI) that will interconnect with the whole of European telecommunications and CATV networks. Supported by the accelerated availability of Euro-ISDN, the EBI shall provide file transfer, e-mail, video, and other multimedia services throughout the European Union. In Japan some 20 companies—including NTT, NEC, Hitachi, and Fujitsu—have established a multimedia observatory to elaborate proposals for a national information network. Taiwan and Hong Kong are likewise planning national multimedia networks.

Telecommunications are vital for the wealth of any country. While the industrialized nations are driving at high speed toward the information age on the aforementioned superhighways, nearly four-fifths of the world's population still lack the most basic access to a telephone. Half of the world's population has never used a telephone. Some 15% of the world's population use more than 85% of the world's telecommunications services. The greater Tokyo area, for example, has more telephones than the whole African continent. Although substantial efforts were made during the last 50 years to improve telecommunications worldwide, unfortunately only limited progress could be obtained in many underdeveloped countries. To improve this situation, the ITU postulated its goal: that by the end of this century each citizen on this globe will have direct access to at least a telephone, be it a private telephone, a community telephone, a "phoneshop," or a payphone. To reach this goal, for example, ITU aims at an average telephone density in the least developed countries of five main lines per 100 persons in urban areas and one main line per 10,000 persons in rural areas.

Should this goal be reached, then the least developed countries would have a telephone density comparable with the Western European countries in the early 1930s!

It is obvious that the financial means to meet this still rather modest goal will not be available from the countries concerned. The competitive situation resulting from the worldwide privatization, liberalization, and globalization of telecommunications, however, will increasingly create a commercial interdependence between the interest of the telecommunications operators and manufacturers of the developed countries with the interest of the operators of the developing countries. In such a common interest, therefore, it is expected that information highways will also be constructed in the developing countries and connected with the GII. With this understanding and to accelerate the pace of telecommunication penetration toward a more equal worldwide distribution, the ITU created "WorldTel" at their first World Telecommunication Development Conference (WT DC-94) in Buenos Aires, Argentina, in March 1994. WorldTel will arrange private financing from large telecommunication operators and manufacturers who will undertake telecommunications investments in underdeveloped countries.

A better global coverage will also be obtained from the mobile telecommunication networks that will emerge with new satellite constellations in LEO and MEO. Those new satellite networks will provide access to the GII at any place on and around the globe, thus filling the telecommunications coverage gaps in geographically remote areas and serving communities with industrial and agricultural activities in difficult accessible rural areas as are prevailing in the developing countries.

As a result of the digitalization, computerization, and deregulation of the telecommunication networks, several operators can now compete for the same subscribers with differentiated and better services. New operators starting from a "green field" situation with a high interest in an earliest network implementation with direct connections to potential customers (bypassing the network of the established national operator) are increasingly using wireless transmission both in the distribution and in the access networks. Figure 10.3, showing an increase in the annual number of new access lines from 50 million to 200

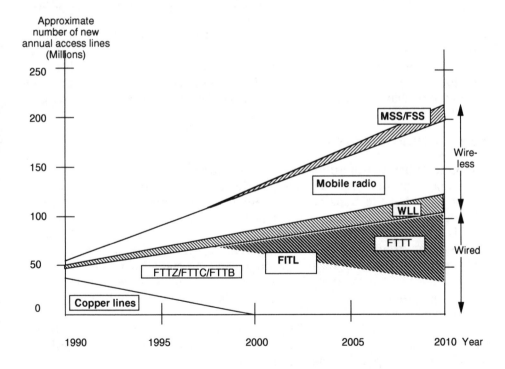

Figure 10.3 Approximate number of annual new access lines worldwide.

million during the 20 years spanning 1990 to 2010 (elaborated by the author from a scenario published in *Telecommunications* in 1993), clearly demonstrates the development from wireline access toward wireless access.

The development over the next five years within each transmission domain shall be briefly described. The future mobile satellite systems operating in LEO and MEO—presenting the major transmission development of the next five years—are covered in more detail because specialty books covering those new systems are not yet available [1–4].

10.2 COPPER LINES

Billions of dollars have been invested in an enormous "copper mine" connecting the majority of the worldwide 600 million telephone subscribers. Millions of dollars will be invested in electronic equipment such as pair-gain and xDSL systems to enhance the transmission capacity of this copper mine and continue its exploitation as a "gold mine" for operators.

Modems being introduced in an end-of-1996 standardized V.34+ version already support 34-Kbps operation for data and voice on copper wire access lines connected with digital exchanges, whereas 56-Kbps modems will be introduced around 1998. Voice compression, for example, from 64 Kbps to 5.3 Kbps, already successfully applied to transmit voice over frame relay, will also ease the digital transmission of voice over copper wire. Internet and multimedia will find wide application through xDSL on copper wire access lines and through broadband modems on coaxial cable CATV networks.

10.3 OPTICAL FIBERS

The optical fiber transmission system of the future will become truly photonic—replacing electronic components by genuine optical components—thus significantly widening the transmission bandwidth. Moreover, the application of EDFAs and soliton transmission will enable the transport of very high capacity signals, for example, 40 to 100 Gbps, without signal regeneration in submarine cable systems around the globe and in terrestrial systems throughout countries and continents. Routing those very-high-capacity signals will take place in cross-connect systems with optical switching devices. While satellite transmission is cost-insensitive to distance, optical fiber transmission will become cost-insensitive to bandwidth.

"Terabit transmission" (1 Tbps = 1,000 Gbps), which will be introduced early next century, was already demonstrated by NEC in mid-1996. Applying WDM with 132 optical signals of 20 Gbps each at a wavelength spacing of 33.3 GHz in the 1,529- to 1,564-nm band, an aggregate 2.64-Tbps signal—thus the equivalent of 40 million telephone channels—could be transmitted over an experimental unrepeatered 120-km link.

Many CATV operators will become full telecommunication operators in multimedia networks, for example, applying WDM for simultaneous transmission of a choice of TV signals and B-ISDN. The high output power of EDFAs will enable CATV operators to serve "ten thousands" of subscribers from one single head-end.

The reduction of the number of switching levels mentioned in Chapter 9 will be supported by the deployment of FITL, which enables an increase of the average length of access lines significantly beyond the current levels of 2 to 5 km in Europe and 5 to 8 km in the United States and Japan. FITL systems in the ultimate *fiber-to-the-terminal* (FTTT) version will evolve into *integrated fiber optic subscribers systems* (IFOSs), just as proposed by NTT for deployment in Japan between 1995 and 2010. Such an IFOS, in fact a multimedia terminal, with a 155-Mbps transmission rate, both upstream and downstream, or 622-Mbps downstream and 155-Mbps upstream, will enable subscribers to receive digital broadcasting, video-on-demand, standard TV, and HDTV channels on top of data transmission and voice and videophone operation. The technologies of "self-healing" and "self-routing" in optical fiber ring networks, widely applied in transport networks, will also be introduced in access networks, thus reducing the effect of access cable and equipment damages [5,6].

10.4 RADIO RELAY

The radio-relay systems of the future will include more intelligent functions such as automatic RF output power regulation as a function of humidity and other propagation deteriorations, and receiver selectivity regulation (of urban systems) as a function of the constantly varying grade of interference from mobile radios. Those and other features will be supported by integrated computing capabilities.

WLL systems, similar to the aforementioned IFOS, will emerge toward *integrated multimedia and radio-relay* (M&R) systems. After all WLL will make much sense if combined as M&R so that the computer capacity required for the aforementioned automatic regulation of the RF output and receiver selectivity can be achieved at minimal additional cost from a PC already integrated in the multimedia terminal. The signal-processing part of the radio-relay equipment thus will be just one of the optional plug-in units of the multimedia terminal (for example, instead of an optical fiber termination unit) that can be connected by a multipair cable with the RF unit of the radio-relay equipment, which in turn can be easily integrated with a small outdoor antenna placed on the balcony or roof top. Such an M&R terminal will be the solution for users who are not yet served with an optical broadband access but still want to escape the traffic jams of the big cities and prefer to *telecommute* or *telework* at home at a time when it is convenient [4].

10.5 MOBILE RADIO

Within less than 10 years after the introduction of cellular operation more than 80 million subscribers worldwide use this facility; analysts predict that by the end of this century there will be over 200 million cellular subscribers (as shown in Figure 9.4), many of them operating a wideband 64-Kbps terminal or the GSM WorldPhone that can be used on all three GSM frequency bands: 900, 1,800, and 1,900 MHz. GSM will evolve to GMM, which stands for Global Multimedia Mobility.

The next goal is the development of a truly global system providing communication "to everyone, everywhere" with a wide range of radio environments covering very small indoor cells with high-capacity, large outdoor terrestrial cells and very large cells with satellite coverage using small convenient lightweight multimode terminals. At WARC-92 the 1,885- to 2,025-MHz and 2,110- to 2,200-MHz frequency bands were allocated for those new services including common subbands for both terrestrial and satellite parts. Many innovative new services and applications are being considered, such as traffic information and control systems, the provision of a globally standard return channel for interactive television, and fixed WLL for developing countries. Standards for two systems are under preparation:

- *Universal Mobile Telecommunications System* (UMTS), by ETSI;
- *International Mobile Telecommunication System* (IMT-2000), by ITU.

UMTS is being developed by a consortium of about 125 telecommunication companies and research organizations as part of a *research for advanced communications in Europe* (RACE) program to define the standards for the future Pan-European PCN. The UMTS standardization process was started in 1991, as ETSI established an UMTS Sub-Technical Committee (SMG5). The purpose of the UMTS is to support global personal communications with worldwide interoperability and roaming, providing a wide range of intelligent network services as identical as possible to those of fixed networks. UMTS will be developed as an open system based on *Open Network Architecture* (ONA) compatible with N- and B-ISDN. First-phase systems are planned to be operational by the year 2002 with bit rates up to 2 Mbps. Operation is planned in the 1,920- to 1,980-MHz and 2,110- to 2,170-MHz bands for the terrestrial services and the 1,980-MHz to 2,010-MHz and 2,170- to 2,200-MHz bands for likewise incorporated satellite services. Full B-ISDN service is planned to be available in 2005.

IMT-2000, initially called FPLMTS (pronounced "flumps" and standing for *Future Public Land Mobile Telecommunications Systems*), is being prepared by Task Group 8/1 (TG 8/1) of the ITU and shares with UMTS the common goal of coming to a global personal telecommunication network, including satellite service, operational at the beginning of the next century.

Paging, the simplest of the mobile radio technologies, will increasingly be digital and two-way, operate several "calling party pays" services, feature online Internet-messaging remote telemetry and supervision, and support international roaming. Paging is likely to survive as a low-cost, small size, easy-to-use, mass-market narrowband personal communications service.

PMR and PAMR will be fully digital and include various intelligent services now already common in cellular networks. PMR will find niche applications in the shadow of cellular radio mainly for security services and PAMR likewise for corporate users [7].

10.6 SATELLITES

10.6.1 General

Despite the tremendous development of cellular radio networks previously mentioned, still less than 20% of the world's land mass and less than 60% of its population have terrestrial cellular coverage. To improve this situation, new personal communications MSS will be developed, implemented, and put into operation within the next five years. An estimated 5 to 15 million subscribers are expected to use the new MSS services at the end of this time period.

Those MSS systems will interwork with terrestrial fixed and cellular radio networks. With MSS, developing countries can leapfrog to wireless technology without the need for substantial investment or improvement of the national PSTN. MSS will complement the cellular radio systems wherever there is no terrestrial cellular coverage and serve international travelers in regions where the terrestrial cellular networks are not compatible with the traveler's home system. Dual-mode and dual-band handsets—eventually in earphone, handglove, or wrist-watch version—will then automatically select the best available cellular or satellite network.

The ITU initiated studies on the aforementioned IMT-2000 system in 1985, covering terrestrial and satellite components. In 1996, ITU separated the two components and created the name *global mobile personal communication by satellite* (GMPCS) for those new satellite services and subdivided the GMPCS systems as follows:

- GEO MSS, for voice and low-speed data mobile personal communications services operating in GEO;

- Little NGEO MSS, for narrowband services, excluding voice, mobile personal communications services operating in non-GEO, also called "little-LEO";
- NGEO MSS, for narrowband, including voice, mobile personal communications services operating in LEO, MEO, and HEO, also called "big-LEO";
- GEO and NGEO FSS, for satellite systems expected to be put into operation during the next five to ten years in order to offer fixed and transportable, multimedia broadband services operating either in geostationary or nongeostationary orbits, also called "super-LEO."

Various GMPCS networks are already in an advanced stage of development and are expected to start operation within the next two to five years. In this chapter, therefore, a brief description of the currently known features is given of those near-future networks that most likely will be operational near the end of this century, such as the MSS networks *Iridium, Globalstar, Odyssey,* and *ICO Global Communications* and the FSS networks *Teledesic* and *Spaceway* [8].

10.6.2 Iridium

IRIDIUM was conceived as a commercial global wireless telecommunications system using handheld telephones in 1987 by engineers from Motorola Satellite Communications. Iridium Inc. was formed by Motorola in 1990 as an international consortium of telecommunications and industrial companies funding and implementing the IRIDIUM system. The name IRIDIUM was chosen because the initial system was based upon a space segment consisting of 77 satellites circling the globe, which is similar to the 77 electrons in the element Iridium. In 1992 the number of satellites was reduced to 66, fortunately without changing the name to "DYSPROSIUM"—the element with ordinal number 66. An artist's rendering of the IRIDIUM system is given in Figure 10.4.

The IRIDIUM system consists of a space segment, a system control segment, fixed gateway stations, and the IRIDIUM terminals, as shown in Figure 10.5 and subsequently briefly described.

The IRIDIUM space segment consists of a LEO constellation of 66 satellites arranged in six orbital planes, each containing 11 satellites traveling at a speed of 26,856 km/h. The satellite altitude is 780 km, which results in an orbital period of approximately 100 min. Each satellite is connected by crosslinks with four adjacent satellites. The satellites have a triangular construction and consist of a communication section, an antenna section, a power section, and a command module.

The communication section contains Ka band crosslink and gateway radio-relay equipment, the subscriber link L-band transponders, and the onboard

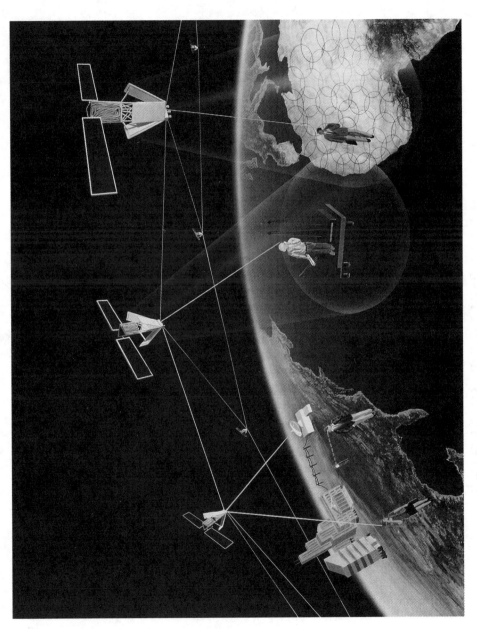

Figure 10.4 Artist's view of the IRIDIUM system (courtesy of IRIDIUM Inc.).

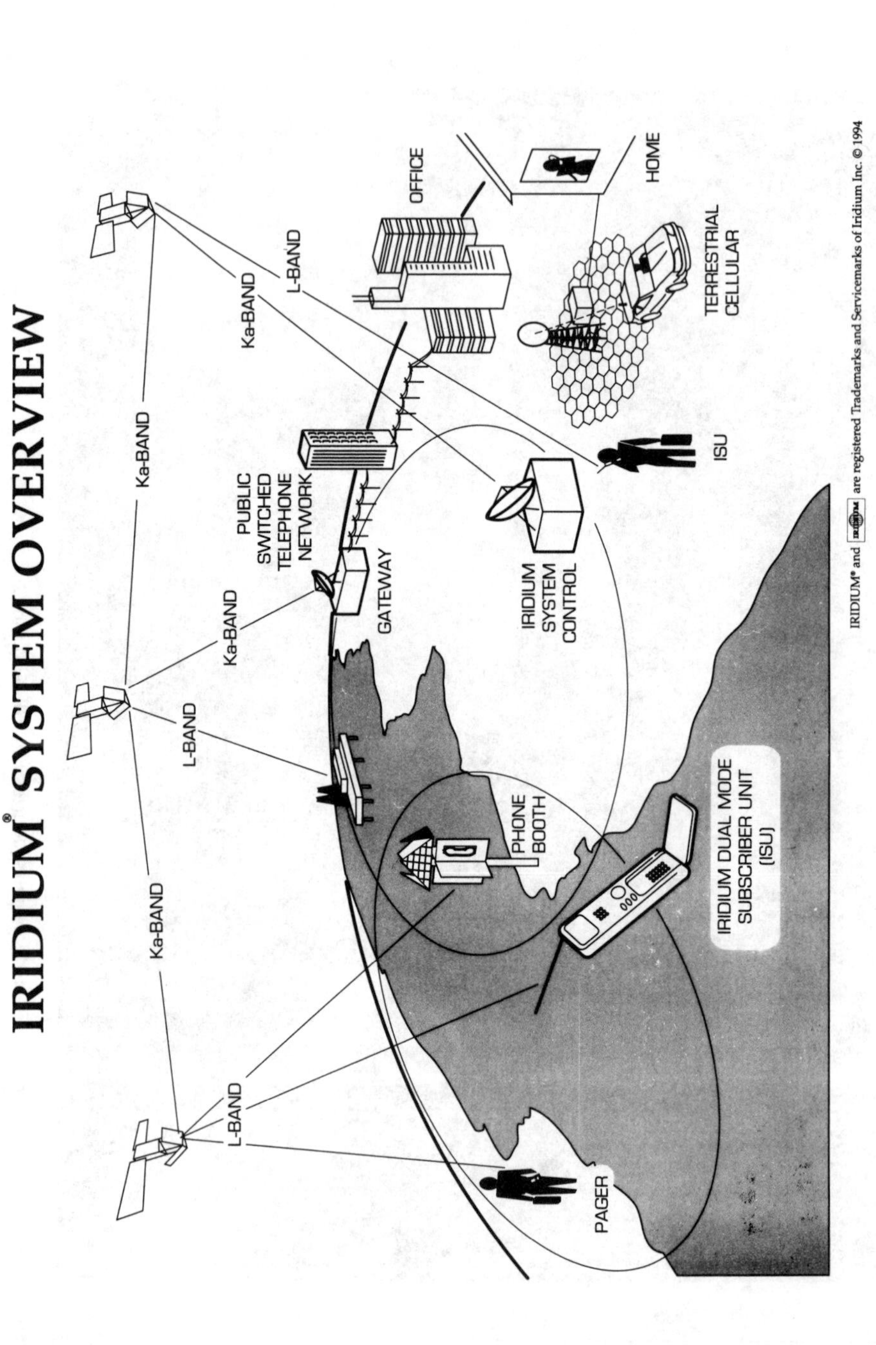

IRIDIUM® SYSTEM OVERVIEW

Ka-BAND

Ka-BAND

Ka-BAND

Ka-BAND

L-BAND

L-BAND

L-BAND

L-BAND

OFFICE

HOME

TERRESTRIAL
CELLULAR

PUBLIC
SWITCHED
TELEPHONE
NETWORK

GATEWAY

IRIDIUM
SYSTEM
CONTROL

ISU

PHONE
BOOTH

IRIDIUM DUAL MODE
SUBSCRIBER UNIT
(ISU)

PAGER

IRIDIUM® and ![logo] are registered Trademarks and Servicemarks of Iridium Inc. © 1994

Figure 10.5 The IRIDIUM system overview (courtesy of IRIDIUM Inc.).

traffic switching. The crosslinks operate in the 23-GHz band in TDD mode with a capacity of 12.5 Mbps. The gateway communication operates in full duplex with 19-GHz band for the downlink and, contrary to usual praxis, on a higher frequency band (29 GHz) for the uplink. The subscriber link transponder operates bidirectionally at 1.6 GHz in FDMA/TDMA mode.

The antenna section contains two fixed crosslink antennas for communication with adjacent satellites in the north-south direction, two moveable crosslink antennas for the east-west direction, four moveable gateway antennas, small secondary link antennas for spacecraft orientation after deployment, and three *main mission antennas* (MMAs). The three identical flat MMAs, at 120-degree distances arranged around the satellite, jointly radiate a ground pattern of up to 48 cells. The 66 satellites cover the Earth with up to 2,150 cells, each with a surface of about 15 million km^2 and each cell simultaneously serving an average of 80 and a maximum of 240 calls. The global throughput thus varies between nominally 172,000 and maximum 500,000 simultaneous calls.

The power section contains two solar panels, the NiCd batteries bridging the frequent sun-eclipse periods, and the voltage regulation equipment. The command module, finally, is to keep the satellite in its exact orbital slot.

The IRIDIUM *system control segment* (SCS) is located in Northern Virginia near Washington, DC with a backup at Rome, Italy. In addition to the TT&C, the SCS also computes and loads the satellites with frequency planning and traffic routing information.

The IRIDIUM gateway stations connect the space segment with the terrestrial PSTN/ISDN. A gateway normally communicates with up to four satellites. Initially 16 gateways are planned worldwide. Each gateway has two or more high-gain 3.048-m parabolic tracking antennas each housed in a 5-m radome to protect the highly sensitive tracking equipment. The colocated antennas are spaced at about 30 km to overcome weather and atmospheric signal fading and blocking. The gateways use mobile switching centers derived from the GSM system, thus implementing various GSM features into the IRIDIUM system.

The IRIDIUM terminals are dual-mode handheld telephones supporting facsimile and data at 2.4 Kbps and alphanumeric paging.

The 66 + 6 satellites will be launched at three different locations: at the Taiyuan Satellite Launch Center in China with two satellites per Long March C rocket; at NASA Vandenberg Airforce Base near Lompoc, California with five satellites per Delta II rocket; and at Baikonur Cosmodrome launch facility of the Krunichev Space Center in Kazakhstan with the largest of the three rockets— the Proton, which has the capacity to place seven IRIDIUM satellites into the *circular transfer orbit* (CTO). From CTO, the IRIDIUM satellites will be moved by the onboard propulsion subsystem (in the command module) to their operational slot.

IRIDIUM is a member of the GSM-MoU Association and is establishing roaming agreements with GSM operators worldwide to provide complementary and value-added global roaming capability to augment their terrestrial wireless offerings. IRIDIUM Inc. expects to serve 2.7 million subscribers by the year 2005 [9,10].

10.6.3 Globalstar

Globalstar is a LEO satellite-based global telecommunication and position location system developed by QUALCOMM and *Space Systems/Loral* (SS/L); both telecommunication companies are based in the United States. QUALCOMM began work on a worldwide satellite-based telecommunication service in 1986, SS/L in 1987. The two companies merged into the *Loral Qualcomm Satellite Services* Inc. (LQSS) in 1991.

The Globalstar network will consist of a space segment, a ground segment, and the terminals. The uplink signals are in the 1.6-GHz band and the downlink signals in the 2.5-GHz band.

The Globalstar space segment consists of 56 satellites equally spaced in eight planes of seven satellites each (including one spare) at an altitude of 1,410 km. The Globalstar constellation is shown in Figure 10.6. The satellite payload is transparent, thus unlike IRIDIUM without crosslinks and on-board traffic processing, all traffic switching happens on the ground and traffic routing is through the existing fixed public networks. A phased-array antenna produces 16 elliptical spot beams that provide continuous multiple satellite coverage, path diversity, and position location.

The ground segment (see Figure 10.7) consists of gateway stations and MSCs. The gateway stations typically spaced at 200 to 400 km are spread over several countries. The MSCs provide the interface with the public-switched fixed and mobile networks of the country. Certain gateways also monitor and control the satellite performance and manage the Globalstar communication networks for call verification and billing, for example.

The Globalstar terminals are handsets, vehicle-mounted mobile units, and fixed stations, for example, Globalstar's fixed multichannel village phone for rural regions without cellular or fixed network. Dual-mode handsets will be able to access both Globalstar and local land-based cellular systems. The terminals will be able to operate with a single satellite in view, although typically two to four satellites will be overhead. The use of CDMA allows each terminal to combine the signals received simultaneously from up to three satellites, thus improving the signal and minimizing call interruptions. The CDMA signal is spread across the entire 16.5-MHz bandwidth to support a total capacity per satellite of 2,808 voice channels at a 4.8-Kbps transmission rate.

Figure 10.6 Globalstar satellite constellation and the dual terminal (courtesy of Globalstar).

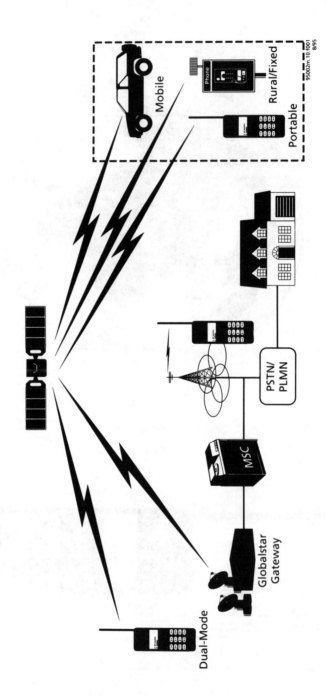

Figure 10.7 Globalstar ground segment (courtesy of Globalstar).

Globalstar service will be compatible with GSM, using various GSM features. The applied position location determines a user's location with an accuracy of 300m.

The Globalstar satellites, like IRIDIUM, will be launched from three countries. The first eight satellites by two McDonnell Douglas Delta II rockets from Vandenberg Airforce Base in the United States. The next 36 by three Zenith-II rockets from Kazakhstan, and the final 12 by one Long March C rocket from Taiyua in China. Launching is planned to start in 1997 to enable commercial operations via an initially 24-satellite constellation in 1998 and full 48-satellite coverage by the end of this century. Globalstar has a cooperation agreement with the GSM-MoU Association [11,12].

10.6.4 Odyssey Telecommunications International

Odyssey is a MEO satellite-based global telecommunication network conceived by TRW Space & Electronics Group (TRW S&EG based in Redondo Beach, California, USA) in 1990. TRW and Teleglobe Inc. (based in Montreal, Canada) formed a joint venture in 1994 to develop, construct, launch, and operate a global telecommunication system under the name Odyssey, since changed to *Odyssey Telecommunications International* (OTI). The OTI network will consist of a constellation of 12 satellites in MEO at an altitude of 10,354 km providing global dual-satellite coverage. OTI, like Globalstar, does not use crosslinks and on-board signal processing and will use the same frequency bands as Globalstar. OTI uses a rigid mounted multiple-beam antenna that produces 37 spot beams uplink and 32 downlink.

The OTI multiple-access scheme uses a combination of CDMA and FDMA. The 16.5-MHz allocated bandwidth is divided into three 4.83-MHz channels that use CDMA to support per satellite 2,300 voice channels at 4.8 Kbps.

OTI, similar to Globalstar, targets the national rather than the international long-distance markets, filling the coverage gaps in geographically remote areas and serving communities with industrial and agricultural activities in difficult-to-access rural areas with their *Odyssey wireless village network* (OWVN). An OWVN typically consists of a solar-powered small Earth station with a PABX connected to a local WLL base station wirelessly serving a community with fixed wireless terminals and public paystations. The OWVN Earth station can interface with existing nearby terrestrial cellular or cordless systems and via an OTI satellite is connected to the national OTI gateway station. OWVNs are expected to be cost effective when the distance between the remote community and the nearest telephone exchange is in the order of hundreds rather than tens of kilometers, as is often the case in Africa and Asia. OTI plans to start commercial operation in 1998 with an initial six satellites in orbit [13].

10.6.5 ICO Global Communications

ICO Global Communications (ICO) is a MEO satellite-based global telecommunication system conceived by INMARSAT. The ICO system consists of a constellation of 12 satellites in two planes each of five operational and one spare satellite in an intermediate circular orbit (hence ICO) at an altitude of 10,355 km. Figure 10.8 shows the ICO system overview. An user segment comprises the terminals located on airplanes, vehicles, and ships as well as the (mainly dual-mode) handheld units and terminals used on semifixed locations.

The space segment is controlled by TT&C Earth stations connected to the *space control center* (SCC). A *network control center* (NCC) maintains the overall management and controls the SCC and one or two system control centers that control the interconnected *satellite access nodes* (SANs). The SANs, each comprising an Earth station with up to five colocated antennas and feeder links operating at 5/7 GHz (subject to frequency allocation), support mobility management (including HLR and VLR) and interconnect with numerous gateways. Up to 12 SANs are planned around the world and will be interconnected through a terrestrial backbone network called P-Net, which will allow calls to be routed through the ground segment of the network to the SAN best able to service the call. ICO also has a cooperation agreement with the GSM-MoU Association [14].

Table 10.1 summarizes the major announced characteristics of those future MSS networks [15].

10.6.6 Teledesic

The satellites for the Teledesic network (backed by Microsoft's Bill Gates and Craig McCaw of McCaw Communications) will be placed in LEO, thus facilitating high-speed interactive online data transmission for small terminals with low EIRP.

The Teledesic network will use 924 satellites. National gateway Earth stations will ensure seamless interface with terrestrial fixed networks and handle two million simultaneous *bandwidth-on-demand* (BOD) connections for a worldwide envisaged 20 million subscribers, mainly Internet-using PC-owners.

The 924 satellites will be located in 21 planes, each with 40 operating and 4 standby satellites. Each operational satellite will support 567 cells of 53 km^2, each capable of handling 1,400 simultaneous 16-Kbps voice channels or 15 T1, respectively, 11 E1 signals or in BOD mode any combination of the cumulative 24-MHz bandwidth. The network will use a packet-switching technology based upon ATM. Each packet will contain a header including address and sequence information, an error-control section, and the payload section with the actual data. Each satellite will be a node in the network and

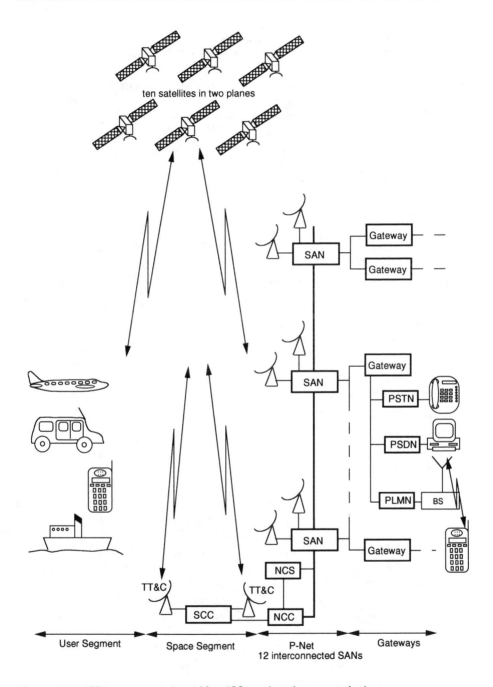

Figure 10.8 ICO system overview (*After*: ICO product documentation).

Table 10.1
Preliminary Data About Future Big LEO Networks

Subject	IRIDIUM	Globalstar	OTI	ICO
Orbit	LEO	LEO	MEO	MEO
Altitude (km)	780	1,410	10,354	10,355
Number of satellites				
Operational	66	48	12	10
In space spares	6	8	2	2
Minimum elevation angle	8.2°	10°	22°	10°
Orbital period	100 min 28 sec	1.5 hr	6 hr	6 hr
Signal delay (ms)	7	18	200	200
Satellite weight				
Dry (kg)	500	320	1,225	
Wet (with fuel, kg)	700	450		2,000
Gateway stations approx.	20	150–200	8	many (12 SANs)
Subscriber link				
uplink (MHz)	1,616–1,626.5	1,610–1,625	1,610–1,625	1,980–2,010
downlink (MHz)	1,616–1,626.5	2,483.5–2,500	2,483.5–2,500	2,170–2,200
Gateway link				
uplink (GHz)	29.1–29.3			5.1–5.25
downlink (GHz)	19.4–19.6			6.925–7.075
RF-power handset (mW)	390	<500		average 250
Multi-access mode	FDMA/TDMA	CDMA	FDMA/CDMA	FDMA/TDMA
Channel modulation	QPSK	QPSK	QPSK	
Voice rate (Kbps)	4.8	1.2–9.6	4.8	
Data rate (Kbps)	2.4	1.2–9.6	1.2–9.6	
Lifetime (years)	5–8	8	12–15	12
Project cost (US$ billion)	2.8	2.0	2.3	2.6
Charge/minute (US$)	3.0	0.35–0.53	0.65	

will have intersatellite links with eight adjacent satellites: two in front and two behind in the same plane and one in each of the four adjacent planes. The network will use a "connectionless" protocol whereby packets of the same connection may follow different paths through the space network and be routed along lines of least delay. To compensate for different delays, the packets can be buffered and resequenced at the destination terminal.

In order to make Teledesic a reality a tremendous amount of technological effort is still required. The remarkable 924 satellites (more than four times the number of telecommunication satellites currently in orbit!) still need to be developed, produced, and launched. With an expected 8 to 20 satellites per launch this means 45 to 115 launches, which at the present rate of 15 annual launches would take three to seven years.

Table 10.2
Preliminary Data About the Future FSS Data Networks Teledesic and Spaceway

Subject	Teledesic	Spaceway
Operator	McCaw Communications	Hughes Communications
Orbit	LEO	GEO
Altitude	696 km	35,786 km
Number of satellites	840 operational	17
	72 spare	
Satellite weight	765 kg	1,700 kg
Ka frequency band	28 GHz	28 GHz
Announced service	BOD 16 Kbps–2 Mbps	similar
	ATM 155 Mbps–1.2 Gbps	
Announced service year	2002	1988
Terminal antenna diameter	25 cm	6.5m
Time to download		
2 Gbytes	18.3 min on 2 Mbps	21.59 hr on 1.5 Mbps
	16 sec on 1.2 Gbps	
Project cost	US$9 billion	US$3.2–6.0 billion
Approximate airtime charge per		
minute/64 Kbps	US$0.05	US$0.02

10.6.7 Spaceway

For the Spaceway network the satellites will be in GEO and thus, due to the high signal delay, gigabytes of information will need to be stored in buffers requiring substantial technical effort to offer the planned high-speed data transmission.

The major data announced about the two FSS data networks are summarized in Table 10.2 [16–18].

10.6.8 Epilogue

These future satellite systems very remarkably were predicted by Arthur C. Clarke in May 1983 at a U.N. General Assembly. Referring to the, at that time, emerging digital wrist watches, Clarke predicted: "The symbols that flicker across those digital displays now merely give time and date. When the zeros flash up at the end of the century, they will do far more than that. They will give direct access to most of the human race through the invisible networks girdling our planet" [19].

References

[1] Dietrich, Mark, "Building the Information Superhighway," *Siemens, telecom report international,* Vol. 17, No. 6, 1994, pp. 12–14.

[2] "Main Highlights of WTDC-94," *Telecommunication,* Vol. 5, 1994, pp. 5–13.

[3] McClelland, Stephen, "Exploiting the Network—An Overview," *Telecommunications,* Vol. 27, No. 11, 1993. pp. A1–A16.

[4] Huurdeman, Anton A., *Radio-Relay Systems,* Norwood, MA: Artech House, 1995.

[5] Shinohara H., I. Yamashita, and T. Miki, "Evolution Scenario for the Integrated Fibre-Optic Subscriber System," *Telecommunications Journal,* Vol. 60, Mar. 1993, pp. 109–118.

[6] Daniels, Guy, "Terabit Transmission," *Communications International,* Vol. 23, Nov. 1996, p. 52.

[7] Donegan, Patrick, "Europe's New Policy Maker," *Mobile Communications International,* Vol. 23, Apr. 1996, pp. 37–41.

[8] Don MacLean, "GMPCS at the WTPF—New challenges for the ITU," *ITU NEWS,* Vol. 8, 1996, pp. 2–7.

[9] *IRIDIUM Today,* Vol. 2, Nos. 3 & 4, 1996.

[10] *IRIDIUM System Networking,* Washington, DC: Iridium, Inc., 1995.

[11] Globalstar Backgrounder, Press kit information from General Partner Globalstar L.P., Arlington, VA.

[12] Globalstar System, "It's for You," Commercial leaflet from Globalstar, San Jose, CA.

[13] Shetty, Vineeta, "To the Far Ends of the Earth," *Communications International,* Vol. 23, No. 10, 1996, pp. 8–12.

[14] ICO System Description, issued by ICO Global Communications, London, Oct. 1995.

[15] Comparetto, Gary. M., and Neal D. Hulkower, "Personal Satcom, More Than Just a Mirage on the Horizon," *Mobile Europe,* Vol. 5, No. 1, 1995, pp. 20–26.

[16] Shankar, Bhawani, "From Narrowband Niche to 'Bandwidth Bull,' " *Telecommunications,* Vol. 29, No. 12, 1995, pp. 44–48.

[17] Shankar, Bhawani, "As Satellites Gather Speed," *Telecommunications,* Vol. 30, No. 7, 1996, pp. 54–57.

[18] Savarnejad, Atoosa, "The Sky's the Limit," *Communications International,* Vol. 23, No. 9, 1996, pp. 31–32.

[19] Clarke, Arthur, C., "Beyond the Global Village," *Developing World Communications,* London: Grosvenor Press International, 1986, pp. 18–29.

Acronyms and Abbreviations

ACA	adaptive channel allocation
ACSE	association control service element
ADSL	asymmetrical digital subscriber line
AGC	automatic gain control
AIM	ATM-inverse multiplexing
AIN	Advanced Intelligent Network
AM-VSB	vestigial side band with amplitude modulation
AMI	alternate mark inversion
AML	added-main-line
AMPS	advanced mobile phone system
ANSI	American National Standards Institute
APC	aeronautical public correspondence
APCO	Association of Public-Safety Communications Officials
APD	avalanche photo diodes
APOC	advanced paging operators code
ARABSAT	Arab Satellite Communications Organization
ARPANET	Advanced Research Project Agency Network
AS	airborne station
ASIC	application specific integrated circuit
ASK	amplitude-shift keying
ASVD	analog simultaneous voice and data
ATM	asynchronous transfer mode
ATPC	automatic transmit power controller
AU	administrative unit
AuC	authentication center
AUSSAT	Australian Satellite Corporation

B-ISDN	broadband-ISDN
BCM	block coding modulation
BECN	backward explicit congestion notification
BER	bit error rate
BOD	bandwidth-on-demand
Bonding	Bandwidth ON Demand INteroperability Group
BSC	base station controller
BTS	base transceiver station
BTTP	broadband-to-the-person
C/R	command/response
CAD/CAM	computer-aided development, respectively, manufacturing
CASE	common application service elements
CATV	cable TV
CBDS	connectionless broadband data service
CC	cross-connecting
CCIR	International Radio Consultative Committee
CCITT	International Telegraph and Telephone Consultative Committee
CCS	common channel signaling
CDDI	copper distributed data interface
CDMA	code-division multiple access
CENTREX	central exchange
CEPT	conférence Européenne des postes et télécommunications
CFDM	companded FDM
CFRP	carbon-fiber-reinforced plastics
CIER	Central Equipment Identity Register
CIF	common intermediate format
CIR	committed information rate
CLNP	connectionless network protocol
CMI	coded mark inversion
CMIP	common management information protocol
CMISE	common management information service element
CMS	circuit multiplication system
CO	central office
COMSAT	Communication Satellite Corporation
CPDP	cellular digital packet data
CPP	calling party pays

CR	conditional replenishment
CSMA/CD	carrier sense multiple access/collision detection
CT	cordless telephony
CTI	computer telephone integration
CTO	circular transfer orbit
CVD	chemical vapor deposition
DAMA	demand assignment multiple access
DASS	demand assignment signaling and switching
DBS	direct-broadcast satellite service
DC-DCPBH	double-channel-double-channel planar buried heterostructure
DCE	data circuit terminating equipment
DCMS	Digital Circuit Multiplication System
DCN	data communication network
DCS	dynamic channel selection
DCT	discrete cosine transform
DE	discard eligibility
DECT	Digitally Enhanced Cordless Telecommunications System
DFS	dispersion-shifted SMF (single-mode fiber)
DIT	directory information tree
DLCI	data link connection identifier data terminal equipment (DTE)
DM	degraded minutes
DM	Dieselhorst-Martin (cable quad)
DMT	discrete multitone
DPCM	differential pulse code modulation
DQDB	distributed queue dual bus
DS	directory services
DS-SS	direct sequence spread spectrum
DSA	data service adapter
DSA	directory system agent
DSF	dispersion-shifted fiber
DSI	digital speech interpolation
DSVD	digital simultaneous voice and data
DT-MF	dual-tone multifrequency
DTH	direct-to-home (TV service)
DTX	discontinuous transmission
DUA	directory user agent

DWDM	dense-WDM (wave division multiplex)
EA	address field extension
EBI	European broadband infrastructure
EDF	energy density factor
EDFA	erbium-doped fiber amplifier
EIA	Electronic Industry Association
EIR	equipment identity register
EIRP	effective isotropic radiated power
ELV	expendable launcher vehicle
ERMES	European Radio Messaging Service
ES	errored second
ETACS	extended TACS (total access communication system)
ETSI	European Telecommunication Standards Institute
EUTELSAT	European Telecommunication Satellite Organization
F/B	front-to-back
FDDI	fiber distributed data interface
FDE	full-duplex Ethernet
FDM	frequency division multiplex
FDMA	frequency division multiple access
FECN	forward explicit congestion notification
FH	frequency hopping
FITL	fiber in the loop
FLAG	fiber-optic link around the globe
FM	frequency-modulated
FOM	figure-of-merit
FPLMTS	future public land mobile telecommunications systems
FSK	frequency-shift keying
FSS	fixed satellite service
FTAM	file transfer, access, and management
FTP	foiled twisted pair
FTTA	fiber to the apartment
FTTB	fiber to the building / or basement / or business
FTTC	fiber to the curb
FTTD	fiber to the desk
FTTF	fiber to the feeder / or floor
FTTH	fiber to the home
FTTK	fiber to the kerb

FTTLA	fiber to the last amplifier
FTTO	fiber to the office
FTTP	fiber to the pedestal
FTTR	fiber to the remote unit
FTTSA	fiber to the service area
FTTT	fiber to the terminal
FTTZ	fiber to the zone
FUNI	frame-UNI (user network interface)
GDRS	general packet radio service
GEO	geostationary Earth orbit
GII	global information infrastructure
GMPCS	global mobile personal communication by satellite
GS	ground station
GSC	ground switching center
GSM	global system for mobile communications
GTO	geostationary transfer orbit
HDB	high-density bipolar
HDCT	hybrid DCT (discrete cosine transform)
HDSL	high-bit-rate digital subscriber line
HDTV	high-definition TV
HEO	highly elliptical orbit
HLL	high-level language
HLR	home location register
HPA	high-power amplifier
HSCSD	high-speed circuit-switched data
HSSI	high-speed serial interface
HT	high tension
IA	implementation agreement
IAD	integrated access device
IBS	international business service
IDD	international direct dialing
IF	intermediate frequency
IFOS	integrated fiber optic subscribers system
IFRB	International Frequency Registration Board
IM	inverse multiplexing
IMCVD	intrinsic microwave heated chemical vapor deposition
IMEI	international mobile station identity

IMT-2000	international mobile telecommunication system
IN	Intelligent Network
INMARSAT	International Mobile Satellite Organization
INTELSAT	International Telecommunications by Satellite Organization
ISDN	integrated services digital network
ISO	International Organization for Standardization
ISPBX	integrated services PBX (private branch exchange).
ITU	International Telecommunication Union
IVR	interactive voice response
JDC	Japanese digital cellular
JPEG	joint photographic expert group
JTACS	Japan-TACS (total access communication system)
JTM	job transfer and manipulation
LAN	local-area network
LAP	link access procedure
LD	laser diode
LED	light-emitting diode
LEO	low Earth orbit
LLC	local link control
LMCS	local multipoint communications system
LNA	low-noise amplifier
LOS	line-of-sight
LT	line termination
M&R	integrated multimedia and radio relay
MA	multiple access
MAC	medium access control
MAN	metropolitan-area network
MC/TDMA	multicarrier-TDMA (time division multiple access)
MCPC	multichannel per carrier
MCS	mobile control station system (Japan)
MCVD	modified chemical vapor deposition
MD	mediation devices
MDF	main distribution frame
MEO	medium Earth orbit
MHS	message handling system
MLSE	maximum likelihood sequence estimation
MMA	main mission antenna

MMC	modulation matched coding
MMDS	multichannel multipoint distribution service
MMF	multi-mode fiber
MMIC	monolithic microwave integrated circuits
MoU	memorandum of understanding
MP	multilink protocol
MPEG	Moving Pictures Expert Group
MQW GRINSCH	multi quantum well graded index separate confinement heterostructure
MS	mobile station
MSC	mobile service switching center
MSS	mobile satellite services
MTA	message transfer agent
MTS	message transfer system
N-ISDN	narrowband-ISDN (integrated services digital network)
NAMPS	narrowband-AMPS (advanced mobile phone system)
NCC	network control center
NII	national information infrastructure
NiMH	nickel metal hybrid
NLT	negative line amplifier
NMF	network management forum
NMT	Nordic mobile telephone
NNI	network node interface
NPCS	narrowband personal communication services
NRZ	non-return-to-zero
NT	network termination
NTACS	narrowband-TACS (total access communication system)
NVOD	near-video-on-demand
O/W	open-wire
OC	optical carrier
OE	opto-electronic
OEIC	optical and electrical integrated circuit
OLP	outside line plant
OMC	operation and maintenance center
OMUX	output multiplexer
ONA	open network architecture
ONT	optical network termination

OS	operation system
OSA	optical subscriber access
OSIRM	open systems interconnection reference model
OTI	Odyssey Telecommunications International
OVD	outside vapor phase deposition
OWVN	Odyssey wireless village network
P-MP	point-to-multipoint
P-P	point-to-point
PABX	private automatic branch exchange
PAC	paging area controller
PACS	personal access communications system
PAD	packet assembly/disassembly
PAS	PanAmSAT
PBX	private branch exchange
PCB	printed circuit board
PCIA	Personal Communications Association Industry
PCM	pulse code modulation
PCMCIA	PCM common interface adapter
PCN	personal communications network
PCS	personal communications services
PCVD	plasma chemical vapor deposition
PDC	personal digital cellular
PDH	plesiochronous digital multiplex
PHS	personal handyphone system
PLL	phase-locked loop
PM	phase modulation
PMD	polarization mode dispersion
PMR	private mobile radio
PNC	paging network controller
POCSAG	(British) Post Office Code Standardization Advisory Group
POH	path overhead
PON	passive optical network
POTS	plain old telephone service
PRA	primary rate access
PSDN	public switched data network
PSK	phase-shift keying
PSTN	public switched telephone network

QAM	quadrature amplitude modulation
QCIF	quarter CIF (common intermediate format)
QPSK	quaternary PSK (phase-shift keying)
RACE	research for advanced communications in Europe
RAM	remote access memory
RBOC	regional Bell operating company
RCB	Radio Regulations Board
RDP	radio distribution point
RFTC	radio from the curb
RIC	receiver identification code
RITL	radio in the loop
RLL	radio in the local loop
ROSE	remote operation service element
RPE	radiation pattern envelope
RSU	remote switching unit
RZ	return to zero
SAN	satellite access node
SASE	specific application service element
SC	service center
SCC	space control center
SCPC	single channel per carrier
SCS	system control segment
SDH	synchronous digital hierarchy
SDLC	synchronous data link control
SDMA	spatial division multiple access
SDSL	symmetrical digital subscriber line
SEA-ME-WE	South-East Asia-Middle East-Western Europe
SES	severely errored second
SIBH	semi-insulated buried heterostructure
SIM	subscriber identity module
SIS	subscriber identity security
SLA	sealed Lead-Acid
SMDS	switched multi-megabit data service
SMF	single-mode fiber
SMR	specialized mobile radio
SMS	short message service
SNA	system network architecture

SNI	subscriber-network interface
SNR	signal-to-noise ratio
SOA	semi-conductor optical amplifier
SOH	section overhead
SOHO	small office home office
SONET	Synchronous Optical Network
SPADE	single-channel-per-carrier PCM multiple-access demand assignment equipment
SPC	stored program control
SPM	self-phase modulation
SS	spread spectrum
SS-TDMA	satellite-switched TDMA (time division multiple access)
STM	synchronous transfer mode
STM-N	synchronous transport module-N
STP	shielded twisted pair
STS	space transport system
STS	synchronous transport signal
SVD	simultaneous voice and data
TA	terminal access
TACS	total access communication system
TASI	time assignment speech interpolation
TAT	transatlantic telephone cable
TCM	time compression multiplex
TCM	trellis coded modulation
TCP/IP	transmission control protocol/Internet protocol
TDD	time division duplex
TDM	time division multiplex
TDMA	time division multiple access
TDP	Telocator data protocol,
TETRA	Trans-European trunked radio
TFTS	Terrestrial Flight Telecommunications System
TH-SS	time hopping SS (spread spectrum)
TIA	Telecommunication Industry Association
TMN	telecommunication management network
TOH	transport overhead
TP	transaction processing
TP-PMD	twisted pair - physical medium dependent

TPDDI	twisted pair distributed data interface.
TSAG	Telecommunications Standardization Advisory Group
TSI	time-slot interchange
TT&C	telemetry tracking and command
TWT	traveling wave tube
TWTA	traveling-wave tube amplifier
UA	user agent
UIM	user identification module
UMTS	universal mobile telecommunications system
UNI	user network interface
USAT	ultra small aperture terminal
UTP	unshielded twisted pair
VAD	vapor axial deposition
VBR-ADPCM	variable bit rate adaptive differential PCM
VC	virtual channel
VCI	virtual channel identifier
VDSL	very-high speed digital subscriber line
VDT	video dial tone
VF	voice-frequency
VHDSL	very high-bit-rate digital subscriber line
VLC	variable length coding
VLR	visitors location register
VLRE	very low rate encoding
VLSI	very large scale integrated circuit
VP	virtual path
VPI	virtual path identifier
VRRA	variable rate reservation access
VSAT	very small aperture terminal
VSWR	voltage standing wave ratio
VT	virtual terminal
WAN	wide-area network
WARC	World Administrative Radio Conference
WDM	wavelength division multiplexing
WLL	wireless local loop
WRC	World Radiocommunication Conference
XPD	cross polarization discrimination
XPIC	cross polarization interference canceler
ZDP	zero-dispersion point

About the Author

Anton Huurdeman was born in Holland. He studied electrical engineering in Amsterdam and international business at Nijenrode Castle (now Nijenrode University), near Amsterdam. After his technical commercial study he apprenticed for one year in England, France, and in Germany, where he then decided to stay.

Starting in 1959 in Ulm at Telefunken, he soon became an export engineer for radiotelephone systems and made his first trips to Iran, Egypt, and India, where he successfully demonstrated the feasibility of radiotelephony between locomotives in the Ghats near Bombay.

He married in 1963 and has two children. In 1964 he changed from AEG-Telefunken to the world of ITT by joining Standard Elektrik Lorenz AG (SEL) in Stuttgart and so found himself 23 years later in the world of Alcatel, without changing companies.

At SEL, after a few years as transmission sales engineer, he became export manager for radio-relay systems, conducting a life as *flying dutchman,* exporting *made in Germany* worldwide.

During the last four years of his career, he was transmission product manager for Alcatel Trade International at the Alcatel Headquarters in Paris. In that capacity he had the opportunity to participate in the process of establishing and integrating the international marketing and sales staff of the newly grouped (previously) CGE and ITT companies; especially by informing them of transmission market opportunities and relevant product availability within the Alcatel group.

As a guideline for Alcatel's marketing and sales staff as well as for the planning and purchasing officials of telecommunication operators, he wrote a summary of transmission in 1991 titled *Transmission: A Choice of Options,* of which some 3,000 copies were eagerly received all over the world and a translation was made into the Russian language. In fact, that summary was the nucleus from which this book emerged. Prior to writing this book, however, he wrote *Radio-Relay Systems,* published by Artech House in 1995.

Index

10Base2, 101
10Base5, 101
10Base-T, 100, 101
24-channel PCM, 58
    digital multiplex hierarchy, 63
    See also Pulse code modulation (PCM)
30-channel PCM, 57–58
    digital multiplex hierarchy, 62
    See also Pulse code modulation (PCM)

Access networks, 1–2
Access transmission systems, 41
Access units (AUs), 104
Adaptive channel allocation (ACA), 266
Adaptive frequency-domain equalizer, 231
Adaptive RF output power control, 231
Adaptive time-domain equalizer, 230
Add/drop multiplexers, 75
Added-main-line (AML) system, 136
Aeronautical public correspondence
    (APC), 277
Air-dielectric coaxial cable, 155–56
ALOHA, 323
Alternate mark inversion (AMI), 58
Amplitude-modulated (AM) system, 209
Amplitude-shift keying (ASK), 91
Analog multiplex, 46–53
    Bell system, 51–53
    CCITT, 49–51
    pregroup translation, 47
    See also Multiplex
Analog-to-digital (a/d) conversion, 57
Analog transmission systems, 53
    major, 56
    from mastergroups, 55

from supergroups, 54
    See also Transmission systems
Antennas (mobile radio), 253–57
    base station, 253–55
    enhancement techniques, 255
    field component diversity, 256
    masthead electronics, 255
    polarization diversity, 255–56
    smart, 256–57
Antennas (radio relay), 219–22
    characteristics, 220–21
    classifications, 220
    F/B ratio, 220
    gain, 220
    half-power beamwidth, 220
    intermodulation, 221
    interport isolation, 220
    parabolic reflector, 222
    reflector, 221–22
    return loss, 220
    RPE, 220
    wind/ice resistance, 221
    XPD, 220
Antennas (satellite), 308–11
    categories, 310
    Earth station, 311
    function of, 308
    requirements, 308–10
    spot, 310–11
Aperture angle, 166–67
Apogee boost motor, 317
Application service elements (ASEs), 97
Application-specific integrated circuits
    (ASICs), 211
ARPANET, 23–24, 99

The Artech House Telecommunications Library

Vinton G. Cerf, Series Editor

For further information on these and other Artech House titles, contact:

Artech House
685 Canton Street
Norwood, MA 02062
617-769-9750
Fax: 617-769-6334
Telex: 951-659
email: artech@artech-house.com
WWW: http://www.artech-house.com

Artech House
Portland House, Stag Place
London SW1E 5XA England
+44 (0) 171-973-8077
Fax: +44 (0) 171-630-0166
Telex: 951-659
email: artech-uk@artech-house.com